Polymeric Micelles: Principles, Perspectives and Practices

Sachin Kumar Singh • Monica Gulati •
Srinivas Mutalik •
Muralikrishnan Dhanasekaran • Kamal Dua
Editors

Polymeric Micelles: Principles, Perspectives and Practices

Springer

Editors
Sachin Kumar Singh
School of Pharmaceutical Sciences
Lovely Professional University
Phagwara, Punjab, India

Monica Gulati
School of Pharmaceutical Sciences
Lovely Professional University
Jalandhar, Punjab, India

Srinivas Mutalik
Department of Pharmaceutics
Manipal College of Pharmaceutical Sciences, Manipal Academy of Higher Education
Manipal, Karnataka, India

Muralikrishnan Dhanasekaran
Harrison School of Pharmacy
Auburn University
Auburn, AL, USA

Kamal Dua
Discipline of Pharmacy
University of Technology Sydney
Sydney, NSW, Australia

ISBN 978-981-99-0360-3 ISBN 978-981-99-0361-0 (eBook)
https://doi.org/10.1007/978-981-99-0361-0

This Springer imprint is published by the registered company Springer Nature Singapore Pte Ltd.
The registered company address is: 152 Beach Road, #21-01/04 Gateway East, Singapore 189721, Singapore

Preface

Polymeric micelles (PMs) are supramolecular configurations formed by amphiphilic block copolymers in aqueous medium. The amphiphilic block copolymers consist of hydrophilic and hydrophobic segments that have self-assembly property and are able to form various self-assembled structures including spherical, lamellar, and vesicles. This depends on their critical packaging parameter and microphase separation process. For the formation of PMs, the critical micelle concentration (CMC) of block copolymers plays a very critical role. This not only ensures their self-assembly but their thermodynamic stability in the solution as well. Above their CMC value, the hydrophilic segment forms the "corona" while the hydrophobic segment forms the "core" of the PMs. In addition, the flexible and functional chemistry of these block copolymers offers the tailoring of physicochemical traits of the PMs for a specific application. This property imparts ease of functionalization to PMs for their utilization in biomedical applications.

In agreement with this, these smart nanocarriers have immense potential to overcome the challenges associated with existing treatment strategies or conventional formulations. This includes poor dissolution rate, lack of site specificity, poor oral bioavailability, poor in vivo stability, and the associated adverse effects. Their nano-size and ease of functionalization have made it possible to deliver therapeutic drugs, nucleic acids, proteins, and phytoconstituents to localized diseased sites to maximize their clinical benefits while limiting unwanted side effects. In addition, PMs could be designed as diagnostic and theranostic nanocarriers to improve the clinical diagnosis of diseases and offer simultaneous treatment monitoring.

This book thoroughly reviews the advancements made in the designing and application of PMs for drug delivery, covering the synthesis of amphiphilic block copolymers and their types, functional chemistry of copolymers required for targeting and sensing, characterization and biomedical applications of PMs.

1. This book provides the possibilities required in the designing of PMs in a range of different drug delivery approaches to the researchers working on this area. This would be the first book to address the molecular parameters of amphiphilic block copolymers required in functionalizing PMs for drug delivery applications.
2. This book presents the recent advances in PMs such as co-delivery, sensing, theranostics, delivery of nucleic acids, proteins, and many more.

3. This book attracts a *range of audiences* including clinical researchers working in the field of drug delivery, diseases diagnosis, undergraduate and postgraduate students from various disciplines such as pharmacy, immunology, pharmacology, pharmacognosy, biotechnology, and health sciences.

Phagwara, Punjab, India Sachin Kumar Singh
Jalandhar, Punjab, India Monica Gulati
Manipal, Karnataka, India Srinivas Mutalik
Auburn, AL, USA Muralikrishnan Dhanasekaran
Sydney, NSW, Australia Kamal Dua

The original version of the book was revised: third editor Dr. Srinivas Mutalik's affiliation has been updated. An correction to this book can be found at https://doi.org/10.1007/978-981-99-0361-0_16

Contents

Editors and Contributors

About the Editors

Sachin Kumar Singh a Professor at School of Pharmaceutical Sciences, Lovely Professional University, India, holds his Master of Pharmacy in Pharmaceutical Analysis from J.S.S. College of Pharmacy, Ooty, Tamil Nadu, India, and holds a Ph.D. degree in Pharmaceutical Sciences from Karpagam University, Coimbatore, India. He has completed one DST-SERB-funded project on the development of SNEDDS of Polypeptide-k for the treatment of diabetes mellitus. Dr. Singh has published more than 200 research papers with cumulative impact of 400 and coauthored 10 book chapters. He has two granted Indian patents and two granted Australian patents. Dr. Singh has reviewed more than 75 research papers upon invitation from publishers of high repute such as Elsevier, American Chemical Society, Bentham Sciences, and Royal Pharmaceutical Society. He has graduated 8 Ph.D. students and 30 Master of Pharmacy students and currently graduating 3 master's and 6 Ph.D. students. Currently, his h-index is 27 with total citations >2300. In January 2020, he received Dr. B.C. Deb memorial award for popularization of science by Indian Science Congress Association at 107th Indian Science Congress held at UAS Bengaluru. He is the first recipient from the field of Pharmaceutical Sciences to receive this award. He is also a recipient of 2014 Bharat Shiksha Ratan Award given by Global Society for Health and Educational Growth, Delhi, India. His core area is improving oral bioavailability of poorly water-soluble drugs using various formulation approaches including nanotechnology, complexation, solid dispersion, and liquisolid technology.

Monica Gulati is the Registrar and Executive Dean at the School of Pharmaceutical Sciences, Lovely Professional University, India, with more than 25 years of teaching, industry, research, and administrative experience. Having completed her B. Pharm., M. Pharm., and Ph.D. from UIPS, Panjab University, Chandigarh, Dr. Gulati served for two years in the R&D department of Panacea Biotec. Shifting later on to academia, she has served for more than 18 years at LPU in various academic and administrative roles. Her research forte is enhancing bio-performance of drug molecules, phytochemicals and probiotics, metabiotics, and other biotherapeutics using conventional and novel dosage forms especially in the area of colon targeting. She holds a keen interest in the quality control parameters of delivery systems. She has more than 150 high-impact publications and 15 invited book chapters to her credit, with an H-index of 28 and RG score of 39.05. She has been granted four Indian patents and three Australian patents. She is on the editorial and review board of more than 25 journals.

Srinivas Mutalik works as Professor and Head of the Department of Pharmaceutics in Manipal College of Pharmaceutical Sciences, Manipal. He has completed his Ph. D. in 2004 from Manipal Academy of Higher Education and 3 years of postdoctoral studies from the University of Queensland, Australia, and was a visiting researcher at A-Star Institute, Singapore, in 2016. Dr. Mutalik has good professional experience in academics and in the pharmaceutical industry at different capacities. Dr Mutalik has published more than 175 papers in reputed journals and has 10 patents to his credit. He has presented research findings at various national and international conferences and delivered several guest lectures. Dr. Mutalik has received several research grants from funding agencies such as DBT, DST, ICMR, AICTE, SERB, BIRAC, VGST, BRNS, and several pharmaceutical companies. He has received several awards like Dr. TMA Gold Medal Award for Outstanding Research, Best Poster/Oral Presentation Awards, Distinguished Alumni Award, APTI Young Pharmacy Teacher Award, VGST Award for Best Research Paper, etc. Dr. Mutalik is supervisor for several Ph.D. and PG students. He is the Life Member of various professional organizations. Dr. Mutalik's research interests include development and evaluation of novel drug delivery systems and nanopharmaceuticals.

Muralikrishnan Dhanasekaran graduated B. Pharm. in 1992 from Annamalai University and obtained M. Pharm. in 1995, and a PhD degree in 1999 from Jadavpur University, India. He received Junior Research Fellowship (JRF) and Senior Research Fellowship (SRF) of the Council of Scientific and Industrial Research (CSIR), Ministry of Science and Technology, Govt. of India. Dr. Dhanasekaran has nearly 30 years of research experience in various synthetic drugs and botanicals including Indian and Western medicinal plants. While in the Harrison School of Pharmacy, he has worked as a principal investigator of several research projects funded by the Alzheimer's Association and pharmaceutical companies. Furthermore, he received several school and university level awards for devoted teaching in the field of Graduate and Pharm. D programs. He has edited 4 books, published 102 scientific research articles in peer-reviewed journals, 40 book chapters, and 200 abstracts in national and international meetings. He has guided 12 M.S. and Ph.D. students as a mentor and has served as a committee member for more than 30 graduate students. For the last many years, Dr. Dhanasekaran has been working as a reviewer in NIH, Alzheimer's Association, and for several SCI journals and grants.

Kamal Dua has research experience of over 12 years working in the field of drug delivery targeting inflammatory diseases. Dr. Dua is also a Node Leader of Drug Delivery Research in the Centre for Inflammation at Centenary Institute/UTS, where the targets identified from the research projects are pursued to develop novel formulations as the first step towards translation into clinics. Dr. Dua research focuses on two complementary areas: drug delivery and immunology, specifically addressing how these disciplines can advance one another helping the community to live longer and healthier. This is evidenced by his extensive publication record in reputed journals. Dr. Dua's research interests focus on harnessing the pharmaceutical potential of modulating critical regulators such as interleukins and microRNAs and developing new and effective drug delivery formulations for the management of inflammation in chronic airway diseases and cancer.

Contributors

Mayur Aalhate Department of Pharmaceutics, National Institute of Pharmaceutical Education and Research (NIPER) Hyderabad, Hyderabad, Telangana, India

Aanawi Tiwari Department of Biotechnology, JIIT Noida, Noida, UP, India

Kalaiselvi Aasaithambi DBT-BUILDER, JSS Academy of Higher Education and Research, Mysuru, Karnataka, India

Ankur Agrawal School of Pharmacy, ITM University, Gwalior, Madhya Pradesh, India

Alaa A. Aljabali Department of Pharmaceutical Sciences, Faculty of Pharmacy, Yarmouk University, Irbid, Jordan

Etikala Amulya Department of Pharmaceutics, National Institute of Pharmaceutical Education and Research (NIPER) Hyderabad, Hyderabad, Telangana, India

Vimal Arora University Institute of Pharma Sciences, Chandigarh University, Mohali, Punjab, India

Sumel Ashique Department of Pharmaceutics, Bharat Institute of Technology (BIT), School of Pharmacy, Meerut, UP, India

Ankit Awasthi School of Pharmaceutical Sciences, Lovely Professional University, Phagwara, Punjab, India

Divya Dhawal Bhandari Panjab University, Chandigarh, India

Shvetank Bhatt Department of Pharmacology, Amity Institute of Pharmacy, Amity University of Madhya Pradesh (AUMP), Gwalior, Madhya Pradesh, India

Bhavna Kumar Faculty of Pharmacy, DIT University, Dehradun, India

Subhasri Bogadi Department of Pharmaceutics, JSS College of Pharmacy, JSS Academy of Higher Education and Research, Ooty, Tamil Nadu, India

Dinesh Kumar Chellappan Department of Life Sciences, School of Pharmacy, International Medical University, Kuala Lumpur, Malaysia

Pavan Kumar Chintamaneni Department of Pharmaceutics, GITAM School of Pharmacy, GITAM Deemed to be University (HYD-Campus), Medak, Telangana, India

Dimple Sethi Chopra Department of Pharmaceutical Sciences and Drug Research, Punjabi University, Patiala, India

Leander Corrie School of Pharmaceutical Sciences, Lovely Professional University, Phagwara, Punjab, India

Kamal Dua Discipline of Pharmacy, University of Technology Sydney, Sydney, NSW, Australia

Harish Dureja Department of Pharmaceutical Sciences, Maharshi Dayanand University, Rohtak, Haryana, India

Debdarshan Dutta Department of Biotechnology, JIIT Noida, Noida, UP, India

Paras Famta Department of Pharmaceutics, National Institute of Pharmaceutical Education and Research (NIPER) Hyderabad, Hyderabad, Telangana, India

Ashish Garg Department of P.G. Studies and Research in Chemistry and Pharmacy, Rani Durgavati University, Jabalpur, MP, India

Monica Gulati School of Pharmaceutical Sciences, Lovely Professional University, Phagwara, Punjab, India

Shivani Gupta Department of Biotechnology, JIIT Noida, Noida, UP, India

Ujala Gupta Department of Pharmaceutics, National Institute of Pharmaceutical Education and Research (NIPER) Hyderabad, Hyderabad, Telangana, India

Afzal Hussain Department of Pharmaceutics, College of Pharmacy, King Saud University, Riyadh, Saudi Arabia

Kaushal K. Jain Industrial Research Laboratory, Department of Pharmacy, Birla Institute of Technology and Science (BITS)—Pilani, Pilani, Rajasthan, India

Naitik Jain Department of Pharmaceutics, National Institute of Pharmaceutical Education and Research (NIPER) Hyderabad, Hyderabad, Telangana, India

Renjil Joshi Shri Rawatpura Sarkar Institute of Pharmacy, Kumhari, Durg, Chhattisgarh, India

Akanksha Yogesh Kadam Industrial Research Laboratory, Department of Pharmacy, Birla Institute of Technology and Science (BITS)—Pilani, Pilani, Rajasthan, India

Deepak N. Kapoor School of Pharmaceutical Sciences, Shoolini University of Biotechnology and Management Sciences, Solan, India

Veera Venkata Satyanarayana Reddy Karri Department of Pharmaceutics, JSS College of Pharmacy, JSS Academy of Higher Education and Research, Ooty, Tamil Nadu, India

Mayank Kashyap Department of Biotechnology, JIIT Noida, Noida, UP, India

Ankur Kaul Division of Cyclotron and Radiopharmaceutical Sciences, Institute of Nuclear Medicine and Allied Sciences (INMAS), Delhi, India

Jaskiran Kaur School of Pharmaceutical Sciences, Lovely Professional University, Phagwara, Punjab, India

Simran Deep Kaur School of Pharmaceutical Sciences, Shoolini University of Biotechnology and Management Sciences, Solan, India

Naveen Khatri College of Pharmacy, Pt. B.D. Sharma University of Health Sciences, Rohtak, Haryana, India

Gowthamarajan Kuppusamy Department of Pharmaceutics, JSS College of Pharmacy, JSS Academy of Higher Education and Research, Ooty, Nilgiris, Tamil Nadu, India

Divya Mahajan Shobhaben Pratapbhai Patel School of Pharmacy & Technology Management, SVKM's Narsee Monjee Institute of Management Studies (Deemed to be University), Mumbai, India

Neeraj Mishra Department of Pharmaceutics, Amity Institute of Pharmacy, Amity University Madhya Pradesh (AUMP), Gwalior, Madhya Pradesh, India

Lavanya Mude Department of Pharmaceutics, JSS College of Pharmacy, JSS Academy of Higher Education and Research, Ooty, Tamil Nadu, India

Shweta Nene Department of Pharmaceutics, National Institute of Pharmaceutical Education and Research (NIPER) Hyderabad, Hyderabad, Telangana, India

Imrankhan Nizam DBT-BUILDER, JSS Academy of Higher Education and Research, Mysuru, Karnataka, India

Divya Pamu Department of Pharmaceutics, JSS College of Pharmacy, JSS Academy of Higher Education and Research, Ooty, Tamil Nadu, India

Madhukiran Parvathaneni Department of Biotechnology, Harrisburg University of Science and Technology, Harrisburg, PA, USA
Arni Medica, South Plainfield, NJ, USA
CRC Pharma LLC, Parsippany, NJ, USA

Deepanshi Pathak Department of Biotechnology, JIIT Noida, Noida, UP, India

Ritika Puri University Institute of Pharma Sciences, Chandigarh University, Mohali, Punjab, India

R. Rachana Department of Biotechnology, JIIT Noida, Noida, UP, India

Saurabh Shah Pharmaceutical Innovation and Translational Research Lab (PITRL), Department of Pharmaceutics, National Institute of Pharmaceutical Education and Research (NIPER) Hyderabad, Hyderabad, Telangana, India

Mansi Sharma Department of Biotechnology, JIIT Noida, Noida, UP, India

Radhika Sharma Department of Biotechnology, JIIT Noida, Noida, UP, India

Saritha Shetty Shobhaben Pratapbhai Patel School of Pharmacy & Technology Management, SVKM's Narsee Monjee Institute of Management Studies (Deemed to be University), Mumbai, India

Rahul Shukla Department of Pharmaceutics, National Institute of Pharmaceutical Education and Research (NIPER)—Raebareli, Lucknow, India

Anupama Sikder Pharmaceutical Innovation and Translational Research Lab (PITRL), Department of Pharmaceutics, National Institute of Pharmaceutical Education and Research (NIPER) Hyderabad, Hyderabad, Telangana, India

Pankaj Kumar Singh Department of Pharmaceutics, National Institute of Pharmaceutical Education and Research (NIPER) Hyderabad, Hyderabad, Telangana, India

Sachin Kumar Singh School of Pharmaceutical Sciences, Lovely Professional University, Phagwara, Punjab, India

Shashi Bala Singh Department of Biological Sciences, National Institute of Pharmaceutical Education and Research (NIPER) Hyderabad, Hyderabad, Telangana, India

Gautam Singhvi Industrial Research Laboratory, Department of Pharmacy, Birla Institute of Technology and Science (BITS)—Pilani, Pilani, Rajasthan, India

Anshita Gupta Soni Shri Rawatpura Sarkar Institute of Pharmacy, Kumhari, Durg, Chhattisgarh, India

Deependra Soni Faculty of Pharmacy, MATS University, Aarng, Raipur, Chhattisgarh, India

Dadi A. Srinivasarao Department of Pharmaceutics, National Institute of Pharmaceutical Education and Research (NIPER) Hyderabad, Hyderabad, Telangana, India

Saurabh Srivastava Pharmaceutical Innovation and Translational Research Lab (PITRL), Department of Pharmaceutics, National Institute of Pharmaceutical Education and Research (NIPER) Hyderabad, Hyderabad, Telangana, India

Divya Suares Shobhaben Pratapbhai Patel School of Pharmacy & Technology Management, SVKM's Narsee Monjee Institute of Management Studies (Deemed to be University), Mumbai, India

Indhumathi Thirugnanasambandham Department of Pharmaceutics, JSS College of Pharmacy, JSS Academy of Higher Education and Research, Ooty, Nilgiris, Tamil Nadu, India

Yashika Tomar Industrial Research Laboratory, Department of Pharmacy, Birla Institute of Technology and Science (BITS)—Pilani, Pilani, Rajasthan, India

Jayesh S. Unde Department of Pharmaceutics, National Institute of Pharmaceutical Education and Research (NIPER)—Raebareli, Lucknow, India

Ganesh Vambhurkar Department of Pharmaceutics, National Institute of Pharmaceutical Education and Research (NIPER) Hyderabad, Hyderabad, Telangana, India

1 Synthesis, Self-Assembly, and Functional Chemistry of Amphiphilic Block Copolymers

Vimal Arora, Divya Dhawal Bhandari, Ritika Puri, Naveen Khatri, and Harish Dureja

Abstract

The polymeric micelles are the colloidal nanostructures composed of amphiphilic copolymers (having two functional blocks, namely hydrophobic and hydrophilic blocks), having a core-shell framework proven as if one of the most efficient carriers for drugs and nucleic acids. The three major pillars for preferring this versatile group of drug carriers are their capability to cater hydrophobic drug candidates, biocompatibility, and in vivo stability. Other functional advantages of these nanostructures are their capability to self-assemble and potential to respond to certain physical, chemical, and biological stimuli like temperature, light, pH, and enzyme. Several efforts have been made to change the polymeric micelles, focusing their longevity or stability and targeting strategies—regarded as important for passive targeting of drugs carried by or dissolved using the micelles. On the other hand, the structural modification in the micellar surface using a variety of ligands has been tried to focus selective targeting or the intracellular drug delivery, thus allowing these nanostructured micelles to respond to a variety of stimuli (both intrinsic and extrinsic) for the predetermined or triggered drug

V. Arora (✉) · R. Puri
University Institute of Pharma Sciences, Chandigarh University, Mohali, Punjab, India

D. D. Bhandari
University Institute of Pharma Sciences, Chandigarh University, Mohali, Punjab, India

Panjab University, Chandigarh, India

N. Khatri
College of Pharmacy, Pt. B.D. Sharma University of Health Sciences, Rohtak, Haryana, India

H. Dureja
Department of Pharmaceutical Sciences, Maharshi Dayanand University, Rohtak, Haryana, India

S. K. Singh et al. (eds.), *Polymeric Micelles: Principles, Perspectives and Practices*,
https://doi.org/10.1007/978-981-99-0361-0_1

release at the target site. Therefore, these nano-carriers may be explored further for designing new generation drug carriers.

Keywords

Nano-carriers · Amphiphilic block copolymers · Drug targeting · Micelles · Polymerization

1.1 Introduction

Polymeric micellar nanostructures, or simply colloidal polymeric chains, comprising amphiphilic copolymers with a core-shell framework have been proven as one of most versatile and efficient carriers for delivery of nucleic acids and drugs. Few of the major properties that lead to preferable utilization and popularity of these novel drug delivery systems include their capability to solubilize poorly soluble drugs/APIs, biocompatibility, better in vitro as well as in vivo stability, along with the ability to accrue in the pathological sites with limited vasculature. Furthermore, via engineering or modifying the surface of these nanostructures using a variety of ligands and introducing moieties with higher cell-penetration capabilities, some additional functionalities can be associated with these micelles to offer specific site targeting and intracellular accumulation, in addition to their loading with contrast agents to offer imaging abilities and stimuli-sensitive drug release. Therefore, in the past decade this drug carrier and delivery system has attracted researchers to explore and integrate these multifunctional smart polymeric micelles in effective drug delivery (Jhaveri and Torchilin 2014).

Polymeric Micelles can be defined as the self-assembled structures (nano-sized) constructed in aqueous solution and comprising amphiphilic blocks. The formation of micellar structures takes place when the block copolymer concentration or strength rises beyond critical micelle concentration (CMC). At or beyond this concentration, block copolymers initiate the linkage of hydrophobic components while reducing their contact or interaction with molecules of water, which eventually forms a vesicular micellar structure (shell-core) (Yoncheva et al. 2012). These nanostructures usually have a core containing a hydrophobic portion of an ionic fragment of nanoparticle, comprising drug molecules within it, which may be loaded using physical, chemical, or electrostatic methods, based on the specific biological functions of the core-forming segment and the drug. On the other hand, the shell part of the micellar structure leads to the formation of the nanoparticle after solvent interaction and thus, which is accounted for its stability in a given liquid medium (Kumar 2021).

In case of amphiphilic block copolymers (ABC), the block of one homopolymer is attached to the block of another type of polymer; these are named so because the block polymers are amphiphilic in nature. The ABCs are utilized in several applications for pharmaceutical drug delivery including designing of sustained release formulations as well as gene delivery. Their therapeutic properties lie with the fact that their chemical composition is unique in terms of the presence of hydrophilic and hydrophobic blocks. These ABCs can be categorized either based on their molecular weight or the solvent environment. Based on solvent environment these are of two types namely:

- Water soluble low molecular weight ABCs.
- Solvent-free ABCs.

Secondly, the polymeric molecules used to form blocks of copolymers may be categorized based on their molecular weight; the molecules whose molecular weight is lesser than 500 are treated as small amphiphilic molecule, while those whose molecular weight is 10–2000 times bigger than the smaller ones are considered as large amphiphilic molecules. These large amphiphilic molecules are the one which play an important role in forming block copolymers. These nanostructured polymeric drug carriers can also be categorized as passive and active targeted polymers in terms of the extent and mechanism of their potential for drug targeting. Further, the active targeted polymeric structures are further sub-categorized as:

- pH sensitive.
- Enzyme sensitive.
- Temperature/heat sensitive.
- Ultra-sound sensitive.
- Light sensitive (Clemons et al. 2018).

1.2 Different Molecular Parameters of Amphiphilic Block Copolymers Affecting Micellar Traits

Amphiphilic block copolymers, which self-assemble into complex structures, serve as a prototypical “bottom-up” system. Whenever two or more chains of repeated segments are joined together, they lead to the formation of a block copolymer, whose stability depends up on the compatibility of blocks; incompatibility causes the polymer to separate into its component parts (Ruzette and Leibler 2005). By using a selected solvent, amphiphilic block copolymers may result in the formation of a variety of micellar structures, in which one block gets dissolved and others remain connected, greatly expanding the range of possible block copolymer morphologies. Although typically block copolymers generate spherical micelles, different forms are readily achievable by careful design of the copolymer composition, selection of the suitable solvent, appropriate production technique, and use of needful additives (Alexandridis and Spontak 1999; Hamley 2005; Boneberg et al. 2008). Micelles may be formed from amphiphilic block copolymers in certain solvents. These micelles may be thought of as compressible soft spheres since they consist of a compact, insoluble shell surrounded by an elastic corona. When deposited as a monolayer onto surfaces, they prefer to form hexagonal ordered arrays, much like hard spheres. Micelles may also act as nano-compartments, into which an inorganic molecule can be loaded in a controlled fashion. After the polymer matrix has been removed using plasma, the inorganic salt may be converted into nanoparticles of the matching noble metal or metal oxide while maintaining the same spatial orientation as the original micelles. There will be clusters of metal or metal oxide distributed in regular hexagonal arrays over the surface. Because the arrays' ultimate arrangement depends heavily on the micellar deposition process, mastering that procedure is essential.

The deposit of block copolymer micelles on two-dimensional surfaces has been studied by many research teams (Yokoyama et al. 2000; Pike et al. 2011; Hahn and Webber 2004). Significant ordering of the micelles was seen in most instances, and factors that reduce the amount of lattice defects have been described, such as the removal rate, loss of solvent, and graphoepitaxy (Hahn and Webber 2004). The order of arrangement in a monolayer of spherical domains of block copolymer melt was studied by Kramer and colleagues while Webber and his team worked on deposition of micelles comprising (cationic block copolymer) from aqueous solutions to a patterned, plain silicon surface, along with the study of the chemistry behind the process (Hahn and Webber 2004, 2003; Segalman et al. 2003a, b). In their experiments, they discovered that micelle deposition onto substrates is significantly affected by both the solvent evaporation rate and the substrate pull out rate. When solvent evaporation predominated throughout the deposition process, the maximum degree of ordering was achieved. Based on literature it was found that the dip-coating approach is the most effective strategy for the uniform distribution of particle arrays over big surfaces. When dispersed in the selected solvent, amphiphilic block copolymers manage to self-organize into the micellar aggregates. Different morphologies, such as spherical, cylindrical, gyroid, and lamellar, may be anticipated based on the percent composition of the constituents, the number of polymeric segments, and their interaction (Lazzari et al. 2006a).

It is essential to remember that the intrinsic molecular structure resulting from the difference in size for both the blocks (hydrophobic and hydrophilic) is responsible for the diversity in block copolymer assemblies. The dimensionless packing parameter (p) is primarily responsible for defining the geometric arrangement of block copolymers in the final copolymer assemblies in aqueous solution. When a volume of hydrophobic blocks (v), the contact area of the head group, a_o, and the hydrophobic block's length, l_c, are all known, p may be calculated as $p = v/a_o l_c$ (Israelachvili 2011). The p theory has been used to predict structures or the geometry of the self-assembled ABCs in the water. It is proportional to the ratio of the molecular volume of hydrophobic blocks to the real volume block copolymers occupied in the resultant assemblies (Kita-Tokarczyk et al. 2005). It has been hypothesized that the structure of polymers in aqueous solution would vary depending on the value of p. Specifically, spherical micelles are generally formed at $p < 1/3$, cylindrical micelles form at a p value ranging between 1/3 and 1/2, and vesicles in the range of p value between 1/2 and 1 (Blanazs et al. 2009; Smart et al. 2008). The self-assembling of these copolymers is affected by both intrinsic (primary and secondary) and extrinsic (temperature, ionic strength, pH) variables (Bae et al. 2003; Solomatin et al. 2003).

These amphiphilic molecules with a hydrophilic head and hydrophobic tail exhibit a concentration-dependent change in their physicochemical properties. These changes are imparted after the formation of micelles, linked with the positioning and combination of amphiphilic polymers in the solution. These micellar structures are dynamic in nature and are in equilibrium with the free monomers, whereas these monomers are regularly switched between inter-micellar solution and micelles. Owing to this phenomenon, these amphiphilic polymers exist as micelle for a fixed time frame known as the "surfactant residence time" (Zana 2005). On the same instance an average count of monomers making micelle is called as the "aggregation number"; micellar structures generally comprise 50–200 monomer units. The diameter of a spherical micelle is almost double the length of a completely extended monomer, which is generally 1–3 nm, and hence micelles are reported to be lying in the colloidal particle size range. The size of these micelles is commonly governed by the geometry and molecular size of the comprising polymer (Rangel-Yagui et al. 2005).

The formation of the micellar structure in the aqueous solution is controlled because of the active contact among the hydrophobic components of the amphiphiles. The driving force for this self-assembling is the significant reduction in the levels of system's free energy. This decrease in free energy is due to elimination of lipophilic segments from the aqueous environment via formation of core of the micelle, stabilized due to exposure of hydrophilic fragments or blocks in water. The free energy change in the process of micellization can be mathematically described as:

$$\Delta G_{\mathrm{mic}} = 1/4 \times RT \ \ \mathrm{In}(C)$$

where R = gas constant, T = temperature of the system, C = critical micelle concentration (Adams et al. 2003).

The active interactions leading to micellization are countered by repulsive interfaces between the heads and an interaction due to the residual alkyl chain and water molecule at the surface of the micelle. One of the major factors affecting the micellization process or self-assembling is the size of hydrophobic component of the polymer (Malmsten 2002). Other factors include the state and type of solvent, polymer concentration, and the temperature. The micelles are treated as micro-heterogeneous because these structures comprise an internal hydrophobic shell and an external hydrophilic surface. This assembling initiates only at or beyond critical micelle concentration (CMS). At lower concentrations, these amphiphilic or surfactant molecules survive separately, and are very small in size, appearing to be in sub-colloidal size range. Below this CMC, amphiphiles undergo adsorption at water-air interface, which further increases with increase in concentration of these amphiphiles. Lastly after achieving CMC, both the interface and bulk phase are saturated with the monomers. Thus, further addition of amphiphile beyond CMC leads to aggregation of these monomers in the bulk phase, and thus reduction in the system-free energy is observed (Zana 2005; Moroi 1992).

Another important parameter associated with the micellar structure is the critical micellization temperature (CMT), which is defined as the temperature at which the aggregation of amphiphilic polymeric structures takes place; below which they exist as unimers or simply the monomers. Thus, the self-assembling or micelle formation may be either via increasing the concentration beyond CMC at a given temperature or via increasing the temperature beyond CMT at any given concentration of amphiphilic block polymers. In the aqueous solution, via attaching a hydrophilic polymer to a hydrophobic or an insoluble polymeric segment leads to initiation of micelle formation of ABCs, leading to changes in their geometry and flow properties, that differs from the properties of either of the parent polymer. One of the major differences between the micelles comprising conventional and polymeric surfactants is that in case of polymeric surfactants there is covalent linkage between the molecules within the core, which does not allow any exchange of monomers between the core and the solution. This in turn offers structural stability and rigidity to these polymeric micellar structures. Therefore, the structures are synthesized because of balance between the covalent bonds holding the blocks together and the intermolecular forces which are reversible in nature and are responsible for their assembling (Hickok et al. 2002). Major factors governing the size of these polymeric micellar structures include:

- Molecular weight of ABCs.
- Aggregation number of the amphiphilic polymers.
- Proportion of hydrophobic and hydrophilic segments.
- Amount of solvent trapped within the core of the micelle.
- Process of their synthesis (Jones and Leroux 1999; Trubetskoy and Torchilin 1996).

The commonly used base materials for the synthesis of these polymeric micelles are polymeric derivatives of:

- Chitosan.
- Polyacrylates.
- Polycaprolactones.
- Polylactides.
- Polyacrylamides.

There are three commonly used approaches for the synthesis of drug-loaded micelles, namely:

- Direct dissolution.
- Solvent evaporation.
- Dialysis.

The direct dissolution technique is the most common and the simplest one, comprising dissolution of the copolymer and API or drug in water. In this case the self-assembling of the drug and copolymer takes place at or above CMC, leading to the formation of drug-loaded micelles. To increase the capacity of drug loading, this technique may be either carried out at higher temperatures or the drug may be added to the system prior to addition of polymer as a thin film. On the other hand, in case of solvent evaporation (which is also known as solution-casting technique) the organic volatile solvents such as ethanol, acetonitrile, methanol, and acetone are utilized for dissolving both the components, namely the drug and the copolymer, resulting in the formation of a thin film of blend of drug and copolymer after due removal of solvent via evaporation. The drug-loaded polymeric micelles can also be prepared via reconstituting these thin films in water or the aqueous buffers or via using the concept of cosolvency (Aliabadi et al. 2007). The composition of the cosolvent (s) may be varied via either altering the type of organic cosolvents (like acetone, tetrahydrofuran, and acetonitrile) or the ratio of organic is to aqueous phase, and the sequence of their addition (processing). The manipulation of the processing may be effectively utilized for improvising the drug loading capacity or process efficiency in the given polymeric nanostructure along with the structural geometry. For example, if the shell-forming polymeric blocks are bulky or lengthy and/or are having higher hydrophobicity, these two above cited techniques are not suitable. In such cases, the micelles have the capability to solubilize the larger amount of poorly hydrophilic drugs. Therefore, in such cases, the third technique dialysis may be used for the synthesis of the supramolecular drug-loaded nanostructures. In this technique, the polymer and the drug solutions prepared in organic solvent(s) are taken in a dialysis bag followed by exchange of solvent with water via dipping the dialysis bag in water, which leads to the assembling of polymers as micelles (Gohy 2005; Jette et al. 2004). Even though it is a highly effective technique, on the contrary, it is a very drawn-out or time-consuming technique (Kulthe et al. 2012).

1.2.1 Self-Assembling of Amphiphilic Block Copolymers

Amphiphilic polymeric molecules exist as surfactants or lipids resulting in intermolecular contacts through two ways, as self-assembled structures (micelles) in solution and adsorbed at the water-air interface or surface. The former involves only amphiphilic polymeric structures or a blended aggregate of low molecular weight polymers. This phenomenon of self-assembling includes creation of a domain of lipophilic and hydrophilic groups. These ABCs assemble to lower down the interfacial tension of hydrophobic blocks resulting in lowering down of interfacial free energy. Further, the increase in the number of ABCs for assembling is complemented with the upsurge in size of the insoluble shell/core, which in turn leads to stretching of the respective blocks comprising the core, because their linkage with the shell-forming blocks must be aligned at their interface to evade thermodynamically adverse phase mixing. In supplement to this, the association of these copolymers also leads to increased density of core-forming components aligned for an extended geometry (Nyrkova and Semenov 2005; Bendejacq and Ponsinet 2008; Hadgiivanova et al. 2011; von Gottberg et al. 1998).

The small molecular surfactants and the ABCs both show comparable characteristic behaviour in a solution, but after getting assembled in the form of micelles, they differ a lot in terms of various attributes. It has also been reported that the ABCs rarely reach thermodynamic equilibrium or simply the Gibbs free energy minima (Nicolai et al. 2010; Stejskal et al. 1992) as these ABCs can be put together in a range of geometrical patterns in the nanoscale range in some selected solvents. The mixing process of micelle offers a simple, and repeatable protocol to synthesize polymeric nanostructures with intended attributes. As the energy barrier for exchange of unimer or a monomer is very high and the strength or amount of free polymeric fragments in the solution is significantly low or is negligible, therefore principally, the exchange or conversion of polymer fragments between the structures is treated as kinetically stalled (Jain and Bates 2004; Patterson et al. 2014). On the other hand the unimer exchange is reported to be taking place at a rational proportion only in case if a solvophobic component block is adequately moveable (lower glass transition temperature, Tg) and having a lower solvophobicity (lower interfacial tension in reference to solvent) (Nicolai et al. 2010; Halperin 2011; Halperin and Alexander 1989; Lund et al. 2006). Such an out-of-equilibrium performance often limits the developments of block copolymer for a variety of applications and challenges the in-depth knowledge of self-assembling of the diblock copolymers in a solution (Hayward and Pochan 2010). On the other hand, the method to surpass such difficulties is the utilization of block-random ABCs; these are defined as the polymers comprising a pure homopolymer block and another copolymer block (Tsitsilianis et al. 2011). This novel approach has directed to advancements in designing a variety of new structures in solutions, where few structures are reaching equilibrium (Wright et al. 2015; Shedge et al. 2014; Dutertre et al. 2012; Lejeune et al. 2010; Borisova et al. 2011, 2012; Bendejacq et al. 2005; Laruelle et al. 2004). An alternate approach can be the copolymer mixing method, in which two polymers are blended together, and the unified structure is having the response, in between the two parent polymers (Wright et al. 2016).

There may be certain supplementary aspects that may be considered, particularly the ones which highlight their chemical structure and related attributes. Let us consider an example of block ionomers, treated as the structures in which the block copolymers comprising neutral and charged components, completely soluble in aqueous environment, when blended with the block ionomers or block copolymers containing an oppositely charged ion, they may lead to the formation of multimolecular self-assemblies known as the poly ion complex (PIC) micelles (Harada and Kataoka 1995, 1999). The formation of PIC is majorly determined by the liberation of oppositely charged ions prompted by ion-pairing among counterion ionomers, evidently leading to elimination of solvent from the insoluble fragments during or after the self-assembling of the copolymers (Cabral et al. 2018). It is important to note that the increase in the size of amphiphilic molecules results in enhancement of both technical and biological applications. Furthermore, it is also proven scientifically that the less soluble monomer among the copolymers has great tendency to act as a good surfactant. Even though the ABCs composed of poly (ethylene oxide)-block-poly(propylene oxide)-block-poly(ethylene oxide) are extensively studied, and the copolymers comprising poly(L-amino acid) and poly(ester) hydrophobic blocks have represented effective efficiency in delivery applications. As every ABC has a distinctive benefit in terms of drug delivery system, therefore it is required to select an appropriate block copolymer for a specific purpose, like prolonged circulation time, modified drug release profile, and introducing targeting moieties. Therefore, these ABCs have been utilized for various applications in pharmaceutical product design and drug delivery including drug solubilization, stabilization, modification of the pharmacokinetic profile of drugs, and suppressing multidrug resistance (Abouzeid et al. 2013).

Numerous efforts have been made to date to modify the polymeric micelles emphasizing up on their longevity and targeting strategies, which are treated as very important for passive targeting of drugs solubilized within the micelles. On the other side, the modification in the micellar surface using ligands has been tried to focus selective targeting and the intracellular drug delivery. Therefore, these modifications aimed to allow micelles to respond to a variety of stimuli (both intrinsic and extrinsic) for the predetermined or triggered drug release at the target site. Now researchers are working on an integrated approach in which different modifications made individually are integrated as multiple components within a micelle. Therefore, an attempt has been made to design micelles with two or more modifications, enabling them to perform important therapeutic functions along with diagnostic functions simultaneously (Torchilin 2006). In simple words these multifaceted micelles owe a variety of distinct attributes and/or functions within a single structure, wherein every single component is functioning faultlessly in coordination with other components; such systems are called as the "multifunctional" micelles (Jhaveri and Torchilin 2014).

1.3 Structural Modifications

There are a variety of copolymers bearing certain characteristic properties or attributes like acceptable biocompatibility, structural strength etc., making them suitable candidates for their use in block polymer synthesis. For example poly (2-hydroxyethyl methacrylate) is an acrylic polymer which is hydrophilic in nature and bearing biocompatibility along with transparent consistency, and effective mechanical properties. On the other side, the pHEMA hydroxy functional groups of this wonderful polymer can be crosslinked with either isocyanates or melamine–formaldehyde resins (Chen and Yang 1994), and epoxides to alter its various characteristic attributes and other properties (Pellice et al. 2007). Another important class of polymers known as polysiloxanes, known for their use as wetting agents and defoamers for better spreading, can be effectively modified by curing with 2-Hydroxyethyl methacrylate leading to the formation of highly flexible and glossy polymeric films (Küçükoğlu et al. 2007; Hill 2002). This class of polymers bears a lower glass transition temperature along with thermal as well as oxidative stability and is also hydrolysable under mildly basic or acidic conditions. Further, the etching of polysiloxane leads to the formation of mesoporous polymeric membranes and materials with controlled pore size (Bruns et al. 2005; Lazzari et al. 2006b).

Furthermore, the method of copolymer preparation or simply the polymerization techniques may have a substantial impact on various structural and functional attributes of the copolymers, for example if the pHEMA and PDMS (class of oxygen permeable, biocompatible polymeric system) based copolymers are prepared using the method of controlled polymerization, their physical properties can be tuned more appropriately as compared to that of via utilizing free radical polymerization. Another such technique is atom transfer radical polymerization (ATRP), which is categorized as a precise free radical polymerization technique, with the ability to control the polydispersity index (PDI) and molecular weight via rescindable transference of halogen atom between the inactive alkyl halide and the active radical in the availability of a transition metal-based catalyst. One of the benefits associated with the ATRP is that it is more tolerant to impurities (available in situ) as compared to that of anionic polymerization whereas the termination reactions in ATRP are not so significant as the concentration of radical species is lower because of higher rate of deactivation (Mühlebach et al. 1998; Bas and Soucek 2012).

There are generally two common raw materials for preparing polymeric micelles, one is biocompatible synthetic polymers and the another is natural macromolecules; these two can also be combined based upon the processing abilities and adaptability of synthetic polymers and the assembling mechanism, structural control as well as functional control of the natural polymers. The assembling process of the copolymers is majorly controlled by the constituting segments, their arrangements, and the load, whereas the stability of these nanostructures controls their performance in the biological surroundings (Kataoka et al. 2001). Furthermore, the assortment of polymers not just deliberate the structural as well as functional attributes of the assembled micelles, but also defines their safety, and it becomes more significant in the conditions where the repeated or frequent administration of micelles is required. Therefore, along with biocompatibility and nontoxicity of biomedical polymers, it is also important that the post-administration (after drug delivery) content of polymers should be such that side effects of these polymers may be easily prevented, like immune response activation (Joint MHLW/EMA n.d.). Therefore, the biopolymers are required to be biodegradable, so that they may be complete disintegrated into the constituting monomers, and safely excreted without any accumulation, and hence preventing any long-term toxicity (Kataoka et al. 2001; Joint MHLW/EMA n.d.).

As far as the functional performance of polymeric micelles is concerned, on one hand its hydrophilic shell reduces their extent of contact with the blood components which leads to elongation of half-life of these nanomedicines in the systemic circulation. On the other hand, it restricts the internalization of products at the target sites. This poor or limited uptake at the target site is a major problem with the product performance which is known as "PEG dilemma"; it is commonly reported in case of the PEGylated nanostructures (Hatakeyama et al. 2013), but it has also been observed in case of some other hydrophilic shell-forming components or polymers (Yu et al. 2012). The best approach to override this problematic situation is the surface alteration of polymeric micellar structures using various ligand molecules, which in turn facilitates the specific targeting and increased cellular uptake (Allen 2002). Furthermore, via these approaches the nanomedicines or the nanostructures having the capabilities for selectively delivery to subcellular targets, enhanced therapeutic efficacy could be reached. The efficacy of ligand-bound micellar structures depends on the designing aspects, like density or abundance of ligands on the micelle surface (Miura et al. 2013; Elias et al. 2013) and flexibility index of spacer placed between the micelle and the ligand, which normally intends to provide mobility to ligands and hence lead to capability of better receptor accessibility (Ishii et al. 2016) (PM-4 ACS).

The attributes of a designated ligand–receptor system comprise the binding affinity (Wiley et al. 2013), which may further be improved via utilization of the multifaceted outcome of different ligands on the nanostructure surface (Liu et al. 2010) which works via increased receptor availability, receptor binding, and biodistribution (Low et al. 2008). The varying expression of receptors also defines the competency of the targeting ligand mechanism. Therefore, while targeting the tumour cells, the parameters involved are the binding affinity, receptor density, and the rate of cellular uptake, because these may directly influence the overall availability and penetration of ligand-bound nanostructures (Danhier et al. 2010). Besides this, the receptor distribution or availability, and varying receptor expression based up on the stage of the cancer (tumour cell type), may also have an impact on the site-targeting efficiency of the ligand-bound nanostructures, and in this situation the ligands capable of suitable cell or cell receptor recognition play an important role (Cabral et al. 2018).

The amphiphilic copolymers constituting polymeric micelles are usually block copolymers (Francis et al. 2003) and based on the structure these are classified as diblock (the A-B type, where A is a lipophobic block and B is a lipophilic block) and triblock copolymers (either comprising two different types of polymers; ABA type or three different polymers ABC type) (Kabanov et al. 2002). Generally, the drug carrier efficiency and applications have been studied using the block copolymers of types A-B and/or A-B-A, due to their close correlation between the micelle properties and structure of polymer (Yokoyama 2011). Thus, the physicochemical attributes of building blocks affect the physical as well as biological properties of the polymeric micelles (Rijcken et al. 2007). One of the most explored class of polymers in the last about one decade is of Pluronics (polyesters) which includes the Polypropylene oxides (PPOs) like poly-amino acids (PAA), poly-lactic acid (PLA), polycaprolactone (PCL), and the copolymers of glycolic acid and lactic acid (Bendejacq et al. 2005; Kabanov et al. 2002). These hydrophobic polymers are used majorly for the core formation owing to their wider range of polarity distribution and structural diversity leading to solubilization of numbers of water insoluble drugs (Kakizawa and Kataoka 2002). Other useful examples of polymer belonging to this class are polystyrene (PS), polystyrene oxide (PSO), polybutylene oxide (PBO), polybutadiene (PB), poly methyl-acrylate oxide (PMA), etc. The stealth properties of the hydrophilic polymeric blocks are generally influenced by the molecular weight and chemical nature which in turn affects the kinetics micellar assembly.

Polyethylene glycol (PEG) is the most frequently used hydrophilic block copolymers, because of its nontoxic nature, whereas poly(N-vinyl-2-pyrrolidone) (PVP) and poly acrylic acid (PAA) are the most used alternatives of PEG (Benahmed et al. 2001; Inoue et al. 1998) (PMC-MDPI). The hydrophilic blocks commonly comprise polyethylene glycol (PEG) and/or polyethylene oxide (PEO) condensate type non-ionic surfactants (due to lower molecular weights, sometimes PEO is not considered as polymer) which are commercially accessible as Tritons, Tweens, Cremophor EL, Soluplus, etc. Other significant examples from this class of hydrophilic blocks are polyvinyl pyrrolidone (PVP), polyvinyl alcohol (PVA), polyvinyl caprolactam (PVCL), poly(N-isopropylacrylamide) (PNIPAM), certain well-known polybases and polyacids (Kuperkar et al. 2022).

1.4 Synthesis of Copolymers

A range of polymers may be prepared by using several distinct polymerization processes or chemical reactions, amongst them the radical polymerization being considered as the most widely used and robust method. The major drawbacks of this technique are larger dispersity and almost no control on molecular mass of the polymer. These drawbacks are addressed by techniques like Nitroxide Mediated Polymerization (NMP; a type of living radical polymerization) (Moad et al. 2005; Rizzardo and Solomon 2012), Reversible Addition Fragmentation Chain Transfer (RAFT) polymerization, and Atom Transfer Reduction Polymerization (ATRP) (Moad et al. 2005; Kato et al. 1995). Among all these techniques, the RAFT polymerization is the only technique which permits the significant level of control in the process of polymerization by including a single supplementary compound. It also allows an easy and simple synthesis of complex polymers (structurally complex) like multi-block copolymers, bottle brush polymers, star polymers, surface coated polymers, and the hyper-branched polymers (Moad et al. 2012). This technique of polymerization is also capable of synthesizing polymers with a wider range of active end groups, which allows for various active processes like coupling or binding with bioactive molecules. For example, the polymers with amine end groups interact with the biological molecules with negative charge and are thus successfully used in numerous drug delivery applications (Fairbanks et al. 2015).

There are some well-established synthetic polymerization procedures, widely used for the synthesis of these polymers and few of these are:

- Anionic polymerization.
- Group-transfer polymerization.
- Ring-opening metathesis polymerization.
- Cationic polymerization.
- Living radical polymerization.
- Active-centre transformation.
- Polymer-analogue reactions (Förster and Antonietti 1998).

Another major concern with these polymers is the toxicity associated with them, majorly influenced by the method of synthesis and the design of the polymers. The chemical compounds mediating the RAFT polymerization are called as the RAFT agents, also named as the chain transfer agents; generally containing a thiocarbonylthio groups associated with substituent "R" and "Z" groups (Moad et al. 2005; Rizzardo and Solomon 2012). In the process of polymerization of the monomer, followed by active extension or elongation of polymeric chain, the radically terminated chain grows through the thiocarbonyl group available in the structure of RAFT agent resulting in the formation of a radical intermediate, which may break apart to release the chain or simply the R group from the chain transfer agent. This released R group radical re-starts the polymerization, resulting in an actively elongating chain of polymer. A polymer chain with thiocarbonylthio group at their terminal is commonly inactive, and in equilibrium (dynamic) with actively, elongating radical polymer chains. This dynamic equilibrium between the actively elongating chains and inactive chains with terminal thiocarbonylthio group delivers equivalent possibility for all the chains to elongate or grow, which in turn results in the formation of polymers with almost narrow molecular weight dispersity (Moad et al. 2012; Chiefari et al. 1998). In this process the identity of R as well as Z groups affects the level of equilibrium met and the extent to which it is maintained, which in turn defines the level of quality of regulation or control in a given polymerization chain reaction. As the kinetic behaviour between the distinct type of monomers differs, a variety of RAFT agents are needed to regulate the polymerization. However, the reactive moieties can be available at either "R" or "Z" and/or both locations; the R groups are commonly chosen due to the instability of the thiocarbonylthio group which results in binding of the synthesized polymer with the Z group site. These RAFT agents are used to synthesize homo or copolymers having a reactive site for the successive association with the small molecules or peptides, drugs, or other bio-macromolecules (Fairbanks et al. 2015).

In the RAFT technique of polymerization only adding a suitable quantity of the RAFT agent to the polymerization reaction mixture is important to attain accurate molecular weight and controlling its dispersity. Therefore, this technique is responsive to redox, radical thermal, and photochemically instigated polymerization (Ham et al. 2012; Tan et al. 2015). Presently, the "PEGylation" (attaching PEG) of the drugs is the most widely used approach to synthesize drug-polymer conjugates with enhanced pharmacokinetic attributes. However, an emerging methodology other than PEG is rising due to oxidative instability of PEGylated complexes and their ability to trigger immune response (Hamad et al. 2008). Otherwise, these RAFT agents may be associated with the drugs or bioactive compounds, for successive direct polymer grafting. For example, Tao et al. synthesized an amine-reactive copolymer via polymerization of poly(HPMA) using thiazolidine-2-thione functionalized chain transfer agent. In this method the polymer was coupled through amide bond formation by incubating it in the presence of the model protein (lysozyme). It led to the formation of a linear homopolymer having a centrally positioned reactive site for successive coupling with a cysteine thiol available on a standard protein. It offers effective masking of the protein surface from certain macromolecules like proteases and the said benefit of this mid-chain conjugation is known as the "umbrella effect". Conjugation of these block copolymers, produced via RAFT technique, offers a limited passive delivery system. These polymers may provide several purposes to influence the delivery of therapeutic payload; some of these functions are:

- Polymer-drug conjugation and grafting.
- Intrinsically bioactive polymers.
- Stimuli-responsive polymers.
- Self-assembled, drug-loaded structures.
- RAFT polymers in nucleic acid delivery vehicles (Fairbanks et al. 2015).

A good deal of consideration has been given to the inclusion of fluorinated groups into synthetic materials, which in turn combines the advantages of both the fluorinated groups and other polymers (Goli et al. 2012). In this direction controlled radical polymerization (CRP) can provide an important tool for preparing a well-definite fluorinated polymer(s) with a variety of structural geometry having a predetermined chain length and lower polydispersities. Generally, the fluorinated block copolymers were prepared (synthetically) previously using CRPs, engaging nitroxide-mediated radical polymerization (NMP) (Bruno 2010), atom transfer radical polymerization (ATRP), and reversible addition fragmentation chain transfer polymerization (RAFT) (Chakrabarty and Singha 2013). The Poly(poly(ethylene glycol) methyl ether methacrylate-co-acrylonitrile) [P(PEGMA)] polymer block prepared by CRPs is a multipurpose polymer group with fine-tuned hydrophobic-hydrophilic balance with a feature of alteration of lengths of both side chain components, namely P(PEGMA) and PEG. Furthermore, P(PEGMA) side chain leads to the introduction of beneficial attributes like water solubility, biocompatibility, and low toxicity (Liu et al. 2001; Wang et al. 2000).

Recently click reactions have drawn significant attention in the segment of synthetic polymer chemistry because of their higher specificity, functional group tolerance, and reaction yields. In another approach of cycloaddition reaction, like CuAAC (copper-assisted cycloaddition), also known as the Cu(I) catalysed Huisgen 1,3-dipolar cycloaddition, the cycloaddition occurs between an alkyne and an azide to provide 1,2,3-triazole, which has arisen as a fruitful tool to prepare multifunctional macromolecular structures, when utilized in combination with techniques like controlled radical polymerization (Kolb et al. 2001; Rostovtsev et al. 2002; Ergin et al. 2007).

Another useful technique for preparing tailor-made polymers is the anionic polymerization for synthesizing polymeric structures with controlled molecular weight along with limited polydispersity. This useful tool offers controlled management of macromolecular structure and leads to the synthesis of simple block copolymers having complex architecture like multi-block polymers, star-block copolymers, star polymers, and graft copolymers (Shipway et al. 2000). Besides this, there are few Y-shaped copolymers (also known as AB2-type or miktoarm), which involves a complex multistep synthesis. The synthesis of Y-shaped copolymer was first explained by Teyssie et al. (Spatz et al. 1998). The Polystyrene-block-[poly-(ethylene oxide)] was synthesized by the process of end-capping of a polystyryl carbanion using a naphthalene derivative, followed by the polymerization of ethylene oxide with the help of naphthalene sodium (radical ion) as initiator. In these days, ATRP and cationic polymers are successfully utilized in the synthesis of miktoarm polymers (Müller et al. 2002); for example ATRP-based synthesis of [PS-(PtBA)^2] Y-shaped block copolymers using a multistep process and the protecting group chemistry. In another process the hydrolysis of tert-butyl groups leads to the formation of Y-shaped block copolymers of PS-(PAA)^2 (Csáki et al. 2002).

In the last one decade a considerable progress in terms of advancement in the synthesis process of block copolymers having varied structural and biological properties as well as other physicochemical attributes like solubility. The solubility of these copolymers varies from solvent to solvent, which in turn depends up on the cohesive energy densities (water is having higher cohesive energy density as compared to silicon oil and the fluorinated solvents). On the other hand, the control or management of functionality is also a very important issue, stabilizes biological interfaces, and optimizes the efficacy of these copolymers. In recent years, various modern methods like living radical polymerization are available that allows the synthesis of newer classes of amphiphilic or multifunctional block copolymers. Such techniques have an advantage of synthesizing the polymers with narrow molecular weight distribution and a determined extent of polymerization (N) which depends only on the molar ratio of monomer [m] to initiator [i]:

$$N = [m]/[i]$$

Another important goal in the synthesis block copolymer is the simplification of reaction conditions to assure efficient production or manufacturing on large scale. There are a variety of polymers available commercially like Kraton (Shell) and Pluronic (BASF Wyandotte) block copolymers; these are otherwise used as emulsifiers and thermoplastic elastomers. Another list of commercially available amphiphilic block copolymers includes polymers based on poly(ethylene oxide) (PEO), polystyrene (PS), and poly(methyl methacrylate) (PMMA). The attributes of the micellar structure like their size, core dimensions, structure, aggregation number, and some other properties are dependent upon concentration, temperature, molecular weight of the blocks, chemical structure, chemistry, solvent-block interactions, and their architecture (Zhao and McCormick 1992; Schrock 1990; Miyamoto et al. 1984).

1.5 Synthetic Advancements in Synthesis of ABCs

Amphiphilic block copolymers can be self-assembled at room temperature using glass hydrophobic blocks using a cosolvent or a solvent switch method (Mai and Eisenberg 2012). This method involves dissolving ABCs in certain organic solvents like DMF, Acetone, or THF, followed by the solvent addition; the solvent should be compatible with only one block; thus there is separation of phases and then self-assembling into nanostructures with a solvophobic shell covered or coated by a solvophilic layer (Alexandridis and Lindman 2000). This method permits only lower polymer concentration (<1%) (Raveendran et al. 2017; Rezaei et al. 2012; Xu et al. 2017; Zhang et al. 1997) and a drawback of it is that the self-assembly occurs post-polymerization; hence it is non-economical and also self-assembly in the solution phase results in diverse morphologies such as micelles, nanofibers, and vesicles (Yadav et al. 2020; Discher and Eisenberg 2002). Advancements have been made to overcome such drawbacks and one among them is polymerization-induced self-assembly (PISA). This method involves one pot approach for both polymerization and self-assembly (Ferguson et al. 2002). The methods that can be used for polymerization in PISA are emulsion and dispersion. The dispersion method requires a monomer and a macromolecule which should be compatible or soluble in same phase either organic or water. Once the polymerization level is reached for the monomer forming core, it starts precipitating out and more monomers will start forming covalent bonds to form stabilized particles. The emulsion method is possible only in the aqueous system, utilizing a water-soluble homopolymer originator molecule and the chain is extended using an immiscible monomer up to a critical level of polymerization. This eventually results in the synthesis of monomeric droplets, stabilized by the homopolymer used as a precursor.

Both polymerization techniques result in the evolution of various morphological forms, which start initially with spherical micelles followed by worm-like or cylinder-shaped nanoparticles then finally the vesicles. The sphere to vesicle transition can be achieved early by modifying some reaction conditions, thus controlling the shape of targeted particles which is an essential feature to increase the drug delivery performance as every distinct shape has its particular in vivo outcome, biodistribution, and post-administration cellular uptake (Zhao et al. 2017a, b; Jindal 2017). The nanoparticles with elongated shape form strong bindings between receptor and ligands in comparison to spherical nanoparticles thus helps in enhancement of specificity and selectivity during endocytosis (Salatin et al. 2015) whereas the sharp edge nanoparticles stay longer in cytoplasm due to disruption of endosome membrane thus shows lesser exocytosis (Chu et al. 2014). The self-assembly of ABCs to form rod-coil copolymers, linear copolymers, and graft copolymers can be achieved by replacing one of block copolymer with rigid segment to form ABCs of distinct morphology. Thus, PISA enhances cellular uptake and bioavailability by giving a range of possible and well-controlled nanostructures along with their distinct optical or opto-electronic attributes, biocompatibility, and stimuli-responsiveness, which offers boundless potential to be the multifaceted materials, especially for drug delivery systems or carriers (Huang et al. 2017).

Along with structure, the size of NPs also matters to be taken by target cells at the site of drug delivery. The ones with <5 nm get eliminated in renal filtration while >200 nm accumulates in the liver and spleen. On the other hand, the ones with <100 nm enhance half-lives of drug due to less adsorption by serum proteins. One of the best methods for PISA synthesis is reversible addition–fragmentation chain transfer (RAFT) (Derry et al. 2016) RAFT dispersion or emulsion can be prepared by thermal, ultrasonic, photochemical, and microwave activations, or using enzymes (Penfold et al. 2019). Other techniques for PISA approach are:

- Ring-opening polymerization (ROP) (Jiang et al. 2019).
- Ring-opening metathesis polymerization PISA (ROMPISA) (Wright et al. 2019).

The physical, chemical, and other attributes or the characteristic properties of an amphiphilic copolymers, like surface tension, modulus (Xu et al. 2007), etc., can be changed via changing the composition, chain length, and structure of copolymer(s). These attributes in turn have an impact on the morphology and critical micelle concentration (CMC) (Yun et al. 2003); for example via increasing the length of a hydrophobic block, a decrease in CMC is generally observed (Remant Bahadur et al. 2007). Similarly via changing the copolymer block attributes, properties like glass transition, crystallinity, thermal degradation, and melting point can be altered that offers a window to control the overall polymer performance (Wang et al. 2005). For example, mechanical properties and swelling index of the block copolymers of either poly(e-caprolactone) or poly(L-lactic acid) and poly(ethylene oxide) (PEO) vary with the type of hydrophobic polymer block component and the composition of the polymer (Bae et al. 2000).

1.5.1 Modification of Amphiphilic Block Copolymers for Drug Targeting Using Functional Moieties

To one side from size and morphological changes, certain functional moieties can be linked into ABCs for a targeted controlled drug delivery. Various research initiatives have been taken to date to use certain functional groups to control the rate of drug delivery and to limit the toxicity of these multifunctional nanostructures. Some of these modifications are:

- Synthesis of redox-responsive 2,2,6,6-tetramethylpiperidine-1-oxyl (TEMPO)-functionalized diblock copolymers of poly(ethylene glycol)-b-poly(2,2,6,6-tetramethylpiperidine-1-oxyl-4-yl) methacrylate-co-N-isopropylacrylamide [PEG-b-P(TMA-co-NIPAM)] using RAFT approach; due to hydrophilic (PEG) and hydrophobic (TEMPO) components, they self-assemble to form nano-vessel drug carriers known as radical nanoparticles (RNPs). On the other hand, with the addition of vitamin C, the TEMPO units decreases and thus reduces the surface hydrophobicity and hence enhances the release of drug at a controlled rate. Furthermore, these TEMPO-based ABCs are reported to be nontoxic in the concentration range up to 500 μg/mL (Shen et al. 2019).
- Another approach reported for introducing modifications is synthesis of amphiphilic copolymers using methoxyl poly(ethylene glycol)-b-poly(L-lactide-co-2-methyl-2-allyloxycarbonyl-propylene carbonate) and allyoxyl poly(ethylene glycol)-b-poly(L-lactide). In this synthesis technique the thiol-ene functional moiety has been used through a radical mediated reaction to modify these two blocks with mercaptans, leading to enhanced biocompatibility (Yue et al. 2010).

1.6 Conclusion

About 33% of the drugs are hydrophobic and half of the drugs in the developmental phase fails in the initial trials because of their poor pharmacokinetics. In such cases, the block copolymer micelles have the potential to solubilize these hydrophobic drugs and lead to alteration or management of both in vitro and in vivo kinetics. As these block copolymer micelles can also be subjected to certain architectural changes leading to inculcation of certain properties which in turn help in the active targeting of the drug (stimuli responsive delivery), these nanostructures may be explored further for designing new generation drug carriers. To accelerate the progress in the development of innovative, controlled, and advanced micelles as the prominent drug delivery vehicles, we must realize the fate of micelles and drug-loaded micellar structures into the cells at the active sites. Therefore, it may be concluded that the superior cellular uptake of micellar structures resulting from distinct and manageable morphology along with higher drug loading capacity may be explored for industrially relevant approaches and formulation techniques for the manufacturing of polymer-based nanostructures for effective and efficient targeted controlled drug delivery.

References

Abouzeid AH, Patel NR, Rachman IM, Senn S, Torchilin VP (2013) Anticancer activity of anti-GLUT1 antibody-targeted polymeric micelles co-loaded with curcumin and doxorubicin. J Drug Target 21(10):994–1000. https://doi.org/10.3109/1061186X.2013.840639

Adams ML, Lavasanifar A, Kwon GS (2003) Amphiphilic block copolymers for drug delivery. J Pharm Sci 92(7):1343–1355. https://doi.org/10.1002/jps.10397

Alexandridis P, Lindman B (2000) Amphiphilic block copolymers. Elsevier, Amsterdam

Alexandridis P, Spontak RJ (1999) Solvent-regulated ordering in block copolymers. Curr Opin Colloid Interface Sci 4(2):130–139. https://doi.org/10.1016/S1359-0294(99)00022-9

Aliabadi HM, Elhasi S, Mahmud A, Gulamhusein R, Mahdipoor P, Lavasanifar A (2007) Encapsulation of hydrophobic drugs in polymeric micelles through co-solvent evaporation: the effect of solvent composition on micellar properties and drug loading. Int J Pharm 329(1–2):158–165. https://doi.org/10.1016/j.ijpharm.2006.08.018

Allen TM (2002) Ligand-targeted therapeutics in anticancer therapy. Nat Rev Cancer 2(10): 750–763. https://doi.org/10.1038/nrc903

Bae YH, Huh KM, Kim Y, Park KH (2000) Biodegradable amphiphilic multiblock copolymers and their implications for biomedical applications. J Control Release 64(1–3):3–13. https://doi.org/10.1016/s0168-3659(99)00126-1

Bae Y, Fukushima S, Harada A, Kataoka K (2003) Design of environment-sensitive supramolecular assemblies for intracellular drug delivery: polymeric micelles that are responsive to intracellular pH change. Angew Chem Int Ed 115(38):4788–4791

Bas S, Soucek MD (2012) Synthesis, characterization and properties of amphiphilic block copolymers of 2-hydroxyethyl methacrylate and polydimethylsiloxane prepared by atom transfer radical polymerization. Polym J 44(11):1087–1097. https://doi.org/10.1038/pj.2012.86

Benahmed A, Ranger M, Leroux JC (2001) Novel polymeric micelles based on the amphiphilic diblock copolymer poly(Nvinyl-2-pyrrolidone)-block-poly(D,L-lactide). Pharm Res 18(3): 323–328. https://doi.org/10.1023/a:1011054930439

Bendejacq DD, Ponsinet V (2008) Double-polyelectrolyte, like-charged amphiphilic diblock copolymers: swollen structures and pH- and salt-dependent lyotropic behavior. J Phys Chem B 112(27):7996–8009. https://doi.org/10.1021/jp712015h
Bendejacq DD, Ponsinet V, Joanicot M (2005) Chemically tuned amphiphilic diblock copolymers dispersed in water: from colloids to soluble macromolecules. Langmuir 21(5):1712–1718. https://doi.org/10.1021/la048983r
Blanazs A, Armes SP, Ryan AJ (2009) Self-assembled block copolymer aggregates: from micelles to vesicles and their biological applications. Macromol Rapid Commun 30(4–5):267–277. https://doi.org/10.1002/marc.200800713
Boneberg J, Habenicht A, Benner D, Leiderer P, Trautvetter M, Pfahler C, Plettl A, Ziemann P (2008) Jumping nanodroplets: a new route towards metallic nano-particles. Appl Phys A 93(2): 415–419. https://doi.org/10.1007/s00339-008-4780-z
Borisova O, Billon L, Zaremski M, Grassl B, Bakaeva Z, Lapp A, Stepanek P, Borisov O (2011) pH-triggered reversible sol–gel transition in aqueous solutions of amphiphilic gradient copolymers. Soft Matter 7(22):10824–10833. https://doi.org/10.1039/c1sm06600e
Borisova O, Billon L, Zaremski M, Grassl B, Bakaeva Z, Lapp A, Stepanek P, Borisov O (2012) Synthesis and pH- and salinity-controlled self-assembly of novel amphiphilic block-gradient copolymers of styrene and acrylic acid. Soft Matter 8(29):7649. https://doi.org/10.1039/c2sm25625h
Bruno A (2010) Controlled radical (Co)polymerization of fluoromonomers. Macromolecules 43(24):10163–10184. https://doi.org/10.1021/ma1019297
Bruns N, Scherble J, Hartmann L, Thomann R, Iván B, Mülhaupt R, Tiller JC (2005) Nanophase separated amphiphilic conetwork coatings and membranes. Macromolecules 38(6):2431–2438. https://doi.org/10.1021/ma047302w
Cabral H, Miyata K, Osada K, Kataoka K (2018) Block copolymer micelles in nanomedicine applications. Chem Rev 118(14):6844–6892. https://doi.org/10.1021/acs.chemrev.8b00199
Chakrabarty A, Singha NK (2013) Tailor-made polyfluoroacrylate and its block copolymer by RAFT polymerization in miniemulsion; improved hydrophobicity in the core-shell block copolymer. J Colloid Interface Sci 408:66–74. https://doi.org/10.1016/j.jcis.2013.07.031
Chen YH, Yang CP (1994) Coemulsion and electrodeposition properties of mixtures of cationic epoxy resin and cationic acrylic resin containing blocked-isocyanate groups. J Appl Polym Sci 51(9):1539–1547. https://doi.org/10.1002/app.1994.070510903
Chiefari J, Chong YK, Ercole F, Krstina J, Jeffery J, Le TPT, Mayadunne RTA, Meijs GF, Moad CL, Moad G, Rizzardo E, Thang SH (1998) Living free-radical polymerization by reversible addition-fragmentation chain transfer: the RAFT process. Macromolecules 31(16):5559–5562. https://doi.org/10.1021/ma9804951
Chu Z, Zhang S, Zhang B, Zhang C, Fang CY, Rehor I, Cigler P, Chang HC, Lin G, Liu R, Li Q (2014) Unambiguous observation of shape effects on cellular fate of nanoparticles. Sci Rep 4: 4495. https://doi.org/10.1038/srep04495
Clemons TD, Singh R, Sorolla A, Chaudhari N, Hubbard A, Iyer KS (2018) Distinction between active and passive targeting of nanoparticles dictate their overall therapeutic efficacy. Langmuir 34(50):15343–15349. https://doi.org/10.1021/acs.langmuir.8b02946
Csáki A, Maubach G, Born D, Reichert J, Fritzsche W (2002) DNA-based molecular nanotechnology. Single Mol 3(5–6):275–280. https://doi.org/10.1002/1438-5171(200211)3:5/6<275::AID-SIMO275>3.0.CO;2-0
Danhier F, Feron O, Préat V (2010) To exploit the tumor microenvironment: passive and active tumor targeting of nanocarriers for anti-cancer drug delivery. J Control Release 148(2): 135–146. https://doi.org/10.1016/j.jconrel.2010.08.027
Derry MJ, Fielding LA, Armes SP (2016) Polymerization-induced self-assembly of block copolymer nanoparticles via RAFT non-aqueous dispersion polymerization. Prog Polym Sci 52:1–18. https://doi.org/10.1016/j.progpolymsci.2015.10.002
Discher DE, Eisenberg A (2002) Polymer vesicles. Science 297(5583):967–973. https://doi.org/10.1126/science.1074972

Dutertre F, Boyron O, Charleux B, Chassenieux C, Colombani O (2012) Transforming frozen self-assemblies of amphiphilic block copolymers into dynamic pH-sensitive micelles. Macromol Rapid Commun 33(9):753–759. https://doi.org/10.1002/marc.201200078

Elias DR, Poloukhtine A, Popik V, Tsourkas A (2013) Effect of ligand density, receptor density, and nanoparticle size on cell targeting. Nanomedicine 9(2):194–201. https://doi.org/10.1016/j.nano.2012.05.015

Ergin M, Kiskan B, Gacal B, Yagci Y (2007) Thermally curable polystyrene via click chemistry. Macromolecules 40(13):4724–4727. https://doi.org/10.1021/ma070549j

Fairbanks BD, Gunatillake PA, Meagher L (2015) Biomedical applications of polymers derived by reversible addition—fragmentation chain-transfer (RAFT). Adv Drug Deliv Rev 91:141–152. https://doi.org/10.1016/j.addr.2015.05.016

Ferguson CJ, Hughes RJ, Pham BTT, Hawkett BS, Gilbert RG, Serelis AK, Such CH (2002) Effective ab initio emulsion polymerization under RAFT control. Macromolecules 35(25): 9243–9245. https://doi.org/10.1021/ma025626j

Förster S, Antonietti M (1998) Amphiphilic block copolymers in structure-controlled nanomaterial hybrids. Adv Mater 10(3):195–217. https://doi.org/10.1002/(SICI)1521-4095(199802)10:3<195::AID-ADMA195>3.0.CO;2-V

Francis MF, Lavoie L, Winnik FM, Leroux JC (2003) Solubilization of cyclosporin A in dextran-g-polyethyleneglycolalkyl ether polymeric micelles. Eur J Pharm Biopharm 56(3):337–346. https://doi.org/10.1016/s0939-6411(03)00111-5

Gohy JF (2005) Block copolymer micelles. Adv Polym Sci 190:65–136. https://doi.org/10.1007/12_048

Goli KK, Rojas OJ, Genzer J (2012) Formation and antifouling properties of amphiphilic coatings on polypropylene fibers. Biomacromolecules 13(11):3769–3779. https://doi.org/10.1021/bm301223b

Hadgiivanova R, Diamant H, Andelman D (2011) Kinetics of surfactant micellization: a free energy approach. J Phys Chem B 115(22):7268–7280. https://doi.org/10.1021/jp1073335

Hahn J, Webber SE (2003) Graphoepitaxy control of the deposition of cationic polymer micelles on SiO2 surfaces. Langmuir 19(8):3098–3102. https://doi.org/10.1021/la0268911

Hahn J, Webber SE (2004) Graphoepitaxial deposition of cationic polymer micelles on patterned SiO2 surfaces. Langmuir 20(4):1489–1494. https://doi.org/10.1021/la035950n

Halperin A (2011) On micellar exchange: the role of the insertion penalty. Macromolecules 44(13): 5072–5074. https://doi.org/10.1021/ma200811x

Halperin A, Alexander S (1989) Polymeric micelles: their relaxation kinetics. Macromolecules 22(5):2403–2412. https://doi.org/10.1021/ma00195a069

Ham M-K, HoYouk J, Kwon Y, Kwark Y et al (2012) Photoinitiated RAFT polymerization of vinyl acetate. J Poly Sci A Poly Chem 50(12):2389–2397. https://doi.org/10.1002/pola.26014

Hamad I, Hunter AC, Szebeni J, Moghimi SM (2008) Poly(ethylene glycol)s generate complement activation products in human serum through increased alternative pathway turnover and a MASP-2-dependent process. Mol Immunol 46(2):225–232. https://doi.org/10.1016/j.molimm.2008.08.276

Hamley IW (2005) Nanoshells and nanotubes from block copolymers. Soft Matter 1(1):36–43. https://doi.org/10.1039/b418226j

Harada A, Kataoka K (1995) Formation of polyion complex micelles in an aqueous milieu from a pair of oppositely charged block copolymers with poly(ethylene glycol) segments. Macromolecules 28(15):5294–5299. https://doi.org/10.1021/ma00119a019

Harada A, Kataoka K (1999) Chain length recognition: core-shell supramolecular assembly from oppositely charged block copolymers. Science 283(5398):65–67. https://doi.org/10.1126/science.283.5398.65

Hatakeyama H, Akita H, Harashima H (2013) The polyethyleneglycol dilemma: advantage and disadvantage of pegylation of liposomes for systemic genes and nucleic acids delivery to tumors. Biol Pharm Bull 36(6):892–899. https://doi.org/10.1248/bpb.b13-00059

Hayward RC, Pochan DJ (2010) Tailored assemblies of block copolymers in solution: it is all about the process. Macromolecules 43(8):3577–3584. https://doi.org/10.1021/ma9026806
Hickok RS, Wedge SA, Hansen AL, Morris KF, Billiot FH, Warner IM (2002) Pulsed field gradient NMR investigation of solubilization equilibria in amino acid and dipeptide terminated micellar and polymeric surfactant solutions. Magn Reson Chem 40(12):755–761. https://doi.org/10.1002/mrc.1099
Hill RM (2002) In: Meyers RA (ed) Encyclopedia of physical science and technology, 3rd edn. Academic, New York, p 14. p. 40
Huang D, Wang Y, Yang F, Shen H, Weng Z, Wu D (2017) Charge-reversible and pH-responsive biodegradable micelles and vesicles from linear-dendritic supramolecular amphiphiles for anticancer drug delivery. Polym Chem 8(43):6675–6687. https://doi.org/10.1039/C7PY01556A
Inoue T, Chen G, Nakamae K, Hoffman AS (1998) An AB block copolymer of oligo(methyl methacrylate) and poly(acrylic acid) for micellar delivery of hydrophobic drags. J Control Release 51(2–3):221–229. https://doi.org/10.1016/s0168-3659(97)00172-7
Ishii T, Miyata K, Anraku Y, Naito M, Yi Y, Jinbo T, Takae S, Fukusato Y, Hori M, Osada K, Kataoka K (2016) Enhanced target recognition of nanoparticles by cocktail pegylation with chains of varying lengths. Chem Commun 52(7):1517–1519. https://doi.org/10.1039/c5cc06661a
Israelachvili J (2011) Intermolecular and surface forces. Academic, New York. https://doi.org/10.1016/C2011-0-05119-0
Jain S, Bates FS (2004) Consequences of nonergodicity in aqueous binary PEO–PB micellar dispersions. Macromolecules 37(4):1511–1523. https://doi.org/10.1021/ma035467j
Jette KK, Law D, Schmitt EA, Kwon GS (2004) Preparation and drug loading of poly(ethylene glycol)-block-poly(-caprolactone) micelles through the evaporation of a cosolvent azeotrope. Pharm Res 21(7):1184–1191. https://doi.org/10.1023/b:pham.0000033005.25698.9c
Jhaveri AM, Torchilin VP (2014) Multifunctional polymeric micelles for delivery of drugs and siRNA. Front Pharmacol 5:article 77. https://doi.org/10.3389/fphar.2014.00077
Jiang J, Zhang X, Fan Z, Du J (2019) Ring-opening polymerization of N-carboxyanhydride-induced self-assembly for fabricating biodegradable polymer vesicles. ACS Macro Lett 8(10): 1216–1221. https://doi.org/10.1021/acsmacrolett.9b00606
Jindal AB (2017) The effect of particle shape on cellular interaction and drug delivery applications of micro- and nanoparticles. Int J Pharm 532(1):450–465. https://doi.org/10.1016/j.ijpharm.2017.09.028
Joint MHLW/EMA (n.d.) Joint MHLW/EMA reflection paper on the development of block copolymer micelle medicinal products. Ema/CHMP/13099/2013
Jones MC, Leroux JC (1999) Polymeric micelles a new generation of colloidal drug carriers. Eur J Pharm Biopharm 48(2):101–111. https://doi.org/10.1016/s0939-6411(99)00039-9
Kabanov AV, Batrakova EV, Alakhov VY (2002) Pluronic block copolymers as novel polymer therapeutics for drug and gene delivery. J Control Release 82(2–3):189–212. https://doi.org/10.1016/s0168-3659(02)00009-3
Kakizawa Y, Kataoka K (2002) Block copolymer micelles for delivery of gene and related compounds. Adv Drug Deliv Rev 54(2):203–222. https://doi.org/10.1016/s0169-409x(02)00017-0
Kataoka K, Harada A, Nagasaki Y (2001) Block copolymer micelles for drug delivery: design, characterization and biological significance. Adv Drug Deliv Rev 47(1):113–131. https://doi.org/10.1016/s0169-409x(00)00124-1
Kato M, Kamigaito M, Sawamoto M, Higashimura T (1995) Polymerization of methyl-methacrylate with the carbon-tetrachloride dichloro-tris(triphenylphosphine)ruthenium (II). Macromolecules 28(5):1721–1723. https://doi.org/10.1021/ma00109a056
Kita-Tokarczyk K, Grumelard J, Haefele T, Meier W (2005) Block copolymer vesicles—using concepts from polymer chemistry to mimic biomembranes. Polymer 46(11):3540–3563. https://doi.org/10.1016/j.polymer.2005.02.083

Kolb HC, Finn MG, Sharpless KB (2001) Click chemistry: diverse chemical function from a few good reactions. Angew Chem Int Ed Engl 40(11):2004–2021. https://doi.org/10.1002/1521-3773(20010601)40:11<2004::AID-ANIE2004>3.0.CO;2-5

Küçükoğlu E, Acar I, İyim TB, Özgümüş S (2007) A novel type of Si-containing acrylic resins: synthesis, characterization, and film properties. J Appl Polym Sci 104(5):3324–3331. https://doi.org/10.1002/app.25942

Kulthe SS, Choudhari YM, Inamdar NN, Mourya V (2012) Polymeric micelles: authoritative aspects for drug delivery. Des Monomers Polym 15(5):465–521. https://doi.org/10.1080/1385772X.2012.688328

Kumar A (2021) A comprehensive review on polymeric micelles: a promising drug delivery carrier. J Anal Pharm Res 10(3):102–107. https://doi.org/10.15406/japlr.2021.10.00372

Kuperkar K, Patel D, Atanase LI, Bahadur P (2022) Amphiphilic block copolymers: their structures, and self-assembly to polymeric micelles and polymersomes as drug delivery vehicles. Polymers 14(21):4702. https://doi.org/10.3390/polym14214702

Laruelle G, François J, Billon L (2004) Self-assembly in water of poly(acrylic acid)-based diblock copolymers synthesized by nitroxide-mediated polymerization. Macromol Rapid Commun 25(21):1839–1844. https://doi.org/10.1002/marc.200400315

Lazzari M, Liu G, Sébastien L (2006a) Block copolymers in nanoscience. Wiley-VCH, Weinheim. https://doi.org/10.1002/9783527610570

Lazzari M, Liu G, Lecommandoux S (2006b) Block copolymers in nanoscience: block copolymers in nanoscience, 391. Wiley-VCH, Weinheim

Lejeune E, Drechsler M, Jestin J, Müller AHE, Chassenieux C, Colombani O (2010) Amphiphilic diblock copolymers with a moderately hydrophobic block: toward dynamic micelles. Macromolecules 43(6):2667–2671. https://doi.org/10.1021/ma902822g

Liu Y, Lee JY, Kang ET, Wang P, Tan KL (2001) Synthesis, characterization and electrochemical transport properties of the poly(ethyleneglycol)-grafted poly(vinylidenefluoride) nanoporous membranes. React Funct Polym 47(3):201–213. https://doi.org/10.1016/S1381-5148(01)00030-X

Liu J, Weller GE, Zern B, Ayyaswamy PS, Eckmann DM, Muzykantov VR, Radhakrishnan R (2010) Computational model for nanocarrier binding to endothelium validated using in vivo, in vitro, and atomic force microscopy experiments. Proc Natl Acad Sci U S A 107(38):16530–16535. https://doi.org/10.1073/pnas.1006611107

Low PS, Henne WA, Doorneweerd DD (2008) Discovery and development of folic-acid-based receptor targeting for imaging and therapy of cancer and inflammatory diseases. Acc Chem Res 41(1):120–129. https://doi.org/10.1021/ar7000815

Lund R, Willner L, Richter D, Dormidontova EE (2006) Equilibrium chain exchange kinetics of diblock copolymer micelles: tuning and logarithmic relaxation. Macromolecules 39(13):4566–4575. https://doi.org/10.1021/ma060328y

Mai Y, Eisenberg A (2012) Self-assembly of block copolymers. Chem Soc Rev 41(18):5969–5985. https://doi.org/10.1039/c2cs35115c

Malmsten M (2002) Micelles. In: Surfactants and polymers in drug delivery. Marcel Dekker, New York, pp 19–50. https://doi.org/10.1201/9780824743758.ch2

Miura Y, Takenaka T, Toh K, Wu S, Nishihara H, Kano MR, Ino Y, Nomoto T, Matsumoto Y, Koyama H, Cabral H, Nishiyama N, Kataoka K (2013) Cyclic RGD-linked polymeric micelles for targeted delivery of platinum anticancer drugs to glioblastoma through the blood–brain tumor barrier. ACS Nano 7(10):8583–8592. https://doi.org/10.1021/nn402662d

Miyamoto M, Sawamoto M, Higashimura T (1984) Living polymerization of isobutyl vinyl ether with hydrogen iodide/iodine initiating system. Macromolecules 17(3):265–268. https://doi.org/10.1021/ma00133a001

Moad G, Rizzardo E, Thang SH (2005) Living radical polymerization by the RAFT process. Aust J Chem 58(6):379–410. https://doi.org/10.1071/CH05072

Moad G, Rizzardo E, Thang SH (2012) Living radical polymerization by the RAFT process—a third update. Aust J Chem 65(8):985–1076. https://doi.org/10.1071/CH12295

Moroi Y (1992) Dissolution of amphiphiles. In: Moroi Y (ed) Micelles. Plenum Press, New York, pp 25–40

Mühlebach A, Gaynor SG, Matyjaszewski K (1998) Synthesis of amphiphilic block copolymers by atom transfer radical polymerization (ATRP). Macromolecules 31(18):6046–6052. https://doi.org/10.1021/ma9804747

Müller J, Sönnichsen C, von Poschinger H, von Plessen G, Klar TA, Feldmann J (2002) Electrically controlled light scattering with single metal nanoparticles. Appl Phys Lett 81(1):171–173. https://doi.org/10.1063/1.1491003

Nicolai T, Colombani O, Chassenieux C (2010) Dynamic polymeric micelles versus frozen nanoparticles formed by block copolymers. Soft Matter 6(14):3111–3118. https://doi.org/10.1039/b925666k

Nyrkova IA, Semenov AN (2005) On the theory of micellization kinetics. Macromol Theory Simulat 14(9):569–585. https://doi.org/10.1002/mats.200500010

Patterson JP, Robin MP, Chassenieux C, Colombani O, O'Reilly RK (2014) The analysis of solution self-assembled polymeric nanomaterials. Chem Soc Rev 43(8):2412–2425. https://doi.org/10.1039/c3cs60454c

Pellice SA, Fasce DP, Williams RJJ (2007) Development of a bilayer coating to improve the adhesion between stainless steel and in situ-polymerized poly(methyl methacrylate). J Appl Polym Sci 105(4):2351–2356. https://doi.org/10.1002/app.26339

Penfold NJW, Yeow J, Boyer C, Armes SP (2019) Emerging trends in polymerization-induced self-assembly. ACS Macro Lett 8(8):1029–1054. https://doi.org/10.1021/acsmacrolett.9b00464

Pike DQ, Müller M, De Pablo JJ (2011) Monte-Carlo simulation of ternary blends of block copolymers and homopolymers. J Chem Phys 135(11):114904. https://doi.org/10.1063/1.3638175

Rangel-Yagui CO, Pessoa A, Tavares LC (2005) Micellar solubilization of drugs. J Pharm Pharm Sci 8(2):147–165

Raveendran R, Mullen KM, Wellard RM, Sharma CP, Hoogenboom R, Dargaville TR (2017) Poly (2-oxazoline) block copolymer nanoparticles for curcumin loading and delivery to cancer cells. Eur Polym J 93:682–694. https://doi.org/10.1016/j.eurpolymj.2017.02.043

Remant Bahadur KC, Bhattarai SR, Aryal S, Khil MS, Dharmaraj N, Kim HY (2007) Novel amphiphilic triblock copolymer based on PPDO, PCL, and PEG: synthesis, characterization, and aqueous dispersion. Colloids Surf A Physicochem Eng Asp 292(1):69–78. https://doi.org/10.1016/j.colsurfa.2006.06.009

Rezaei SJT, Nabid MR, Niknejad H, Entezami AA (2012) Folate-decorated thermoresponsive micelles based on star-shaped amphiphilic block copolymers for efficient intracellular release of anticancer drugs. Int J Pharm 437(1–2):70–79. https://doi.org/10.1016/j.ijpharm.2012.07.069

Rijcken CJF, Soga O, Hennink WE, van Nostrum CFV (2007) Triggered destabilisation of polymeric micelles and vesicles by changing polymers polarity: an attractive tool for drug delivery. J Control Release 120(3):131–148. https://doi.org/10.1016/j.jconrel.2007.03.023

Rizzardo E, Solomon DH (2012) On the origins of nitroxide mediated polymerization (NMP) and reversible addition-fragmentation chain transfer (RAFT). Aust J Chem 65(8):945–969. https://doi.org/10.1071/CH12194

Rostovtsev VV, Green LG, Fokin VV, Sharpless KB (2002) A stepwise huisgen cycloaddition process: copper(I)-catalyzed regioselective ligation of azides and terminal alkynes. Angew Chem Int Ed Engl 41(14):2596–2599. https://doi.org/10.1002/1521-3773(20020715)41:14<2596::AID-ANIE2596>3.0.CO;2-4

Ruzette AV, Leibler L (2005) Block copolymers in tomorrow's plastics. Nat Mater 4(1):19–31. https://doi.org/10.1038/nmat1295

Salatin S, Maleki Dizaj S, Yari Khosroushahi A (2015) Effect of the surface modification, size, and shape on cellular uptake of nanoparticles. Cell Biol Int 39(8):881–890. https://doi.org/10.1002/cbin.10459

Schrock RR (1990) Living ring-opening metathesis polymerization catalyzed by well-characterized transition-metal alkylidene complexes. Acc Chem Res 23(5):158–165. https://doi.org/10.1021/ar00173a007

Segalman RA, Hexemer A, Kramer EJ (2003a) Edge effects on the order and freezing of a 2D array of block copolymer spheres. Phys Rev Lett 91(19):196101. https://doi.org/10.1103/PhysRevLett.91.196101

Segalman RA, Hexemer A, Hayward RC, Kramer EJ (2003b) Ordering and melting of block copolymer spherical domains in 2 and 3 dimensions. Macromolecules 36(9):3272–3288. https://doi.org/10.1021/ma021367m

Shedge A, Colombani O, Nicolai T, Chassenieux C (2014) Charge dependent dynamics of transient networks and hydrogels formed by self-assembled pH-sensitive triblock copolyelectrolytes. Macromolecules 47(7):2439–2444. https://doi.org/10.1021/ma500318m

Shen X, Cao S, Zhang Q, Zhang J, Wang J, Ye Z (2019) Amphiphilic TEMPO-functionalized block copolymers: synthesis, self-assembly and redox-responsive disassembly behavior, and potential application in triggered drug delivery. ACS Appl Poly Mater 1(9):2282–2290. https://doi.org/10.1021/acsapm.9b00293

Shipway AN, Katz E, Willner I (2000) Nanoparticle arrays on surfaces for electronic, optical, and sensor applications. ChemPhysChem 1(1):18–52. https://doi.org/10.1002/1439-7641(20000804)1:1<18::AID-CPHC18>3.0.CO;2-L

Smart T, Lomas H, Massignani M, Flores-Merino MV, Perez LR, Battaglia G (2008) Block copolymer nanostructures. Nano Today 3(3–4):38–46. https://doi.org/10.1016/S1748-0132(08)70043-4

Solomatin SV, Bronich TK, Bargar TW, Eisenberg A, Kabanov VA, Kabanov AV (2003) Environmentally responsive nanoparticles from block ionomer complexes: effects of pH and ionic strength. Langmuir 19(19):8069–8076. https://doi.org/10.1021/la030015l

Spatz JP, Moessmer S, Moeller M, Herzog T, Plettl A, Ziemann P (1998) Functional nanostructures by organized macromolecular-metallic hybrid systems. J Lumin 76–77:168–173. https://doi.org/10.1016/S0022-2313(97)00144-0

Stejskal J, Hlavatá D, Sikora A, Konňák Č, Pleštil J, Kratochvíl P (1992) Equilibrium and non-equilibrium copolymer micelles: polystyrene-block-poly (ethylene-co-propylene) in decane and in diisopropylether. Polymer 33(17):3675–3685. https://doi.org/10.1016/0032-3861(92)90655-G

Tan J, Rao X, Yang J, Zeng Z (2015) Monodisperse highly cross-linked "living" microspheres prepared via photoinitiated RAFT dispersion polymerization. RSC Adv 5(24):18922–18931. https://doi.org/10.1039/C4RA15224G

Torchilin VP (2006) Multifunctional nanocarriers. Adv Drug Deliv Rev 58:1532–1555. https://doi.org/10.1016/j.addr.2006.09.009

Trubetskoy VS, Torchilin VP (1996) Polyethylene glycol-based micelles as carriers for delivery of therapeutic and diagnostic agents. STP Pharma Sci 6:79–86

Tsitsilianis C, Gotzamanis G, Iatridi Z (2011) Design of "smart" segmented polymers by incorporating random copolymers as building blocks. Eur Polym J 47(4):497–510. https://doi.org/10.1016/j.eurpolymj.2010.10.005

von Gottberg FK, Smith KA, Hatton TA (1998) Dynamics of self-assembled surfactant systems. J Chem Phys 108(5):2232–2244. https://doi.org/10.1063/1.475604

Wang P, Tan KL, Kang ET (2000) Surface modification of poly(tetrafluoroethylene) films via grafting of poly(ethylene glycol) for reduction in protein adsorption. J Biomater Sci Polym Ed 11(2):169–186. https://doi.org/10.1163/156856200743634

Wang SF, Lu LC, Gruetzmacher JA, Currier BL, Yaszemski MJ (2005) A biodegradable and cross-linkable multiblock copolymer consisting of poly(propylene fumarate) and poly(e-caprolactone): synthesis, characterization, and physical properties. Macromolecules 38(17):7358–7370. https://doi.org/10.1021/ma050884c

Wiley DT, Webster P, Gale A, Davis ME (2013) Transcytosis and brain uptake of transferrin-containing nanoparticles by tuning avidity to transferrin receptor. Proc Natl Acad Sci U S A 110(21):8662–8667. https://doi.org/10.1073/pnas.1307152110

Wright DB, Patterson JP, Pitto-Barry A, Cotanda P, Chassenieux C, Colombani O, O'Reilly RK (2015) Tuning the aggregation behavior of pH-responsive micelles by copolymerization. Polym Chem 6(14):2761–2768. https://doi.org/10.1039/C4PY01782J

Wright DB, Patterson JP, Gianneschi NC, Chassenieux C, Colombani O, O'Reilly RK (2016) Blending block copolymer micelles in solution; obstacles of blending. Polym Chem 7(8): 1577–1583. https://doi.org/10.1039/C5PY02006A

Wright DB, Thompson MP, Touve MA, Carlini AS, Gianneschi NC (2019) Enzyme-responsive polymer nanoparticles via ring-opening metathesis polymerization-induced self-assembly. Macromol Rapid Commun 40(2):e1800467. https://doi.org/10.1002/marc.201800467

Xu J, Bohnsack DA, Mackay ME, Wooley KL (2007) Unusual mechanical performance of amphiphilic crosslinked polymer networks. J Am Chem Soc 129(3):506–507. https://doi.org/10.1021/ja067986i

Xu F, Li H, Luo YL, Tang W (2017) Redox-responsive self-assembly micelles from poly (N-acryloylmorpholine-block-2-acryloyloxyethyl ferrocenecarboxylate) amphiphilic block copolymers as drug release carriers. ACS Appl Mater Interfaces 9(6):5181–5192. https://doi.org/10.1021/acsami.6b16017

Yadav S, Sharma AK, Kumar P (2020) Nanoscale self-assembly for therapeutic delivery. Front Bioeng Biotechnol 8:127. https://doi.org/10.3389/fbioe.2020.00127

Yokoyama M (2011) Clinical applications of polymeric micelle carrier systems in chemotherapy and image diagnosis of solid tumors. J Exp Clin Med 3(4):151–158. https://doi.org/10.1016/j.jecm.2011.06.002

Yokoyama H, Mates TE, Kramer EJ (2000) Structure of asymmetric diblock copolymers in thin films. Macromolecules 33(5):1888–1898. https://doi.org/10.1021/ma9912047

Yoncheva K, Calleja P, Agüeros M, Petrov P, Miladinova I, Tsvetanov C, Irache JM (2012) Stabilized micelles as delivery vehicles for paclitaxel. Int J Pharm 436(1–2):258–264. https://doi.org/10.1016/j.ijpharm.2012.06.030

Yu M, Huang S, Yu KJ, Clyne AM (2012) Dextran and polymer polyethylene glycol (PEG) coating reduce both 5 and 30 nm iron oxide nanoparticle cytotoxicity in 2D and 3-D cell culture. Int J Mol Sci 13(5):5554–5570. https://doi.org/10.3390/ijms13055554

Yue J, Li X, Mo G, Wang R, Huang Y, Jing X (2010) Modular functionalization of amphiphilic block copolymers via radical-mediated thiol-ene reaction. Macromolecules 43(23):9645–9654. https://doi.org/10.1021/ma101960d

Yun J, Faust R, Szilágyi LS, Kéki S, Zsuga M (2003) Effect of architecture on the micellar properties of amphiphilic block copolymers: comparison of AB linear diblock, A1A2B, and A2B heteroarm star block copolymers. Macromolecules 36(5):1717–1723. https://doi.org/10.1021/ma021618r

Zana R (2005) Introduction to surfactants and surfactant self-assemblies. In: Zana R (ed) Dynamics of surfactant self-assemblies micelles, microemulsions, vesicles, and lyotropic phases. Taylor & Francis, Philadelphia, PA, pp 1–35. https://doi.org/10.1201/9781420028225.ch1

Zhang L, Shen H, Eisenberg A (1997) Phase separation behavior and crew-cut micelle formation of polystyrene-b-poly(acrylic acid) copolymers in solutions. Macromolecules 30(4):1001–1011. https://doi.org/10.1021/ma961413g

Zhao XK, McCormick LD (1992) Oriented crystal particles of semiconductor PbS on Langmuir monolayer surfaces. Appl Phys Lett 61(7):849–851. https://doi.org/10.1063/1.107765

Zhao Y, Wang Y, Ran F, Cui Y, Liu C, Zhao Q, Gao Y, Wang D, Wang S (2017a) A comparison between sphere and rod nanoparticles regarding their in vivo biological behavior and pharmacokinetics. Sci Rep 7:4131

Zhao J, Lu H, Wong S, Lu M, Xiao P, Stenzel MH (2017b) Influence of nanoparticle shapes on cellular uptake of paclitaxel loaded nanoparticles in 2D and 3-D cancer models. Polym Chem 8(21):3317–3326. https://doi.org/10.1039/C7PY00385D

Advances in Polymer Optimization for Enhanced Drug Delivery

2

Mayank Kashyap, Deepanshi Pathak, Shivani Gupta, Aanawi Tiwari, Mansi Sharma, Debdarshan Dutta, Radhika Sharma, Saritha Shetty, and R. Rachana

Abstract

Nanocarriers are being used to deliver those active pharmaceutical ingredients (APIs) to the site of action that are toxic, biodegradable, and have less bioavailability. One type of nanocarrier that can meet the aforementioned criteria in the form of therapeutic formulations is polymeric micelles. They are revolutionary drug delivery nanocarriers that can be developed using various types of polymers to achieve specific features such as target specificity and controlled release for the drugs. There are different categories of polymers used to prepare such micelles such as graft polymers, diblock polymers, and triblock polymers. Diblock polymers such as polymer made up of poly(ε-caprolactone) [PCL] and poly (ethylene glycol) [PEG]; graft copolymers such as polymer made up of gelatin and hyaluronic acid; and triblock copolymers such as polymer made of polyethylene oxide PEO-b-PB-b-PEO (PB, polybutadiene; PEO, polyethylene oxide) are being employed in designing and developing improved polymeric micelles which may respond better in the harsh inter/intracellular environments, for example: in changed ionic concentrations, in presence of several proteins and enzymes, pH changes and even at higher temperatures. In the present chapter, various types of polymers, which are used to prepare conventional, mucoadhesive, pH-sensitive and temperature-sensitive polymeric micelles, are discussed. The major focus of this chapter is to review various polymers used worldwide, which are used to

M. Kashyap · D. Pathak · S. Gupta · A. Tiwari · M. Sharma · D. Dutta · R. Sharma · R. Rachana (✉)
Department of Biotechnology, JIIT Noida, Noida, UP, India
e-mail: rachana.dr@iitbombay.org

S. Shetty
Shobhaben Pratapbhai Patel School of Pharmacy and Technology Management, SVKM's Narsee Monjee Institute of Management Studies (Deemed to be University), Mumbai, India

S. K. Singh et al. (eds.), *Polymeric Micelles: Principles, Perspectives and Practices*,
https://doi.org/10.1007/978-981-99-0361-0_2

obtain specialized micelles for specific applications, with desired characteristics for drug formulations.

Keywords

Micelles · pH-sensitive polymeric micelles · Temperature-sensitive polymers · Targeted drug delivery · Conventional polymeric micelles · Mucoadhesive polymeric micelles · pH-sensitive polymeric micelles · Temperature-sensitive polymeric micelles

2.1 Introduction

Getting a drug delivered at the desirable site involves intricate analysis about the nature of the drug, its carriers, and all other various factors that may affect its bioavailability. Numerous studies have been conducted and multiple advancements have been made in this research area of drug delivery. One of the strategies utilized to overcome the hurdles mentioned above is choosing an appropriate carrier such as nanoparticles, liposomes, micelles, or microemulsions (Natarajan et al. 2014). The drug carriers are chosen on the basis of their hydrophilicity and lipophilicity of the material used, its interactions with API (Active Pharmaceutical Ingredient) and its release, degradative pattern at the target, etc. In the selection of drug carriers, the nature of API and its pharmacokinetic (PK) play a major role (Irvine and Dane 2020). Sometimes, if a drug is hydrophobic in nature and possesses low molecular mass, it gets rapidly circulated in the body, due to large volume distribution and so does not reach the target site in desired concentrations. Furthermore, its increased exposure and localization in other healthy tissues result in various side effects and toxicity. In such cases, unfavorable solubility is the main obstacle to be handled to get the above complexities resolved (Zuegg and Cooper 2012).

At present, polymeric micelles (PMs) (Fig. 2.1) are among the most potent drug delivery carriers which have been proven to overcome a number of aforementioned difficulties. They are tiny, colloidal dispersions having a diameter in nanorange. Their diameter is determined by the head group's structure of the molecule used and the accompanying alkyl chain's length (Hanafy et al. 2018). These molecules are mostly amphiphilic copolymers, i.e., they have a hydrophilic tail to withstand the aqueous environment and a hydrophobic head for drug encapsulation (Fig. 2.1). This improves their capacity to load the drug and the stability, bioavailability, and longevity of the drug molecule in the circulation (Williams et al. 2013).

A major problem associated with micelles is that, when they get diluted (in the body fluids), they become unstable and release the drug prematurely. The most successful strategy to avoid such incidents is choosing the right material for the micelles' preparation. Scientists all over the world are currently working with various "polymers-based drug carriers" to provide them with unique properties. The present manuscript discusses various polymers used to prepare polymeric micelles to analyze their importance in drug delivery (Liu et al. 2019).

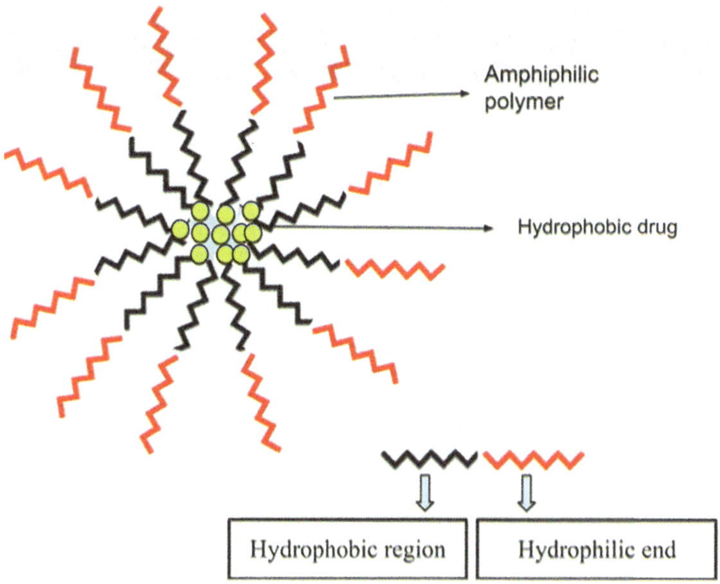

Fig. 2.1 Standard structure of a polymeric micelle

2.2 Types of Polymers Used

For the preparation of these micelles, three categories of polymers, such as diblock copolymer, triblock copolymer, and graft copolymer, are used (Torchilin 2006). Among diblock and triblock copolymers, hydrophilic blocks of polyethylene oxide or PEG polymers are most commonly used. Amphiphilic diblock or triblock copolymers are found to form spherical micelles in aqueous media, depending on the length of hydrophilic (longer) and hydrophobic (smaller) blocks. The length of the hydrophilic and hydrophobic polymeric chains affects the shape and size of the copolymers formed and that can exist in water as either single molecules (unimers) or in other diverse shapes. If the chain of hydrophilic blocks is too long, copolymers exit as single molecules, whereas, if the hydrophobic blocks are long, they take on diverse shapes (Kulthe et al. 2012).

2.2.1 Diblock Copolymer

Diblock copolymers are composed of two different types of monomers (e.g., A and B). They are created by attaching monomers in such a way (Fig. 2.2) that the chains made up of each monomer join to form a single polymer chain (Sharma et al. 2014). When these copolymers are assembled in huge amounts, a polymeric structure

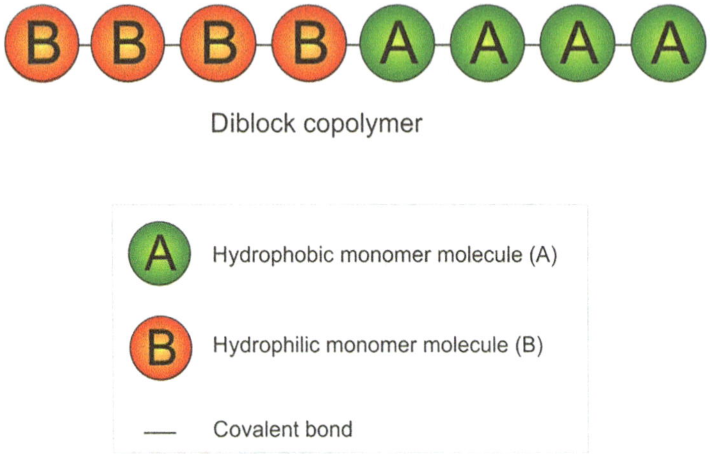

Fig. 2.2 Diblock copolymer assembly

known as a polymeric melt is formed, which further melts at a particular temperature called the transition temperature (Matsen 2020).

Letchford et al. (2008) have studied water-soluble diblock copolymers MePEG-b-PCL [Methoxy poly(ethylene glycol)-block-poly(ε-caprolactone)] with different MePEG and PCL block lengths. The length of the PCL block was found to be determined by the ratio of the weight of caprolactone to the reaction mix.

Further, chlorin-cored, 4-armed star-shaped mPEG-b-PCL diblock copolymer micelles were made by Peng et al. (2008). It was found to get self-assembled into nanoscaled micellar form and acted as photosensitizing compounds for PDT (Photodynamic Therapy) and also encapsulated hydrophobic therapeutics like paclitaxel. These micelles could significantly improve paclitaxel cytotoxicity for MCF-7 cells after irradiation.

This theory further suggests that the micelle-based delivery system can be used as a potential dual carrier for the symbiotic association of PDT and cancer treatment via chemotherapy (Peng et al. 2008).

2.2.2 Graft Copolymers

Graft copolymers are made up of one or more homopolymer blocks grafted onto the main chain made up of another polymer as branches (Fig. 2.3). One to several side chains may be present on the main chain of the polymer used. When these polymers are dissolved in a certain solvent, they either form a unimolecular or a multimolecular micelle (Williams et al. 2016).

The backbone length, arm length, and grafting density of these polymers can be changed systematically to alter the quality of the micelles (Cheng et al. 2012a, b; Williams et al. 2013). These aids can produce a wide range of nanostructures. The

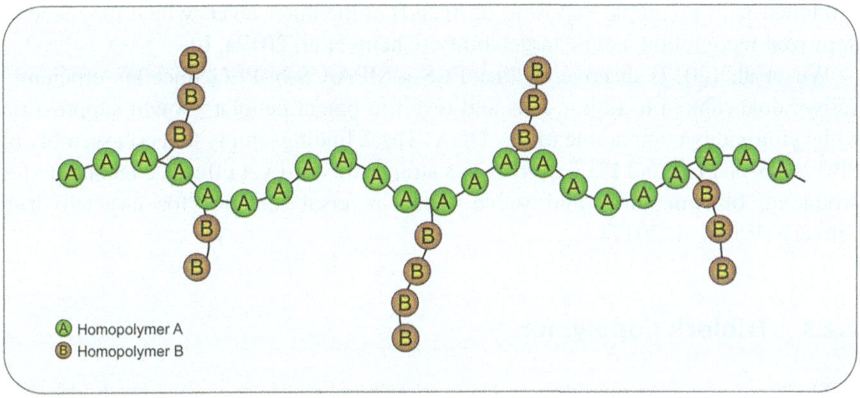

Fig. 2.3 Graft copolymer showing a linear chain of hydrophobic A polymer and branched chains of hydrophilic B polymer

hydrophobic backbone of linear graft copolymers, which is prone to collapse, is protected from adverse solvent interactions by the hydrophilic side arms, forming a core-shell structure (Achilleos and Vamvakaki 2010).

A novel and distinctive type of polymer called a cyclic graft polymer has a linear grafted side chain and cyclic backbone. Since cyclic polymers have a more constrained conformation than equivalent linear polymers, they have smaller hydrodynamic volumes and gyratory radii (Morgese et al. 2018). Cyclic polymers have recently been found to be better than the linear polymers when utilized as potential delivery systems for the delivery of genes or drugs.

Steric hindrance from the grafted side chains regulates their physical characteristics and conformations. According to research by Cortez et al. (2015), cyclic poly(ethylene imine) has a higher transfection effectiveness than linear poly (ethylene imine). Further investigation revealed that cyclic polymer-based micelles are more trustworthy for drug administration because they have higher ionic strength and durability than linear polymers (Cortez et al. 2015).

Cheng et al. (2012a, b) could prepare biodegradable micelles from graft copolymer PCL-g-PHEMA [poly(3-caprolactone)-g-poly(2-hydroxyethyl methacrylate)] in order to release DOX (doxorubicin) effectively, in cancer cells. The micelles provided numerous distinctive advantages as a drug carrier as their backbone length is unrestricted. The characteristics of the micelle prepared, such as surface properties, drug permeability, and size, were elegantly controlled by working on the density of graft and length of graft as well as of the backbone. These polymeric micelles are stable in nature and possess lower CMC than block copolymer micelles. In contrast to usual biodegradable block copolymer micelles, they could demonstrate a very rapid release of DOX in response to a mild acidic pH that replicates the environment of endo/lysosomal compartments, attaining enhanced antitumor action. It was also noted that the existence of several hydrophilic grafts per macromolecule promotes the recombination of high-density targeting ligands when they are combined with favorable exposure of ligands at the outer layer, which may lead to improved recognition and/or targetability (Cheng et al. 2012a, b).

Wu et al. (2017) discovered that PBS-g-MPA/CS-bi-PEG micelles efficiently deliver doxorubicin to tumor cells and result in potent cellular growth suppression with cytotoxicity comparable to free DOX. These findings imply that co-assembly of PBS-g-CS-bi-PEG and PBS-g-MPA is a straightforward yet effective technique for producing biocompatible and stable charge-reversal micelles for targeted drug delivery (Wu et al. 2017).

2.2.3 Triblock Copolymer

Triblock polymers are prepared by using three blocks of homopolymers (hydrophobic and hydrophilic polymers) to form three-block copolymers. Depending on their structure and architecture, there are two types of triblock copolymers: symmetric and asymmetric triblock copolymers. In addition, symmetric triblock copolymers can be classified as either ABA triblock copolymers or BAB triblock copolymers depending on their structure (Fig. 2.4). Two different types of polymers combine to form symmetric triblock copolymers: A (a hydrophilic polymer, for reference) and B (a hydrophobic polymer, for reference). On the other hand, asymmetric triblock copolymers are made up of three distinct polymer types, A, B, and C, respectively.

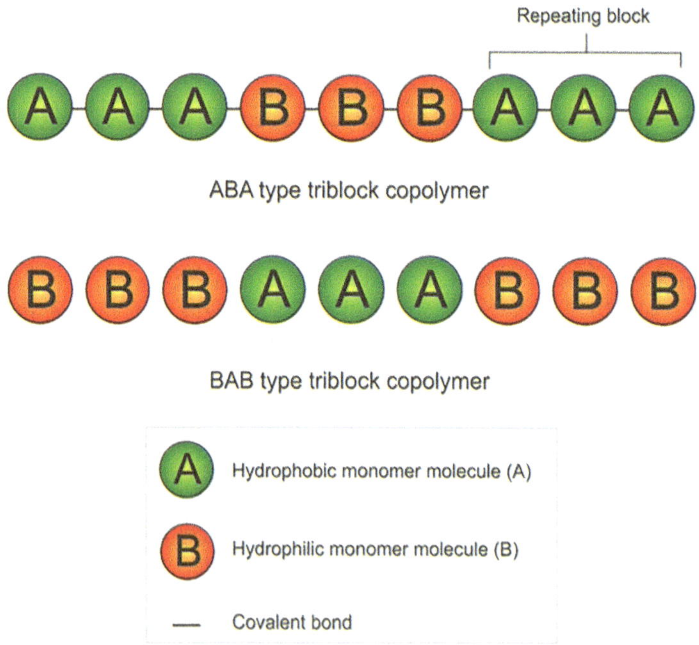

Fig. 2.4 Symmetric triblock copolymer

Asymmetric triblock polymers consist of three polymers, two of which are hydrophobic and one of which is hydrophilic. As a result, they are also known as ABC triblock copolymers. Two of the three blocks, either A, B, or C, depending on how polymers are used, could either be hydrophobic or hydrophilic (Hoang et al. 2017).

According to Jeong et al. (2000), the speed with which drugs are released from these types of drug delivery systems (DDS) is determined by the characteristics of the drug, the drug-core interaction, and the amount of drug placed in the core (Jeong et al. 2000). Guo et al. (2010) prepared a DDS where superparamagnetic iron oxide nanoparticles (SPION) and a hydrophobic drug were successfully combined with the micelles made of PEG-poly(diethylaminoethyl methacrylate)-poly(glycerol monomethacrylate) (PEG-PDEAEMA) or PEG-PDMAEMA-poly(glycerol monomethacrylate) (PEG-PDMAEMA-PGMA) (Guo et al. 2010).

2.2.3.1 Symmetric Triblock Copolymers for Micelle Preparation

Micelles with the aforementioned features can be made from either BAB types of symmetric polymers (where B indicates hydrophilic polymer and A specifies triblock copolymers) or ABA types of symmetric polymers (Fig. 2.4), wherein there is a hydrophobic polymer A and a hydrophilic polymer B. BAB triblock copolymers can produce micelles with a two-block hydrophilic shell and an A block-based hydrophobic core. On the other hand, ABA copolymers can produce a micelle that resembles a flower by foliating the B block and forming a foliate shell around it (Hoang et al. 2017).

Ortel and his colleagues published a series of symmetric PEO-b-PB-b-PEO (PB, polybutadiene; PEO, polyethylene oxide) triblock copolymers in 2012 that were based on pluronic-type triblock copolymers. Low CMC for these polymers was obtained by a large hydrophilic-hydrophobic contrast between the PB and PEO blocks (Ortel et al. 2012).

2.2.3.2 Asymmetric Triblock Copolymers for Micelle Preparation

The hydrophilic-hydrophobic ratio of an ABC copolymer can be changed to change the DDS properties of asymmetric triblock copolymers, just as symmetric triblock copolymers (Endres et al. 2011; Liu et al. 2002; Tang et al. 2003). The most frequent building block for surface layers in asymmetric copolymers is PEG (Jee et al. 2012).

A functional block can be added in place of the third block to help generate a variety of nanostructures, including environmentally sensitive polymer vesicles, micelles resembling flowers, and star-shaped micelles (Endres et al. 2011; Guo et al. 2010). The hydrophilic (A block for reference) and hydrophobic (B block for reference) portions of the ABC triblock copolymer can be used to create the star-shaped micelles (B block and C block for reference) (Jee et al. 2012; Kubowicz et al. 2005). Each block of the ABC triblock copolymers, which form micelles that resemble stars, has a specific feature, such as being pH or thermally sensitive (Lim et al. 2016; Sugihara et al. 2004). Nevertheless, the micelles prepared using this method are similar to those formed from BAB copolymers. Even star-like micelles made of ABC triblock copolymers are utilized to load various sorts of medicines using various block configurations (Endres et al. 2011). It was suggested that one can employ ABC triblock copolymers, in which A and C are hydrophobic blocks and B is the hydrophilic block, to create a flower-like micelle (Van Butsele et al. 2009).

Using a specific kind of asymmetric triblock copolymer poly(styrene-acrylic acid-ethylene glycol) (PS-PAA-PEG), Bastakoti et al. (2013) created multifunctional core-shell-corona polymeric micelles. The micelles formed by PS-PAA-PEG self-assembly in aqueous solutions were found to have an anionic PAA shell and a PS core and hydrophilic and neutral PEG corona. The PS core was loaded with the phenoxazine dye Nile Red (NR) using the dialysis technique. The presence of many carboxylic groups on the pH-sensitive PAA shell increased the interaction of the drug with the micelles, which allowed a rapid release of cisplatin at pH 5.0 (slightly acidic) but not at physiological pH. The PAA shell was selectively calcified by the calcium phosphate (CaP), which further boosted the fluorescence intensity of Nile Red, improving detection sensitivity and efficacy. It also acted as a barrier for diffusion, causing controlled release of the drug cisplatin. Micelles could not aggregate because of the steric protection offered by the PEG shell. HepG2 cells absorbed the polymeric NR micelles and localized them in their cytoplasm and nucleus. They were nontoxic, exhibited 90% viability at polymer concentrations up to 500 μg/mL, and showed high biocompatibility. A dose-dependent cytotoxic effect was observed for the cisplatin-loaded micelles on HepG2 cells (Bastakoti et al. 2013).

According to research by Epps et al. (2004) addition of a third block into the network phases of PI-b-PS-b-PEO (PI, polyisoprene) enhanced the network phases of triblock copolymer (ABC) as opposed to diblock copolymer AB (Epps et al. 2004). Meng et al. (2006) used PS-b-PMMA-b-PEO [PMMA, poly(methyl methacrylate)] as a template to create ordered mesoporous carbon from an asymmetric triblock copolymer (Meng et al. 2006).

2.3 Role of Various Polymers in Achieving Target-Specific Drug Release

PMs have been criticized for their drawback they carry, releasing the drug prematurely into the bloodstream; this also reduces the amount of drug at the targeted location. Thus, to avoid early drug release and to reduce its interaction of the drug with nontargeted tissues during circulation, ideally the drug should continue to be integrated into PMs, while in circulation. This can be obtained by stabilizing the micelle or by enhancing the interactions of the medicines in the core of the micelles by hydrogen bonds. The construction of stimuli-responsive PMs is a very efficient method for ensuring tissue-targeted medication administration and sufficient release (Zhou et al. 2018). Numerous different triggers have been utilized to deliver the drug to the target site, as the diseased sites mostly have a different physiological state such as higher temperature, altered redox potential, upregulated enzyme or pH (Lee and Thompson 2017).

Sometimes such stimuli are not available at diseased sites; in those cases, external triggers have been utilized for the drug release at the site of action such as light,

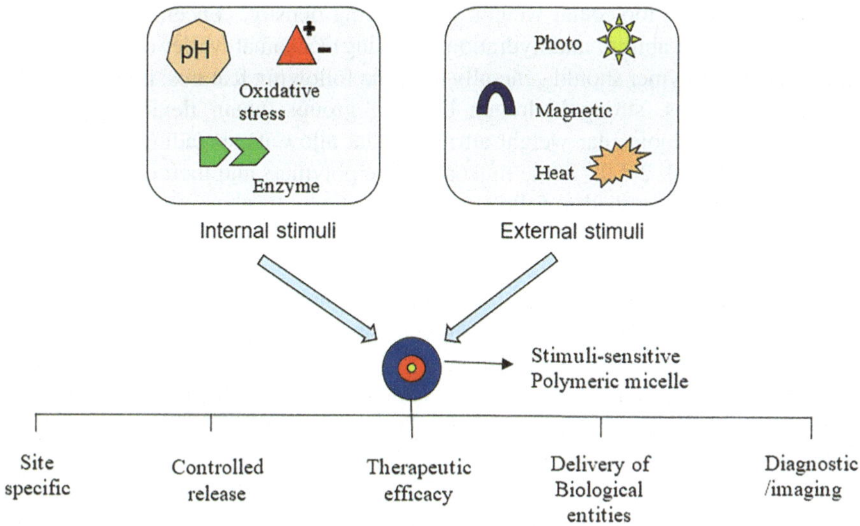

Fig. 2.5 Stimuli-sensitive polymeric micelles

temperature, and magnetic field (Fig. 2.5) for diagnostic and therapeutic purposes (Oerlemans et al. 2010). These PMs, which are referred to as "smart" nanocarriers, are utilized to transport drugs and/or imaging agents for the diseased sites. In the following section polymers used in such customized micelles are discussed.

Figure 2.5 explains how a change in the environment can trigger the release of API at the site of action.

2.3.1 Polymers Used for the Micelles Sensitive to Internal Triggers

The rate at which pharmaceuticals are released from micelles is influenced by the partition coefficient, micelle stability, drug diffusion rate, and copolymer biodegradation rate (Oerlemans et al. 2010). Many of these features are affected/influenced by the physiological parameters, which might differ at the site of action in comparison to the normal tissues. The concentration of the drug in the micelles, the hydrophobic polymer's length, its molecular weight, its physicochemical properties, and its placement within the micelles are some other parameters that also affect the release (Ghezzi et al. 2021). By manipulating the composition of polymers, the desired properties of micelles can be obtained. The following section discusses different polymers used to prepare micelles which can stimulate the drug release by internal triggers.

2.3.1.1 Polymers Used to Prepare Mucoadhesive Micelles

Mucoadhesive polymeric micelles are employed to deliver medication to mucosal cavities. The mucoadhesive qualities of polymers are influenced by various elements, including molecular weight, crosslinking density, concentration, charge, hydrogen bonding ability, and hydration (swelling) (Salamat-Miller et al. 2005). A mucoadhesive polymer should generally have the following features: potent cationic or anionic charges, strong hydrogen bonding groups, chain flexibility, surface energy, and high molecular weight attributes that allow its spreading on the layer of mucus (Lee et al. 2000). Some mucoadhesive polymers and their contribution to drug delivery are described as follows.

Chitosan

Chitosan is a polysaccharide made from chitin. Chitin has a molecular mass ranging from 300 to 1000 kDa, and it depends upon its source. It is a copolymer of D-glucosamine and N-acetyl-D-glucosamine. Chitosan is a linear, semi-crystalline polymer with at least 60% of the glucosamine residues, which are deacetylated (this equals a 60% deacetylation degree). Chitin can be deacetylated chemically in highly alkaline circumstances or enzymatically in the presence of certain enzymes such as chitin deacetylase. Chitosan is a polysaccharide that has numerous applications as it has various beneficial properties including antibacterial activity, adaptability, non-toxicity, and biodegradability (Gonçalves et al. 2021).

Despite being well known for its mucoadhesive qualities, chitosan is only effective under certain circumstances. It has weak mucoadhesive bonds and low water solubility at neutral and basic pHs. In order to overcome these drawbacks, chitosan has been experimented and has been chemically modified in several ways.

One of its derivatives is glycol chitosan, where the hydrophilic (-OH and $-NH_2$) are found inside the micelles and hydrophobic groups are at the outer shells. This enables glycol chitosan's mucoadhesive characteristics to be retained during self-assembly (Sogias et al. 2008). Another derivative of chitosan, quaternary ammonium palmitoyl glycol chitosan, was found to self-assemble into micelles with a high drug loading capacity. Hydrophobic therapeutics like ranitidine and griseofulvin were found to be transported more easily by this derivative than the hydrophilic pharmaceuticals like propofol and griseofulvin, albeit to a lesser extent, via intestinal and blood-brain barriers (Ways et al. 2018).

Gelatin

Gelatin is a biopolymer that is produced by thermally denaturalizing collagen obtained from animal bones and skin. 4-hydroxyproline, glycine, and proline residues are present in significant amounts in gelatin. Its powder is practically tasteless, colorless, and translucent (Deshmukh et al. 2017). Gelatin has been used as an excipient with the drugs since long in order to make hard capsules and softgels as it ensures appropriate dissolving rates of the active pharmaceutical ingredients (API) and also is economic and simpler to produce. Additionally, it aids in shielding delicate substances from contaminants such as air, light, microbial development, and others. Similar to chitosan, gelatin has also been derivatized to enhance its properties in order to make the hydrophilic gelatin amphiphilic. Selestin et al. (2018) conjugated it with oleylamine, utilizing genipin as a crosslinking agent. In a water-based medium, the amphiphilic GOC (gelatin-oleylamine conjugate) formed micelles by self-assembling. Target triple negative breast cancer cells (TNBC-type cells) was the goal for this preparation (MDA-MB-231). It was discovered that the cells could stop the cell cycle by going into the G2/M phase, which causes cellular death, and internalization of drug Catechin/CT-bound GOC nanocarriers (CT-GOC) (Selestin et al. 2018).

Hyaluronic Acid (HA)

HA is a naturally occurring polysaccharide that comprises D-glucuronic acid and N-acetyl-D-glucosamine with a high affinity for water. It is a biocompatible, biodegradable, nonimmunogenic, and nontoxic polymer (Larraneta et al. 2018). It has been widely used in the biomedical area, cosmetics and beyond, as it has capacity to alter cellular activity. Hyaluronic acid has been chemically and physically crosslinked to get the desired properties. According to reports, it has a high degree of natural affinity for cell-specific surface markers like RHAMM and CD44. HA receptors are present on malignant tumor cells in abundance, which enhances their binding and uptake to the cancer cells (Spadea et al. 2019).

Like other polymers it has also been modified by various methods such as crosslinking it to acid-octadecanoic acid (HA-C18) and FA-HA-C18 amphiphilic graft copolymers which are used for physical encapsulation with PTX (Paclitaxel). Liu et al. (2011) created a number of innovative dual-targeting micelle-based drug delivery systems of HA. His newly developed nanomicellar systems exhibit exceptional performance traits, such as high drug absorption, small particle size, high drug loading capacity, prolonged drug release, dual-targeting capability and effective cellular uptake via cancer cells (Liu et al. 2011).

Hyaluronic acid-based micelles loaded with corticosteroid were created by Bongiovì et al. (2017) for ocular surface application with improved corneal penetration of the APIs. In another study by Bongiovì et al. (2018) ethylenediamine (EDA), hyaluronic acid (HA), polyethylene glycol (PEG), hexadecyl chains (C16), or l-carnitine were used as the starting materials to create HA-EDA-C16, HA-EDA-C16-CRN and HA-EDA-C16-PEG, micelles. These nanocarriers demonstrated ideal mucoadhesive and particle size characteristics.

DTDS micelles using ALN-HA-C18, loaded with Curcumin (CUR), were designed by Spadea et al. (2019). This delivery system had an appropriate particle size with high drug loading capacity, which promotes sustained release, and a high binding affinity with hydroxyapatite. DTDS was able to deliver more CUR into osteosarcoma tissue, which further improved in vivo antitumor activity of CUR with low systemic toxicity. The study further suggests that the DTDS could be useful for bone applications (Spadea et al. 2019).

2.3.1.2 Polymers Used to Prepare Enzymatically Triggerred Polymeric Micelles

Keeping the fact in mind that various enzymes might be present/expressed, differently at the target site, researchers successfully designed and synthesized well-defined enzyme-responsive amphiphilic linear-dendritic diblock copolymers bis-MPA-Gn-b-PHEG and G-b-PHEG. These enzymatic polymers were based on a hydrophilic linear block of poly(hydroxyethyl L-glutamine) (PHEG) and a hydrophobic dendron derived from cysteamine or an aliphatic polyester of 2,2′-dimethylolpropionic acid (bis-MPA) via a combination of the ring-opening polymerization of α-amino acid N-carboxyanhydrides (NCA) and the "dendron-first" strategy. Due to their amphiphilic properties, these copolymers form either the micelles or the nanospherical vesicles in the aqueous phase. It was observed that at higher enzyme concentrations, it results in faster disassembly rates. These copolymers are suggested to be extremely useful in the field of medication administration since they are enzyme sensitive and biocompatible (Qian et al. 2019).

In another example, Park et al. (2021) designed amphiphilic block copolymers with quinone propionic acid (QPA) groups. These could be depolymerized by NAD(P)H:quinone oxidoreductase-1 (NQO1), an overexpressed enzyme in certain cancers via cascade cyclization of the polycaprolactone (PCL) backbone. It was found that the micelles exhibited enzyme-specific behaviors like NQO1-responsive disassembly and drug release. Furthermore, when compared to control micelles, QPA-PM-DOX demonstrated NQO1-responsive intracellular drug release behavior; further it has enhanced anticancer effects in the tumor mouse model. The findings suggest that enzyme-responsive polymeric micelles have a high potential as carriers for tumor-specific drug delivery and release (Park et al. 2021).

2.3.1.3 Polymers Used to Prepare Redox Triggered Polymeric Micelles

Most of the time, various diseases are initiated due to imbalanced oxidative stress, and the level of oxidative stress might be different at the diseased site in comparison to the normal tissue; therefore, Gupta et al. (2012) described the use of reactive oxygen species responsive PMs that will release entrapped hydrophobic drug cargo in pro-inflammatory, oxidative environments. It was found that the poly(PS74-b-DMA310) micelles prepared by them were responsive to various chemical moieties such as peroxynitrite, H_2O_2 and SIN-1, which indicates that the multiple reactive species could contribute to get the drug release at the inflammation site (Gupta et al. 2012).

By combining living anionic ring-opening polymerization and ATRP (atom transfer radical polymerization), Tang et al. (2017) demonstrated that PPS [poly(propylene sulfide)]-PNIPAm block copolymers could be successfully synthesized. The synthesized polymer was used to develop a dual stimuli-responsive drug delivery platform for improved therapy. It was found to be effective at changing temperature responses and the overproduction of ROS in cancer cells, releasing the encapsulated cargos inside the cells in vitro that were producing ROS.

This method offers fresh perspectives to create oxidation and thermally responsive polymeric vesicles for the loading of drug like DOX to deliver it to a site with reactive oxygen species overproduction such as cancer cells (Tang et al. 2017).

2.3.1.4 Polymers Used to Prepare pH-Sensitive Polymeric Micelles

The stimuli-responsive polymers that can detect minute changes in the microenvironment pH can be used to make pH-sensitive nanocarriers. pH change can further initiate variation in the size, shape, and hydrophobicity of the polymer. Since primary and metastatic tumors have lower pH values than healthy tissue, the acidic pH has been regarded as the internal stimulus and best trigger for the selective release of anticancer medications among all applied stimuli, after temperature, enzyme activity, etc. (Karimi et al. 2016).

Poly(Styrene-Alt-Maleic Anhydride) Furfurylamine (PHSM/f)

Styrene maleic anhydride (SMA or SMAnh) is a synthetic polymer, having styrene and maleic anhydride as its constituents. These polymers are resistant to heat, transparent by nature, have high dimensional stability, and exhibit the unique reactivity of anhydride groups. The latter characteristic causes SMA to dissolve in alkaline (water-based) solutions and get dispersed (Hill et al. 1985).

Lee et al. (2005) had designed DDS using PHSM/f polymer at a concentration of 10 mg/mL of DOX, where micellar viability came out to be between 70% and 80%. At pH 6.8, PHSM showed increased cytotoxicity and efficacy of DOX. Due to the high local concentration, DOX enters the cell without being blocked, increasing the likelihood of cell death (Lee et al. 2005).

Poly(Ethylene Glycol)-Poly(Aspartate-Hydrazone Adriamycin) [PEG-p(Asp-Hyd-Adr)]

PEG-p(Asp-Hyd-Adr) micelles were created by Bae et al. in 2005. By monitoring the intracellular pH drop in lysosomes and endosomes (pH 5–6) they demonstrated that these micelles store drugs in a stable state under typical conditions. It was shown that they could release pharmaceuticals at a selective pH. Studies carried out in vivo and in vitro further disclosed that these micelles exhibit properties, such as the capacity to release medications when the intracellular pH changes, the capacity to infiltrate tumors, and the capacity to treat tumors with a very low level of toxicity (Bae et al. 2005a).

Polyacrylic Acid (PAAc)

Poly(acrylic acid/carbomer) $(C_3H_4O_2)_n$ is a polymer derived from acrylic acid $(C_3H_4O_2)$. It is a biocompatible substance with a long history of use as a pH-responsive drug carrier (Ohara et al. 2000). In a water solution with a pH of 7, PAAc remains as an anionic polymer. Such types of polymers cause many of their side chains to lose their protons and change to negatively charged ions. Polyelectrolyte forms of PAAc, such as partially or completely deprotonated PAAcs, have the capacity to absorb and hold water as well as swell several times their initial volume. The equilibrium between hydrogen bonds and electrostatic repulsion forces dictates how hydrophobic or hydrophilic PAAc behaves at different pH levels. Since it has pendant carboxylic groups, it has the ability to accept (at low pH) and release protons (at high pH) (Tran and Lee 2021).

Xue et al. (2009) demonstrated micelle generation from the PAA block copolymer PAAc-b-PDLLA by using a dialysis approach. They obtained a PDI of 0.125, a monosize distribution, with a mean size of 187 nm. The zeta potentials of the micelles dropped as the pH rose from 1.5 to 12 whereas the diameters of the micelles gradually increased as the pH rose from 3 to 5. These micelles also demonstrated decreased cytotoxicity. Drugs that are hydrophobically released from micelles showed pH dependence for this release. It is anticipated that pH-responsive drug delivery systems using PAAc-bPDLLA micelles would have a bright future, especially for targeted administration to the intestine in the GI tract (Xue et al. 2009).

Poly(N,N′-Dimethyl Aminoethyl Methacrylate) (PDMAEMA)

PDMAEMA is one of the significant polymers that exhibits rapid conformational modifications in response to variations in ionic strength, pH, and temperature (Lee et al. 2011). The water-soluble cationic polymer PDMAEMA can bind to DNA via electrostatic interactions. Van de Wetering et al. (2000) created complexes that are positively charged with a size of around 0.2 μ at a plasmid/polymer ratio higher than 2 (w/w). PDMAEMA is somewhat cytotoxic, just like other cationic polymers. PDMAEMA copolymers with MMA (Methyl Methacrylate), N-vinyl pyrrolidone (NVP) or ethoxy triethylene glycol methacrylate (triEGMA), were also used as transfection agents (Van de Wetering et al. 2000).

The P(DMAEMAco-MMA)-b-PPEGMA amphiphilic triblock copolymers were created by Cheng et al. (2012a, b). They displayed remarkable stimuli responsiveness and reversible self-assembly processes. Additionally, these researchers created multiblock copolymers that are pH-responsive and temperature sensitive. They had significant potential for both medicine delivery and environmental protection because they could release chemicals, dyes, and some specialized hydrophobic anticancer medications, wherever required. Organic-inorganic hybrid nanoparticles can also be created using stimuli-responsive PDMAEMA-b-PNIPAM micelles as templates (Hu et al. 2019).

Poly(4-Vinylpyridine) (PVP)

PVP is a polymer having pendant pyridine groups due to which these polymers become useful for several purposes, including modification of surfaces, immobilizing particles or atoms, electrochemical sensors, creating antibacterial surfaces, creating pH-sensitive systems, and 3D molecular level ordering systems (Mavronasou et al. 2022).

It has also derivatized to improvise its functions; for example, Mishra et al. (2021) synthesized PNVP-b-P4VP micelle with outstanding stability and pH sensitivity by the combination of RAFT (reversible addition fragmentation chain transfer) and ATRP (atom transfer radical polymerization) technologies, using an amide-based difunctional chloroamide-xanthate iniferter [S-(3-azidopropyl propanamide)-(O-ethyl xanthate)] having both a xanthate group (RAFT-CTA) and an activated chlorine group for ATRP initiator. They could change their shapes at different pH such as at pH > 7.4, spherical shaped micelles were maintained, at pH 6.9, they were rearranged, and at pH 3.5, they were completely dissolved. When the pH drops, the nitrogen atoms in P4VP protonate into pyridinium ions. The high concentration of DOX (doxorubicin) was confirmed in the micelles by a UV-Vis spectrometer, and it was believed to be the result of strong hydrogen bonding between the hydroxyl group of DOX and the nitrogen atoms of P4VP. Sustained drug release of DOX was obtained for the pH values between 6.4 and 6.9. The entire amount of DOX was observed to be fully liberated in 12 h at a pH of 3.5 (Mishra et al. 2021).

Poly(Histidine) PHIS

Poly(histidine) is a potential pH-sensitive polymer with outstanding biodegradability, biocompatibility, and fusogenic activity. Due to its imidazole ring's ability to undergo protonation and deprotonation, PHIS exhibits an amphoteric characteristic (Wu et al. 2013). Additionally, when the pH falls below 6, PHIS becomes hydrophilic and highly positively charged, which causes the endosomal membrane to rupture and makes it easier for endosomes to escape by the "proton sponge effect" (Oh et al. 2008). PHIS is very sensitive to ambient pH, which might compromise its core's stability. In order to create stable micelles, PHIS was frequently coupled with more hydrophobic polymers (Kim et al. 2005).

In 2012, Liu and his coworkers created pH-sensitive PEO-b-PHIS-b-PLLA triblock terpolymers [PLLA is poly(L-lactide)]. Using the amino end-functionalized PEO as the macroinitiator of the ROP (ring opening polymerization) of DNP (2,4-dinitrophenyl groups) -protected HIS-NCA (histidine-N-carboxy anhydride), the PEO-b-P-[HIS(DNP)] copolymer was initially created by the synthetic process. The terminal hydroxyl group of the newly synthesized PLLA was then converted to the carboxyl group by reaction with succinic anhydride, coupled to the terminal amine group of the PEO-b-P[HIS(DNP)], and PHIS was deprotected with a thiol. The resulting terpolymer has its dispersity indices as high as 1.87–1.94, indicating coupling is incomplete and the existence of PEO-b-PHIS precursors (Liu et al. 2012).

In a work by Wu et al. (2013), PHIS-PEG and DSPE-PEG (1:1, m/m) polymers were combined to create pH-sensitive PMs. At tumor pH (above 6), the mixed micelle was found to be stable and was shown to be endocytosed as an intact micelle. When the pH decreases to around 5.5 in an endosome, the destabilization of micelles is observed due to phase separation in the micelle core and dissociation of the ionized PHIS-PEG molecules. An instability of micelles was seen at pH 5.5 on which phase separation occurred in the core of the micelle and dissociation of the ionized PHIS-PEG molecules. According to an in vitro cytotoxicity assay with these micelles, the PCT-loaded mixed micelles demonstrated pH-dependent cytotoxicity, which was most prominent at lower pH levels. The conjugation of tumor-specific mAb 2C5 to pH-sensitive mixed micelles further enhanced the tumor cell killing as 2C5 could mediate specific recognition and internalization. The combination of two virtues, the active site drug targeting and pH-triggered drug release, could show the potential to significantly increase intracellular drug concentration to efficiently kill tumor cells while avoiding the development of drug resistance (Wu et al. 2013).

Jia et al. (2017) constructed a pH-sensitive micelle of poly(ethylene glycol)-poly (D,L-lactide)-poly(L-histidine) (mPEG-PLA-PHIS) for the intracellular delivery of DOX from the copolymer. Micelles loaded with DOX showed uniform distribution, high encapsulation efficiency, and small particle size. Release of DOX-loaded micelles occurred in a pH-dependent manner and had a high level of toxicity against the MCF-7/ADR cells. These copolymer micelles have been shown to be an effective nanocarrier for the intracellular delivery of DOX in order to reverse tumor MDR (multidrug resistance) (Jia et al. 2017).

Poly(β-Amino Ester) (PβAE/PAE)

Poly(β-amino ester) is a polymer created by Michael addition reaction of an acrylate and an amine that possesses characteristics of tertiary amines and esters, such as biodegradability and pH responsiveness. PβAEs can be synthesized into distinct formulations either alone or in combination with other materials (nano chaperones) for natural complex mimicry and for delivering anticancer and antimicrobials (Liu et al. 2019).

Iqbal and his colleagues (2022) synthesized PβAE-447 and various parameters (colloidal stability, zeta potential and size) were characterized to enhance its robustness. It was found that early endosomal escape was facilitated by PβAE-447s exhibition of its high buffering ability at an acidic pH. After being stored for 4 months, lyophilized nanoparticles managed to maintain the proper size, zeta potential, and transfection activity (Iqbal et al. 2022).

Min et al. (2010) prepared pH-responsive methoxypoly(ethylene glycol)-block-poly(-amino ester) (MPEG-PAE) micelles to target the acidic pH of tumors. Due to its nonantigenic, nontoxic, and nonimmunogenic properties, the MPEG provides a protective shell to the drug-loaded system, maintains its stability during longer periods of drug circulation, and extends the system's bloodstream half-life. It was noted that MPEG-PAE micelles encapsulating the fluorescence dye TRITC could efficiently deliver the drug CPT (camptothecin) to TRITC MDA-MB231 human breast tumor bearing mice in a pH-responsive manner. The targeting ability of these micelles was found to be 11 times greater than that of MPEG-PLLA micelles without pH responsiveness to the tumor tissue. It was found to have boosted therapeutic efficacy with minimal adverse effects in comparison to free CPT (Min et al. 2010).

2.3.1.5 Polymers Used to Prepare Heat-Sensitive Polymeric Micelles

Temperature, which can be either an external or internal stimulation, is among the most studied stimuli for delivery of drugs. The thermo-responsive blocks in the polymers that make up the thermo-sensitive PMs experience a dramatic change in physical characteristics in reaction to a change in temperature, which destabilizes micelles and initiates drug release (Zhang et al. 2017).

The first thermoresponsive polymer made from N-isopropylacrylamide was PNIMPAm and is available commercially (Schild 1992). It is the most used thermosensitive polymer used for creating thermosensitive micelles. PNIMPAm is water soluble at room temperature, but precipitates in water above 32 °C due to the phase transition of a polymer to a hydrophobic state, which causes turbidity due to aggregation. PNIMPAm has been used for 3D extracellular matrices (ECM) materials, drug delivery vehicles, and thermoresponsive tissue culture dishes because of this unique thermoresponsive property (Ohya et al. 2004).

Thermosensitive behavior of the special polymers is exhibited due to the existence of hydrophobic groups, such as methyl, ethyl, and propyl groups. The lower critical solution temperature (LCST) and upper critical solution temperature (UCST) are two additional important parameters which these polymers exhibit (Singh et al. 2007). The LCST, also called as "cloud point," is the temperature at which the polymeric monophasic system separates into two phases: hydrophobic and insoluble. The UCST is the temperature at which all of the constituent parts of a mixture become miscible. Since UCST systems require high temperatures, which is unfavorable for biomolecules and medicines which are heat-labile, therefore, an LCST system is usually recommended for drug delivery technologies (James et al. 2014).

Temperature below the LCST initiates the formation of hydrogen bonds between the polymeric chains and water-based molecules, thus making the water-soluble polymer, whereas when the temperature is above the critical solution these hydrogen bonds break due to heat, making the polymer insoluble. A desirable value of LCST, lying between 37 °C and 42 °C, could be achieved by adding suitable hydrophobic or hydrophilic comonomer. This temperature range is lower than hyperthermia and greater than the body's normal temperature (Nizardo et al. 2022). As an example, poly(N-isopropylacrylamide) (PNIPAAm) have an LCST of 32 °C and is widely used as thermo-responsive polymer for creating thermosensitive PMs as the temperature is close to body's physiological temperature. For better targeting and enhanced drug release, random copolymerization could be done using different monomers to adjust the LCST of PNIPAAm. For instance, a smart thermos-responsive micelle was created by Panja et al. (2016). They used two unique four-arm star-shaped block copolymers, PE-PCL-b-PNIPAAm-FA and PE-PCL-b-poly (N-vinylcaprolactam)-FA. By varying the block length of the thermo-responsive segment, they also produced several temperature-responsive polymers with specified LCST (ranging from 30 °C to 39 °C) (Panja et al. 2016).

Yang et al. (2014) formed a comb-like polymer with a hydrophobic backbone of polyacrylate (PA) with mPEG blocks and thermolabile PNIPAAm grafted chains (mPEG-b-PA-g-PNIPAAm), loaded with camptothecin (CPT). At an LCST ranging between 40 °C and 45 °C, the micelles showed a temperature-responsive phase transition from hydrophilic to hydrophobic. Without any initial burst release, the release of CPT from the micelles was constant and enhanced above the LCST. The CPT-loaded thermo-responsive micelles showed a hazardous nature to the cancerous cells, while they did not affect the healthy cells (Yang et al. 2014).

2.3.2 Polymeric Micelles Sensitive to External Stimulus

2.3.2.1 Photosensitive Polymeric Micelles

Photosensitive micelles are prepared, typically using the chromophores such as pyrene, cinnamoyl, nitrobenzyl azobenzene groups, or spirobenzopyran by including or by conjugating them to the polymeric structure. When exposed to light, the nanostructure of the PMs that contain chemical compounds alters releasing the payloads (Zhou et al. 2018).

Azobenzene and derivatives are highly researched upon light-responsive chromophores. When exposed to UV light, azobenzene can photoisomerize the nitrogen double bond from trans to cis, which can then be converted back to trans under visible light (Ren et al. 2022). However, some chromophores like 2-diazo-1,2-naphthoquinone (DNQ) undergo irreversible cleavage in response to light, changing hydrophobic to hydrophilic groups (Zhou et al. 2018).

Alemayehu et al. (2020) described creation of photosensitive polypropylene glycol (PPG) polymer that comprises of bifunctional uracil-containing end groups (BU-PPG), which solve many issues associated with existing nanocarriers. BU-PPG self-assembles spontaneously, into spherical shaped micelles which are nanosized when suspended in water. They were found to have the capability to encapsulate DOX at high concentration. Compared to non-irradiation micelles, the DOX-loaded irradiated micelles considerably reduced cell viability and increased the fraction of apoptotic cells (Alemayehu et al. 2020).

Moleavin et al. (2010) created and characterized amphiphilic azo-polysiloxanes with quaternary ammonium groups. According to the photochromic experiments, the maximal amount of cis-isomer content can be attained within 20–30 s of UV exposure. All amphiphilic polysiloxanes had the ability to produce micelles, with CCA (concentration of aggregation) values falling between 10^3 and 10^2 g/L. Different circumstances with regard to the aggregate's morphology were found to depend on the concentration and amine structure. In water, individual micelles as well as primary aggregates were observed at low concentrations. Individual micelles or micellar clusters were formed during the micellar aggregation process depending on the ternary amine structure utilized in the quarterization reaction. If the amine had a lengthy hydrocarbon segment, the multi-micellar clusters formation occurred after the primary micellar association process (Moleavin et al. 2010).

2.3.2.2 Magnetic-Sensitive Polymeric Micelles

An external magnetic field (MF) can be used to target tumors with MF-responsive PMs and to stimulate rapid release of drug, temperature was raised alternating the magnetic field. Typically, they combine a magnetically active component with therapeutic payloads (Senapati et al. 2018). The iron oxide nanoparticles such as magnetite (Fe_3O_4) and maghemite γ-(Fe_2O_3) are widely used as magnetic material used for magnetosensitive PMs (Arora 2012).

In a study, Wang et al. (2012) described the use of magnetic nanoparticles loaded with super magnetic iron oxide nanoparticles (SPIONs) coated with chitosan and polyethyleneimine (PEI) to prepare CP-mag-micelles for the transport of plasmid DNA. CP-mag-micelles demonstrated much greater transfection efficiency than PEI or lipofectamine (Wang et al. 2012). Varshosaz and his colleagues (2013) formulated folate-conjugated dextran (FA-DEX)/retinoic acid (RA) magnetic micelle as an anticancer drug for the dual targeted delivery (DOX) for breast cancer encapsulating OA-coated (Oleic acid) magnetic Fe_3O_4. Hydrothermal technique was successfully used to create OA-coated MNPs (Varshosaz et al. 2013).

Li et al. (2012) and his colleagues formulated magnetomicelles by combining fluorine amphiphilic poly(HFMA-g-PEGMA) block copolymer with oleic acid containing Fe_3O_4 particles. TEM confirmed the presence of Fe_3O_4 clusters enclosed in polymeric micelles. These magnetomicelles were stable in both water and sodium chloride solution. The magnetomicelles' compatibility with cells was demonstrated using the MTT cytotoxicity experiment. These magnetomicelles were proposed to show potential in in vivo MRI and medication delivery as they allowed controlled release of the 5-FU. According to in vitro and in vivo studies it was observed that magnetomicelles have enhanced opposite effects in the spleen and liver because of the high T2 relaxivity (134.27/mM s) (Li et al. 2012).

Wei et al. (2017) proposed a clickable and imageable multiblock polyurethane (MPU) system with switchable tumor targeting and triggered drug release capabilities for accurate tumor therapy and targeted MR imaging. In this case PEG and PCL made the MPUs' soft segments, whereas L-lysine ethyl ester diisocyanate (LDI) was in the hard segments. An L-cysteine-derived versatile chain extender (Cys-PA) was used to endow polymers with a number of reduction-cleavable disulfide linkages in the backbone and clickable alkyne sites on the side chains, eliminating the need for multiple chain extenders and difficult synthetic procedures for the construction of multifunctional MPUs (Wei et al. 2017).

Table 2.1 summarizes all the polymers that have been discussed in this chapter explaining their types, constituents, uses, and limitations.

2.4 Conclusions and Future Directions

In the light of recent advancements in drug delivery techniques, the development of polymeric micelles is one of the most promising technologies being explored. They have superior encapsulation capabilities, amphiphilicity, kinetic stability, biodegradability, and lower toxicity, as compared to other nanocarriers. PMs have several applications and have been used for many years. Surfactant optimization, multidrug loading capacity, insoluble drug conjugates, delivery of biological entities like siRNA, antibodies, etc., and the development of stimuli-sensitive polymers are currently the main research areas. PMs have demonstrated notable in vivo and in vitro successes. It is evident from the above discussion that in order to prevent off-target delivery and to ensure controlled release, research is now being done on stimuli-sensitive micellar optimization, in response to internal or environmental cues. In addition, PMs have been widely explored for their application for cancer therapeutics, but now they are also being investigated for various other applications such as for neuroendocrine and metabolic problems. It is also evident from the above description that there is a lot of potential in studying polymer chemistry and their modification procedures because it carries a lot of potential to modulate the properties of the polymers to obtain the desired goal for the miceller DDS. Currently, extremely intriguing studies focusing on targeted delivery and regulated release in response to short-term and long-term internal stresses and environmental circumstances are being conducted using novel tools and modifications. There are notable examples of their therapeutic effectiveness, but their detailed uses, effective administration methods, and molecular interactions in the bloodstream have not been fully investigated. To develop a next-generation drug delivery approach centered on micellar carriers, further descriptive studies based on the stability and molecular

Table 2.1 Different types of polymeris micelles, polymers used and their advantages

Type of micelle	Property of polymer	Polymer name	Advantages	References
Triblock polymeric micelles	pH-sensitive polymeric micelles	PHSM f	Enhanced cytotoxicity, improved efficiency of drug release	Lee et al. (2005)
Triblock polymeric micelles		Poly(ethylene glycol)-poly(aspartate-hydrazone adriamycin) [PEG-p(Asp-Hyd-Adr)]	Tumor-infiltrating permeability, effective antitumor activity with extremely low toxicity	Bae et al. (2005b)
Graft polymeric micelles		Polyacrylic acid (PAAc)	Biocompatible, pH-responsive drug carriers	Xue et al. (2009)
Triblock polymeric micelles		Poly(N,N′-dimethyl aminoethyl methacrylate) (PDMAEMA)	Rapid conformational changes during changes in the temperature, pH, and ionic strength	
Diblock polymeric micelles		Poly(4-vinylpyridine) (PVP)	Antibacterial activity, pH-sensitive	Mishra et al. (2021)
Graft polymeric micelles		Poly(histidine) (PHIS)	Excellent biodegradability, biocompatibility, and fusogenic activity	Wu et al. (2013)
Diblock polymeric micelles		Poly(β-amino ester)	pH responsiveness and biodegradability	Min et al. (2010)
Triblock polymeric micelles	Thermosensitive polymeric micelles	Poly(N-isopropyl acrylamide) (PNIMPAM)	Lowers critical temperature to human body temperatures, excellent candidate for drug delivery applications	Ohya et al. (2004)

(Note: The limitations column in the table above is optional)

interaction of these polymers, intermediate colloids, and nature of the loaded entity are needed.

References

Achilleos DS, Vamvakaki M (2010) End-grafted polymer chains onto inorganic nano-objects. Materials 3(3):1981–2026

Alemayehu YA, Fan WL, Ilhami FB, Chiu CW, Lee DJ, Cheng CC (2020) Photosensitive supramolecular micelle-mediated cellular uptake of anticancer drugs enhances the efficiency of chemotherapy. Int J Mol Sci 21(13):4677

Arora S (2012) Superparamagnetic iron oxide nanoparticles: magnetic nanoplatforms as drug carriers. Int J Nanomedicine 7:3445

Bae Y, Nishiyama N, Fukushima S, Koyama H, Yasuhiro M, Kataoka K (2005a) Preparation and biological characterization of polymeric micelle drug carriers with intracellular pH-triggered drug release property: tumor permeability, controlled subcellular drug distribution, and enhanced in vivo antitumor efficacy. Bioconjug Chem 16(1):122–130

Bae Y, Jang WD, Nishiyama N, Fukushima S, Kataoka K (2005b) Multifunctional polymeric micelles with folate-mediated cancer cell targeting and pH-triggered drug releasing properties for active intracellular drug delivery. Mol Biosyst 1(3):242–250

Bastakoti BP, Wu KCW, Inoue M, Yusa SI, Nakashima K, Yamauchi Y (2013) Multifunctional Core-Shell-Corona-type polymeric micelles for anticancer drug-delivery and imaging. Chem A Eur J 19(15):4812–4817

Bongiovì F, Di Prima G, Palumbo FS, Licciardi M, Pitarresi G, Giammona G (2017) Hyaluronic acid-based micelles as ocular platform to modulate the loading, release, and corneal permeation of corticosteroids. Macromol Biosci 17(12):1700261

Bongiovì F, Fiorica C, Palumbo FS, Di Prima G, Giammona G, Pitarresi G (2018) Imatinib-loaded micelles of hyaluronic acid derivatives for potential treatment of neovascular ocular diseases. Mol Pharm 15(11):5031–5045

Cheng C, Convertine AJ, Stayton PS, Bryers JD (2012a) Multifunctional triblock copolymers for intracellular messenger RNA delivery. Biomaterials 33(28):6868–6876

Cheng R, Wang X, Chen W, Meng F, Deng C, Liu H, Zhong Z (2012b) Biodegradable poly (ε-caprolactone)-g-poly (2-hydroxyethyl methacrylate) graft copolymer micelles as superior nano-carriers for "smart" doxorubicin release. J Mater Chem 22(23):11730–11738

Cortez MA, Godbey WT, Fang Y, Payne ME, Cafferty BJ, Kosakowska KA, Grayson SM (2015) The synthesis of cyclic poly (ethylene imine) and exact linear analogues: an evaluation of gene delivery comparing polymer architectures. J Am Chem Soc 137(20):6541–6549

Deshmukh K, Ahamed MB, Deshmukh RR, Pasha SK, Bhagat PR, Chidambaram K (2017) Biopolymer composites with high dielectric performance: interface engineering. In: Biopolymer composites in electronics. Elsevier, Amsterdam, pp 27–128

Endres TK, Beck-Broichsitter M, Samsonova O, Renette T, Kissel TH (2011) Self-assembled biodegradable amphiphilic PEG–PCL–lPEI triblock copolymers at the borderline between micelles and nanoparticles designed for drug and gene delivery. Biomaterials 32(30): 7721–7731

Epps TH, Cochran EW, Bailey TS, Waletzko RS, Hardy CM, Bates FS (2004) Ordered network phases in linear poly (isoprene-b-styrene-b-ethylene oxide) triblock copolymers. Macromolecules 37(22):8325–8341

Ghezzi M, Pescina S, Padula C, Santi P, Del Favero E, Cantù L, Nicoli S (2021) Polymeric micelles in drug delivery: an insight of the techniques for their characterization and assessment in biorelevant conditions. J Control Release 332:312–336

Gonçalves C, Ferreira N, Lourenço L (2021) Production of low molecular weight chitosan and chitooligosaccharides (COS): a review. Polymers 13(15):2466

Guo M, Yan Y, Liu X, Yan H, Liu K, Zhang H, Cao Y (2010) Multilayer nanoparticles with a magnetite core and a polycation inner shell as pH-responsive carriers for drug delivery. Nanoscale 2(3):434–441
Gupta MK, Meyer TA, Nelson CE, Duvall CL (2012) Poly (PS-b-DMA) micelles for reactive oxygen species triggered drug release. J Control Release 162(3):591–598
Hanafy NA, El-Kemary M, Leporatti S (2018) Micelles structure development as a strategy to improve smart cancer therapy. Cancers 10(7):238
Hill DJ, O'Donnell JH, O'Sullivan PW (1985) Analysis of the mechanism of copolymerization of styrene and maleic anhydride. Macromolecules 18(1):9–17
Hoang NH, Lim C, Sim T, Oh KT (2017) Triblock copolymers for nano-sized drug delivery systems. J Pharm Investig 47(1):27–35
Hu L, Wang Y, Yin Q, Du K, Yin Q (2019) Multiple morphologies of a poly (methyl methacrylate)-block-poly (N, N-dimethyl aminoethyl methacrylate) copolymer with pH-responsiveness and thermoresponsiveness. J Appl Polym Sci 136(38):47972
Iqbal S, Martins AF, Sohail M, Zhao J, Deng Q, Li M, Zhao Z (2022) Synthesis and characterization of poly (β-amino Ester) and applied PEGylated and non-PEGylated poly (β-amino ester)/ plasmid DNA nanoparticles for efficient gene delivery. Front Pharmacol 13:722
Irvine DJ, Dane EL (2020) Enhancing cancer immunotherapy with nanomedicine. Nat Rev Immunol 20(5):321–334
James HP, John R, Alex A, Anoop K (2014) Smart polymers for the controlled delivery of drugs—a concise overview. Acta Pharm Sin B 4(2):120–127
Jee JP, McCoy A, Mecozzi S (2012) Encapsulation and release of amphotericin B from an ABC triblock fluorous copolymer. Pharm Res 29(1):69–82
Jeong H, Tombor B, Albert R, Oltvai ZN, Barabási AL (2000) The large-scale organization of metabolic networks. Nature 407(6804):651–654
Jia N, Ye Y, Wang Q, Zhao X, Hu H, Chen D, Qiao M (2017) Preparation and evaluation of poly (L-histidine) based pH-sensitive micelles for intracellular delivery of doxorubicin against MCF-7/ADR cells. Asian J Pharma Sci 12(5):433–441
Karimi M, Eslami M, Sahandi-Zangabad P, Mirab F, Farajisafiloo N, Shafaei Z et al (2016) pH-sensitive stimulus-responsive nanocarriers for targeted delivery of therapeutic agents. Wiley Interdiscip Rev Nanomed Nanobiotechnol 8(5):696–716
Kim GM, Bae YH, Jo WH (2005) pH-induced micelle formation of poly (histidine-co-phenylalanine)-block-poly (ethylene glycol) in aqueous media. Macromol Biosci 5(11):1118–1124
Kubowicz S, Baussard JF, Lutz JF, Thünemann AF, von Berlepsch H, Laschewsky A (2005) Multicompartment micelles formed by self-assembly of linear ABC triblock copolymers in aqueous medium. Angew Chem Int Ed 44(33):5262–5265
Kulthe SS, Choudhari YM, Inamdar NN, Mourya V (2012) Polymeric micelles: authoritative aspects for drug delivery. Des Monomer Polym 15(5):465–521
Larraneta E, Henry M, Irwin NJ, Trotter J, Perminova AA, Donnelly RF (2018) Synthesis and characterization of hyaluronic acid hydrogels crosslinked using a solvent-free process for potential biomedical applications. Carbohydr Polym 181:1194–1205
Lee Y, Thompson DH (2017) Stimuli-responsive liposomes for drug delivery. Wiley Interdiscip Rev Nanomed Nanobiotechnol 9(5):e1450
Lee JW, Park JH, Robinson JR (2000) Bioadhesive-based dosage forms: the next generation. J Pharm Sci 89(7):850–866
Lee ES, Na K, Bae YH (2005) Doxorubicin loaded pH-sensitive polymeric micelles for reversal of resistant MCF-7 tumor. J Control Release 103(2):405–418
Lee H, Son SH, Sharma R, Won YY (2011) A discussion of the pH-dependent protonation behaviors of poly (2-(dimethylamino) ethyl methacrylate)(PDMAEMA) and poly (ethylenimine-ran-2-ethyl-2-oxazoline)(P (EI-r-EOz)). J Phys Chem B 115(5):844–860
Letchford K, Liggins R, Burt H (2008) Solubilization of hydrophobic drugs by methoxy poly (ethylene glycol)-block-polycaprolactone diblock copolymer micelles: theoretical and experimental data and correlations. J Pharm Sci 97(3):1179–1190

Li X, Li H, Liu G, Deng Z, Wu S, Li P et al (2012) Magnetite-loaded fluorine-containing polymeric micelles for magnetic resonance imaging and drug delivery. Biomaterials 33(10):3013–3024
Lim C, Youn YS, Lee KS, Hoang NH, Sim T, Lee ES, Oh KT (2016) Development of a robust pH-sensitive polyelectrolyte ionomer complex for anticancer nanocarriers. Int J Nanomedicine 11:703
Liu S, Weaver JV, Tang Y, Billingham NC, Armes SP, Tribe K (2002) Synthesis of shell cross-linked micelles with pH-responsive cores using ABC triblock copolymers. Macromolecules 35(16):6121–6131
Liu Y, Sun J, Cao W, Yang J, Lian H, Li X et al (2011) Dual targeting folate-conjugated hyaluronic acid polymeric micelles for paclitaxel delivery. Int J Pharm 421(1):160–169
Liu R, He B, Li D, Lai Y, Tang JZ, Gu Z (2012) Synthesis and characterization of poly (ethylene glycol)-b-poly (l-histidine)-b-poly (l-lactide) with pH-sensitivity. Polymer 53(7):1473–1482
Liu Y, Li Y, Keskin D, Shi L (2019) Poly (β-amino esters): synthesis, formulations, and their biomedical applications. Adv Healthc Mater 8(2):1801359
Matsen MW (2020) Field theoretic approach for block polymer melts: SCFT and FTS. J Chem Phys 152(11):110901
Mavronasou K, Zamboulis A, Klonos P, Kyritsis A, Bikiaris DN, Papadakis R, Deligkiozi I (2022) Poly (vinyl pyridine) and its quaternized derivatives: understanding their solvation and solid state properties. Polymers 14(4):804
Meng Y, Gu D, Zhang F, Shi Y, Cheng L, Feng D et al (2006) A family of highly ordered mesoporous polymer resin and carbon structures from organic—organic self-assembly. Chem Mater 18(18):4447–4464
Min KH, Kim JH, Bae SM, Shin H, Kim MS, Park S et al (2010) Tumoral acidic pH-responsive MPEG-poly (β-amino ester) polymeric micelles for cancer targeting therapy. J Control Release 144(2):259–266
Mishra AK, Lim J, Lee J, Park S, Seo Y, Hwang H, Kim JK (2021) Control drug release behavior by highly stable and pH sensitive poly (N-vinylpyrrolidone)-block-poly (4-vinylpyridine) copolymer micelles. Polymer 213:123329
Moleavin I, Grama S, Cârlescu I, Scutaru D, Hurduc N (2010) Photosensitive micelles based on polysiloxanes containing azobenzene moieties. Polym Bull 65(1):69–81
Morgese G, Cavalli E, Rosenboom JG, Zenobi-Wong M, Benetti EM (2018) Cyclic polymer grafts that lubricate and protect damaged cartilage. Angew Chem 130(6):1637–1642
Natarajan JV, Nugraha C, Ng XW, Venkatraman S (2014) Sustained-release from nanocarriers: a review. J Control Release 193:122–138
Nizardo NM, Alimin DF, Lestari ML (2022) Synthesis and characterization of dual-responsive poly (N-vinylcaprolactam-co-N-methylolacrylamide) nanogels. Des Monomers Polym 25(1): 155–164
Oerlemans C, Bult W, Bos M, Storm G, Nijsen JFW, Hennink WE (2010) Polymeric micelles in anticancer therapy: targeting, imaging and triggered release. Pharm Res 27(12):2569–2589
Oh KT, Lee ES, Kim D, Bae YH (2008) L-histidine-based pH-sensitive anticancer drug carrier micelle: reconstitution and brief evaluation of its systemic toxicity. Int J Pharm 358(1–2): 177–183
Ohara T, Sato T, Shimizu N, Prescher G, Schwind H, Weiberg O et al (2000) Acrylic acid and derivatives. In: Ullmann's encyclopedia of industrial chemistry. Wiley-VCH Verlag, Weinheim
Ohya S, Nakayama Y, Matsuda T (2004) In vivo evaluation of poly (N-isopropylacrylamide) (PNIPAM)-grafted gelatin as an in situ-formable scaffold. J Artif Organs 7(4):181–186
Ortel E, Fischer A, Chuenchom L, Polte J, Emmerling F, Smarsly B, Kraehnert R (2012) New triblock copolymer templates, PEO-PB-PEO, for the synthesis of Titania films with controlled Mesopore size, wall thickness, and bimodal porosity. Small 8(2):298–309
Panja S, Dey G, Bharti R, Kumari K, Maiti TK, Mandal M, Chattopadhyay S (2016) Tailor-made temperature-sensitive micelle for targeted and on-demand release of anticancer drugs. ACS Appl Mater Interfaces 8(19):12063–12074

Park J, Jo S, Lee YM, Saravanakumar G, Lee J, Park D, Kim WJ (2021) Enzyme-triggered disassembly of polymeric micelles by controlled depolymerization via cascade cyclization for anticancer drug delivery. ACS Appl Mater Interfaces 13(7):8060–8070

Peng CL, Shieh MJ, Tsai MH, Chang CC, Lai PS (2008) Self-assembled star-shaped chlorin-core poly (ε-caprolactone)-poly (ethylene glycol) diblock copolymer micelles for dual chemo-photodynamic therapies. Biomaterials 29(26):3599–3608

Qian Y, You D, Lin F, Wei J, Wang Y, Bi Y (2019) Enzyme triggered disassembly of amphiphilic linear-dendritic block copolymer micelles based on poly [N-(2-hydroxyethyl-l-glutamine)]. Polym Chem 10(1):94–105

Ren H, Yang P, Yu H (2022) Recent progress in azopyridine-containing supramolecular assembly: from photoresponsive liquid crystals to light-driven devices. Molecules 27(13):3977

Salamat-Miller N, Chittchang M, Johnston TP (2005) The use of mucoadhesive polymers in buccal drug delivery. Adv Drug Deliv Rev 57(11):1666–1691

Schild HG (1992) Poly (N-isopropylacrylamide): experiment, theory and application. Prog Polym Sci 17(2):163–249

Selestin I, Thangam R, Fathima NN (2018) Polymeric micelle of a gelatin-oleylamine conjugate: a prominent drug delivery carrier for treating triple negative breast cancer cells. ACS Appl Bio Mater 1(5):1725–1734

Senapati S, Mahanta AK, Kumar S, Maiti P (2018) Controlled drug delivery vehicles for cancer treatment and their performance. Signal Transduct Target Ther 3(1):1–19

Sharma S, Dua A, Malik A (2014) Polyaspartic acid based superabsorbent polymers. Eur Polym J 59:363–376

Singh S, Webster DC, Singh J (2007) Thermosensitive polymers: synthesis, characterization, and delivery of proteins. Int J Pharm 341(1–2):68–77

Sogias IA, Williams AC, Khutoryanskiy VV (2008) Why is chitosan mucoadhesive? Biomacromolecules 9(7):1837–1842

Spadea A, Rios de la Rosa JM, Tirella A, Ashford MB, Williams KJ, Stratford IJ, Tirelli N, Mehibel M (2019) Evaluating the efficiency of hyaluronic acid for tumor targeting via CD44. Mol Pharm 16(6):2481–2493

Sugihara S, Kanaoka S, Aoshima S (2004) Stimuli-responsive ABC triblock copolymers by sequential living cationic copolymerization: multistage self-assemblies through micellization to open association. J Polym Sci A Polym Chem 42(11):2601–2611

Tang Y, Liu SY, Armes SP, Billingham NC (2003) Solubilization and controlled release of a hydrophobic drug using novel micelle-forming ABC triblock copolymers. Biomacromolecules 4(6):1636–1645

Tang M, Hu P, Zheng Q, Tirelli N, Yang X, Wang Z et al (2017) Polymeric micelles with dual thermal and reactive oxygen species (ROS)-responsiveness for inflammatory cancer cell delivery. J Nanobiotechnol 15(1):1–11

Torchilin VP (2006) Multifunctional nanocarriers. Adv Drug Deliv Rev 58(14):1532–1555

Tran VA, Lee SW (2021) pH-triggered degradation and release of doxorubicin from zeolitic imidazolate framework-8 (ZIF8) decorated with polyacrylic acid. RSC Adv 11(16):9222–9234

Van Butsele K, Cajot S, Van Vlierberghe S, Dubruel P, Passirani C, Benoit JP et al (2009) pH-responsive flower-type micelles formed by a biotinylated poly (2-vinylpyridine)-block-poly (ethylene oxide)-block-poly (ε-caprolactone) triblock copolymer. Adv Funct Mater 19(9):1416–1425

Van de Wetering P, Schuurmans-Nieuwenbroek NME, Van Steenbergen MJ, Crommelin DJA, Hennink WE (2000) Copolymers of 2-(dimethylamino) ethyl methacrylate with ethoxytriethylene glycol methacrylate or N-vinyl-pyrrolidone as gene transfer agents. J Control Release 64(1–3):193–203

Varshosaz J, Sadeghi-Aliabadi H, Ghasemi S, Behdadfar B (2013) Use of magnetic folate-dextran-retinoic acid micelles for dual targeting of doxorubicin in breast cancer. Biomed Res Int 2013: 680712

Wang C, Ravi S, Martinez GV, Chinnasamy V, Raulji P, Howell M et al (2012) Dual-purpose magnetic micelles for MRI and gene delivery. J Control Release 163(1):82–92
Ways TM, Lau WM, Khutoryanskiy VV (2018) Chitosan and its derivatives for application in mucoadhesive drug delivery systems. Polymers 10(3):267
Wei J, Shuai X, Wang R, He X, Li Y, Ding M et al (2017) Clickable and imageable multiblock polymer micelles with magnetically guided and PEG-switched targeting and release properties for precise tumor theranosis. Biomaterials 145:138–153
Williams HD, Trevaskis NL, Charman SA, Shanker RM, Charman WN, Pouton CW, Porter CJ (2013) Strategies to address low drug solubility in discovery and development. Pharmacol Rev 65(1):315–499
Williams RJ, Pitto-Barry A, Kirby N, Dove AP, O'Reilly RK (2016) Cyclic graft copolymer unimolecular micelles: effects of cyclization on particle morphology and thermoresponsive behavior. Macromolecules 49(7):2802–2813
Wu H, Zhu L, Torchilin VP (2013) pH-sensitive poly (histidine)-PEG/DSPE-PEG co-polymer micelles for cytosolic drug delivery. Biomaterials 34(4):1213–1222
Wu S, Zheng L, Li C, Xiao Y, Huo S, Zhang B (2017) Grafted copolymer micelles with pH triggered charge reversibility for efficient doxorubicin delivery. J Polym Sci A Polym Chem 55(12):2036–2046
Xue YN, Huang ZZ, Zhang JT, Liu M, Zhang M, Huang SW, Zhuo RX (2009) Synthesis and self-assembly of amphiphilic poly (acrylic acid-b-DL-lactide) to form micelles for pH-responsive drug delivery. Polymer 50(15):3706–3713
Yang XL, Luo YL, Xu F, Chen YS (2014) Thermosensitive mPEG-b-PA-g-PNIPAM comb block copolymer micelles: effect of hydrophilic chain length and camptothecin release behavior. Pharm Res 31(2):291–304
Zhang Q, Weber C, Schubert US, Hoogenboom R (2017) Thermoresponsive polymers with lower critical solution temperature: from fundamental aspects and measuring techniques to recommended turbidimetry conditions. Mater Horiz 4(2):109–116
Zhou Q, Zhang L, Yang T, Wu H (2018) Stimuli-responsive polymeric micelles for drug delivery and cancer therapy. Int J Nanomedicine 13:2921
Zuegg J, Cooper AM (2012) Drug-likeness and increased hydrophobicity of commercially available compound libraries for drug screening. Curr Top Med Chem 12(14):1500–1513

Dynamics of Micelle Formation

3

Anupama Sikder, Saurabh Shah, Shashi Bala Singh, and Saurabh Srivastava

Abstract

Developing a novel drug delivery system is critical in managing and treating a broader spectrum of diseases. The intensified demand for an effective targeting carrier has led to rapid development and a better understanding of the principles of forming polymeric micelles. Micelles exist in a state of dynamic equilibrium. Henceforth, the dynamics involved in the formation of micelles have become obligatory to understand. This chapter deals with the underlying parameters influencing micelle development and the simulation methods. The advent of analytical tools such as fluorescence resonance energy transfer (FRET) and dissipative particle dynamics (DPD) will help in ascertaining the micellar dynamics in a much more effective manner. Various theories and energetics involved in micelle formation have been deliberated.

Keywords

Micelle formation dynamics · Thermodynamics of micelles · Molecular dynamics · FRET · DPD · Theories of micelle formation

A. Sikder · S. Shah · S. Srivastava (✉)
Pharmaceutical Innovation and Translational Research Lab (PITRL), Department of Pharmaceutics, National Institute of Pharmaceutical Education and Research (NIPER) Hyderabad, Hyderabad, Telangana, India
e-mail: saurabh@niperhyd.ac.in

S. B. Singh
Department of Biological Sciences, National Institute of Pharmaceutical Education and Research (NIPER) Hyderabad, Hyderabad, Telangana, India

S. K. Singh et al. (eds.), *Polymeric Micelles: Principles, Perspectives and Practices*,
https://doi.org/10.1007/978-981-99-0361-0_3

3.1 Introduction

Polymeric micelles are amphiphilic assemblies that depict a particle size in nanoscale dimensions and spherical morphology (Sasidharan and Nakashima 2014). The hydrophobic core of micelles facilitates carrying drug or contrast agents for the treatment and diagnosis of various diseases like cancer, neurological disorders, topical diseases, etc. (Karabasz et al. 2020; Krishnan and Mitragotri 2020; Domínguez et al. 2014). The hydrophilic shell is responsible for the stability of micelles and promotes their water solubility, aiding intravenous delivery (Lu and Park 2013). The massive advancements in polymer science have opened novel avenues for the fabrication of micelles depicting properties like target specificity, enhanced loading capability, protection against the reticulo-endothelial system, better therapeutic activity, enhanced stability in vitro and in vivo, and ease of surface modification with aptamers and ligands (Mollazadeh et al. 2021; Li et al. 2019; Li et al. 2016; Maghsoudnia et al. 2020).

Due to their versatility, diverse drug-loaded micelles have been studied in preclinical and clinical trials (Cabral and Kataoka 2014). The architectural composition of micelles decides the molecular weight, critical micelle concentration (CMC), stability, drug encapsulation, dissociation time, and site specificity (Qiu and Bae 2006). Hence, it becomes indispensable to understand the micellar architecture and the dynamics behind its formation. Various shapes of micelles have been depicted in Fig. 3.1.

The temperature at which micelles assemble in an aqueous medium is known as kraft temperature (T_k) (Chu and Feng 2012). Despite some insights into micelles basic architecture, their specific fabrication mechanisms are still unclear, including drug encapsulation by the micelles and discrete phases of core and corona

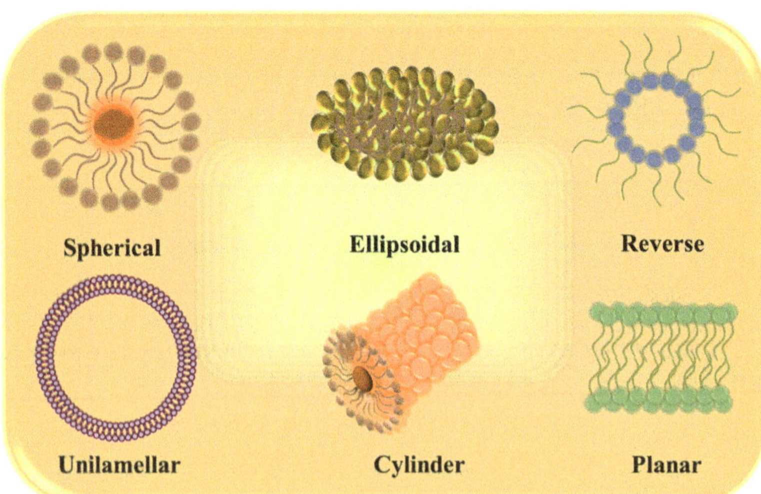

Fig. 3.1 Diagrammatic representation depicting the different shapes of micelles

development during micellar growth (Zhang et al. 2017). These amphiphilic assemblies are in a state of dynamic equilibrium and constantly dissociating and reorganizing, which correlates with their stability (Zhang and Waymouth 2017). Various theories have been proposed to overcome the lacunas in thoroughly understanding the kinetics involved in the formation of micelles. This chapter discusses the associated theories and factors persuading micelle formation dynamics. The related thermodynamics responsible for the micelle's assembling has been conveyed. Various evaluation techniques for evaluating micelle formation have been covered.

3.2 Factors Influencing Micelle Formation Dynamics

3.2.1 Amphiphilic Block Copolymer

The amphiphilic block copolymers (ACs) are extensively used in the fabrication of polymeric micelles because of their inherent properties like tunable structure, chemistry, physicochemical properties, target specificity, diagnosis, and efficient delivery of hydrophobic drugs (Gu et al. 2014). These ACs are composed of hydrophilic and hydrophobic fragments joined via a covalent bond, which self-assemble in a specified solvent to generate polymeric micelles. Various techniques have been employed in designing ACs, including nitroxide-mediated and radical polymerization (Hawker et al. 2001). These techniques offer several advantages like chemical compatibility, controlled degree of polymerization, imparting less polydispersity index, eliminating impurities from the fabricated copolymers, etc., and are preferred to conventional preparation methods. The standard polymeric backbones widely utilized in manufacturing ACs include chitosan, polylactide, polycaprolactone, and polyethyleneimine (Perin et al. 2021). The ACs manufactured from more than one type of monomer will generate moieties with opposing affinities in an aqueous medium (Schmidt 2018). Based on the number of monomers, various copolymers can be manufactured, tabulated in Table 3.1.

The critical packing parameter (Cpp) of the block copolymers determines the nanocarriers' morphology based on the type of aggregation formed by the surfactants. Along with Cpp, other factors that influence the morphology of micelles include degree of polymerization (DP) and volume fraction of the amphiphilic blocks (f) (Jiao et al. 2020).

C_{pp} can be calculated using the following equation,

$$Cpp = V/Al$$

where V is the volume of the lipophilic tail, A is the surface area of the hydrophilic head, l is the effective chain length.

The different values of C_{pp} which leads to the formation of diverse morphology of micelles are depicted pictographically in Fig. 3.2 (Lombardo et al. 2015). The hydrophilic-lipophilic balance (HLB) of block copolymers regulates their solubility

Table 3.1 Types of block copolymers

S. no	Type of copolymers	Characteristics	Structure	References
1	Di-block	Made up of two monomers possessing dissimilar chemical properties	A-A-A-A-B-B-B-B	(Garnier and Laschewsky 2006)
2	Tri-block	Three monomers		(Altintas et al. 2015)
	ABA type	Here two monomers are chemically equivalent	A-A-A-B-B-B-A-A-A	(Guo et al. 2021)
	ABC type	All three monomers are chemically diverse	A-A-A-B-B-B-C-C-C	(Gupta et al. 2017)
3	Graft	These are segmented polymers w-ith a backbone chain of one monomer and various branches of another, portraying different chemical compositions	A-A-A-A-A-A-A-A-A-A-A / | B B \ | B B	(Thakur et al. 2013)

in an aqueous solvent which is determined by hydrophilic volume fraction (Bahadur et al. 2015). Block copolymers depicting a hydrophilic volume fraction of 25%, 35%, and 45% will form inverted micelles, polymeric vesicles, and spherical micelles, respectively (Won et al. 2002).

3.2.2 Micelle Size and CMC

The surface tension of the micelles is observed to change dynamically below the CMC, so studying the relationship between CMC and the formation of micelles becomes critical (Church et al. 2021). Micelles are self-assembled moieties composed of amphiphilic copolymers that depict a particle size range of 10–100 nm (Zhou et al. 2016). The CMC is the concentration above which the surfactant forms micelles (Shi et al. 2011). The factors affecting the micellar size and CMC include temperature, hydrophobic chain length, the behavior of the polar group, and the surfactant counterions (Di Michele et al. 2011). The value of CMC decreases logarithmically with an increase in the size of the hydrophobic chain (Zhao and Zheng 2011). A lower CMC depicts enhanced micellar stability. The conventional micelles have a CMC value in the range of 10^{-3}–10^{-4} M. In contrast, polymeric micelles have a lower CMC of 10^{-6}–10^{-7} M (Mandal et al. 2017). Yang et al. prepared redox-sensitive micelles for liver cancer. This study selected poly-(N-ε-carbobenzyloxy-L-lysine) (PZLL) as the hydrophobic chain. With an increase in the DP of PZLL blocks from 10 to 25, a marked decrease in the CMC values from 0.0145 to 0.0072 mg/mL was observed (Yang et al. 2020).

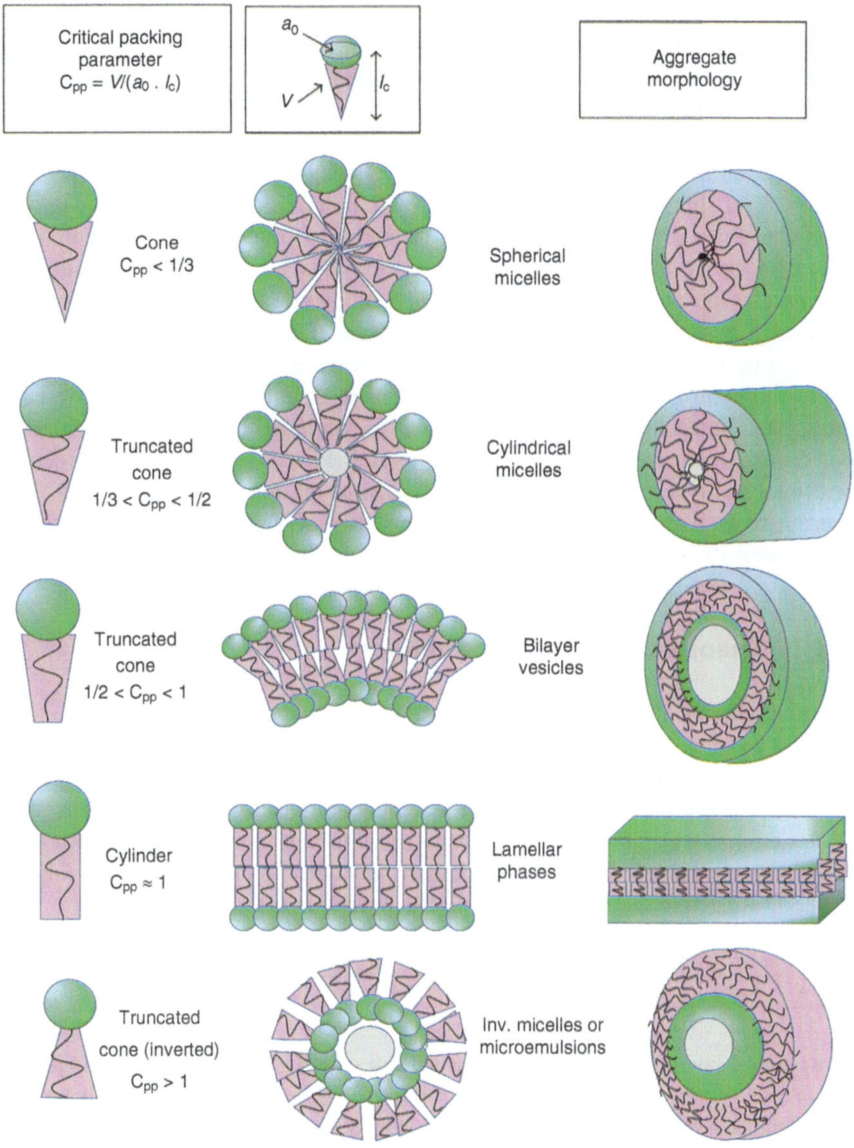

Fig. 3.2 Various self-assembled structures predicted from critical packing parameter C_{pp}. (Reproduced with permission from Lombardo et al. (Lombardo et al. 2015) Copyright © 2015 published by Hindawi Publishing Corporation)

3.2.3 Hydrogen Bonding

Hydrophobic interactions are the driving force for micelle formation (Maibaum et al. 2004). The self-assembling of the ACs into micelles is mediated by various interaction parameters like electrostatic interactions, coordinate bonds, and hydrogen bonding between the polymers (Li et al. 2020). For small hydrophobic molecules, water tends to reorganize itself around the hydrophobic solute without losing its hydrogen bond, which attributes to its low solubility (Schmid 2001) while in the presence of a bulky hydrophobic moiety, it becomes challenging to maintain the hydrogen bond interactions geometrically (Krieg et al. 2016). Techniques like Monte Carlo simulations are used to study the relationship between the hydrogen bond dynamics of the solvent and the formation of micelles from amphiphiles. Heinzelmann et al. performed Monte Carlo simulations using a solution of amphiphiles in water to determine hydrogen bonding in water and micellar dynamics between a non-micellized and a micellized state. The results depicted a gradual decrease in the number of hydrogen bonds with the growth of micelles, while the number of hydrogen bonds increased with an increase in the exposed surface area of the solute (Heinzelmann et al. 2012). Hu et al. depicted that the complexation of polydimethylsiloxane-graft-poly(ethylene oxide) (PDMS-g-PEO) and poly(acrylic acid)-block-polyacrylonitrile (PAA-b-PAN) led to the formation of micelles. This was due to the hydrogen bonding between the ether oxygen and the carboxyl groups of PEO and PAA (Hu et al. 2010).

3.3 Theories of Micelle Formation

The hydrophobic effect is responsible for amphiphilic molecules propensity to form micelles in an aqueous solution, as discussed in the previous section. The stoichiometric constraint that prevents the demixing of polar and hydrophobic groups is the considerable gap between the micellar assembly and macroscopic phase separation (Boles et al. 2016). This section discusses the different mechanisms governing the formation of micelles. The structure-property relationship and computer simulation methods utilized in the dynamics of micelle formation are also covered.

3.3.1 Law of Mass Action

Mass action law applies to the micellar system which is considered a homogenous one-phase system (Chen et al. 2005). In this method, aggregation of surfactants is regarded as a reversible chemical reaction and is represented by the following equation:

$$\sum_i nB \leftrightarrow M$$

wherein M represents micelle, n and B are the aggregation numbers.

The above equation is a fast reaction because of the short half-lives of surfactant and micelles (Rusanov 1993). Amphiphiles tend to form aggregates of various sizes and shapes. In this law, the aggregation number portrays the role of the stoichiometric coefficient. The aggregation number is directly proportional to the surfactant concentration; however, an insignificant change in aggregation number is observed in the CMC region (Maibaum et al. 2004).

3.3.2 Flory-Huggin's Interaction Parameters

Paul Flory and Maurice Huggins introduced Flory-Huggin's interaction parameters to study interaction effects in their lattice model of polymeric solutions (Russell et al. 2014). It is a dimensionless quantity and is represented by χ. This parameter is employed to demonstrate thermodynamic properties like swelling ability and solubility equilibrium of amphiphilic micelles. The interaction parameter also postulates the amount of inconsistency between the subsequent amphiphilic blocks. The degree of microphase segregation is governed by the segregation product represented as χN. This parameter takes into consideration the enthalpy of the solvent and polymer mixture and is expressed as

$$\chi = (\delta_1 - \delta_2)^2 \frac{V}{RT}$$

here, V is the molar volume of the solvent, δ_1 and δ_2 are solubility parameters of the solvent and polymer.

According to this theory, two moieties depicting $\chi < 0.5$ is said to be soluble, whereas $\chi > 0.5$ is designated as phase separation. A group of scientists determined the solubility of eleven drugs in micelles and found that the solubility decreased with an increase in χ (Forrest et al. 2006).

3.3.3 Molecular Dynamics

Molecular dynamics (MD) is an emerging tool for investigating atomistic and mesoscale dynamics (Lau et al. 2018). MD was used for the first time to determine the self-assembling nature of surfactants. The observations depicted that the surfactants formed micelles in an aqueous medium, whereas reverse micelles were obtained in a lipophilic medium (Klein and Shinoda 2008; Seif and Montazeri 2022). The MD simulations compute and compare numerous equilibrium properties of the surfactants starting from their target phase with the experimentally available raw data (Farafonov et al. 2020). MD can also interpret the dynamic properties like stability parameters and rheology of the formed micelles (Liu et al. 2019). The more prominent role of MD simulations is to determine the aggregation of surfactant molecules into micelles by investigating their surface structure and hydrophobic nature (Jafari et al. 2018). The results obtained from MD simulations depicted that the increase in the number of amphiphilic groups led to an increase in the aggregation of micelles (Crespo et al. 2020). Luna et al. employed 1-alkyl-3-methylimidazolium cations in an aqueous mixture of ionic liquids to determine the

CMC and the formation of micelles using MD simulations. The simulation results were consistent with those from the experimental evidence (Vicent-Luna et al. 2017).

3.3.4 Quantitative Structure-Property Relationship

The former discussed approach is inadequate for larger data sets and is a time-consuming method. Simulations in MD encounter the necessity of enormous cost and computational power (Liu et al. 2017). Hence, statistical methods like quantitative structure-property relationship (QSPR) could be employed to estimate drug and polymer compatibility in micelles (Upadhya et al. 2021). QSPR is extensively utilized in toxicology and medicinal chemistry to determine the biocompatibility and efficacy of small drugs (Cern et al. 2012). It involves the statistical analysis of large data sets, forecasting drug solubilization in micelles, drug loading in micelles, and cheminformatics study of micelles. Wu et al. utilized QSPR modeling and genetic function approximation for determining the influence of polymer configuration on the drug loading capacity of doxorubicin-loaded polymeric micelles. The results depicted that QSPR modeling method was an optimized and good fitting approach to predict drug loading capabilities in micelles (Wu et al. 2015).

3.4 Thermodynamics of Micelle Formation

Surfactants are composed of amphiphilic structures of hydrophilic and hydrophobic groups. Hydrophobic chains repel the surrounding water molecules due to their diminished affinity. This causes an increase in the surface free energy, which further increases the Gibbs free energy at the micelle-water interface. Surfactants tend to reduce their surface free energy to attain an equilibrium in two ways, i.e., by concentrating at the surface via aligning the hydrophobic chains toward the air, known as adsorption (Reeve et al. 2021). The other scenario includes the aggregation of surfactants with hydrophobic chains oriented toward the interior while hydrophilic chains face the exterior with a continuous aqueous phase. This process is known as micellization (Jover et al. 2021). Micelle formation is governed by a string of factors like temperature, CMC, ionizing groups, aggregation number, hydrophobic and hydrophilic chain length and volume, continuous phase viscosity, etc. (Sikder et al. 2022). Standard thermodynamic parameters could help in understanding the dynamics of the process undergone by surfactant molecules, i.e., either adsorption or micelle formation. Generally, for ionized surfactants, the degree of dissociation of the counter ion plays an essential role in governing micelle dynamics. Xi et al. employed the postulates of the density functional theory to study thermodynamics, alteration in the microstructure, and solubilization of block co-polymeric micelles. Inhomogeneous statistical associating fluid theory was used to contemplate the structure formation of block copolymers in dilute and concentrated solutions. They unveiled that hexagonal, lamellar, and cubic mesophasic micelles existed with an increase in concentration compared to spherical micelles at relatively dilute concentrations. Temperature negatively influenced CMC, while aggregation number positively correlated with the length of the hydrophobe (Xi et al. 2019).

3.5 Techniques to Evaluate Micelle Formation Dynamics

3.5.1 Isothermal Titration Calorimetry

Isothermal titration calorimetry considers the change in heat flow at a constant temperature which is correlated with the measurement of assembly in micelles. Kadam and colleagues visualized the exothermic heat of Tetronic® T904 produced in a sigmoidal manner. Concentration below CMC was distinguished by solid enthalpy released, resulting in the disassembly of micelles to the generation of monomers. A decrease in the enthalpy chain with an increasing concentration near CMC demonstrated micelles' formation. However, a significant drawback of this method is that the CMC at temperatures lower than 25 °C and higher than 45 °C cannot be accurately estimated. At lower temperatures, feeble enthalpy effects are observed. In contrast, at higher temperatures, an association of surfactant monomers to micelles is much higher and corresponds to pseudo-CMC, indicating erroneous thermodynamic results for micellar formation (Kadam et al. 2010; Tso et al. 2020; Sarolia et al. 2022).

3.5.2 Förster Resonance Energy Transfer

Förster resonance energy transfer (FRET) includes the exchange of energy from an excited molecule termed as a donor to another molecule termed as acceptor. The advantage of this technique includes the measurement of interaction of nearby dipoles depending on Förster distance. Bos and coworkers employed FRET to devise a model to study the underlying macromolecular exchange rates in complex coacervate micelles among oppositely charged ions. They showed that the exchange rate was strongly influenced by the ionic strength of the aqueous phase. The understanding of thermodynamics was enriched by the structural association and dissociation of micellar assembly. FRET is useful for studying micellar dynamics at lower concentrations compared to other techniques preferably with ionic interactions governing expulsion and insertion characteristics (Bos et al. 2021).

3.5.3 Conductometry

The alteration in conduction of surfactant molecules and assembled micellar structures could result in the quantification of thermodynamic parameters governing micelle formation. Thoppil et al. evaluated Gibb's free energy, enthalpy and entropy of micelle formation using conductometric principles. In the pre-micellar region, the increase in conductivity was observed with increasing surfactant concentration attributed to the increasing amount of free ions. During micelle formation, the alteration in slope led to the reduction of ionic moieties. Further, the decrease in the slope of the post-micellar zone results in the effective loss of conducting ions owing to the presence of counter ions on the surface. This further results in size increment and lower micellar mobility (Thoppil et al. 2020). The calculation of Gibb's free energy, enthalpy and entropy of micellization was done using the application of pseudo-phase order equation respectively given by,

$$\Delta G_m{}^{\circ} = (1+\beta)RT \ \ln X_{\mathrm{CMC}}$$

$$\Delta H_m{}^{\circ} = -(1+\beta)RT^2 \ {\partial \ln X_{\mathrm{CMC}}}/{\partial T}$$

$$\Delta S_{\mathrm{m}}{}^{\circ} = {\Delta H_{\mathrm{m}}{}^{\circ} - \Delta G_{\mathrm{m}}{}^{\circ}}/{T}$$

wherein β—degree of association of counter ion, ΔG_{m}—change in Gibb's free energy, ΔH_{m}—change in enthalpy, and ΔS_{m}—change in entropy of micelle formation.

3.5.4 Dissipative Particle Dynamics

Dissipative particle dynamics (DPD) is a stochastic mesoscopic simulation process that delineates clusters of fragments collectively moving in a Lagrangian manner when subjected to random, dissipative forces and conservative forces (Barnes et al. 2019). DPD utilizes a group of subtle intermingling particles which are employed to simulate a fluid medium. Individual bead will symbolize several atoms (Ramezani and Shamsara 2016). The dynamics behind these beads is governed by Newton's second law and is given as

$$\frac{\mathrm{d}r_i}{\mathrm{d}t} = v_i, m_i\frac{\mathrm{d}v_i}{\mathrm{d}t} = f_i$$

herein, m_i is the mass of the bead which is assumed to be unity, r_i, v_i, f_i are position vector, velocity, and total force on the particle i. Masses of all beads is fixed as 1 DPD unit.

The conventional DPD cannot work in certain cases like non-isothermal and two-phase fluid systems due to the lack of necessary degrees of freedom (Dzwinel and Yuen 2000). To overcome this problem, extensions of DPD could be used like Smoothed DPD (Müller et al. 2015), Energy Conserving DPD (Li et al. 2014), Fluid Particle Model (Dzwinel et al. 2002), and Many-Body DPD (Arienti et al. 2011). Wang and his coworkers performed DPD simulations to achieve the delivery of multiple cargos using a single micellar structure. The results revealed that different payloads occupied their peculiar preferred locations on the basis of their structure (Wang et al. 2022).

3.6 Conclusion

Micelles are self-assembling aggregates designed using amphiphilic copolymers. These nanocarriers are gaining highlights among researchers because of their ability to improve the bioavailability of hydrophobic drugs, imparting target specificity with sustained release, improving therapeutic efficiency, responding to stimuli, and reducing toxicity. In this chapter, we have conveyed the fundamental dynamics behind the micelle formation and the underlying theories. The micelles which are fabricated from ACs are the perfect nanocarriers for drug delivery in the management and diagnosis of different disease conditions. The role of ACs, types of block copolymers, and how the value of packing parameters will influence the morphology of micelles are discussed. The effect of CMC and hydrogen bonding in micelle formation along with its solubility and stability is also conveyed through this chapter. Aggregation number determines the size and morphology of the micelles and is explained by the mass action law. The solubility of amphiphilic polymers is further explained by the Flory-Huggin's interaction parameter by taking enthalpy of solvent into consideration.

Micelles are believed to be in a dynamic equilibrium with frequently deforming and reforming. The MD simulation studies could be a useful tool to study the behavior of active pharmaceutical ingredients and payloads in a solvent medium. However, the unmet needs associated with this technique include the inability to determine the values for large data sets and is an expensive method. QSPR modeling was introduced to compute larger data sets in lesser time and cost.

The evaluation methods including dynamic light scattering, differential scanning calorimetry, etc. will determine the morphology, size, and CMC of micelles. However, the detailed dynamics of micelle formation could be only determined via molecular level simulations like DPD. The variations in heat flow, conducting property of surfactants, and energy transfer between moieties should be studied to determine the formation of micelles which is also deliberated in this chapter.

Acknowledgments The authors would like to acknowledge the research funding support by the Department of Pharmaceuticals (DoP), Ministry of Chemicals and Fertilizers, Govt. of India to National Institute of Pharmaceutical Education and Research (NIPER), Hyderabad, INDIA.

Declaration of Interest The authors declare that there is no conflict of interest.

References

Sasidharan M, Nakashima K (2014) Core-shell-corona polymeric micelles as a versatile template for synthesis of inorganic hollow nanospheres. Acc Chem Res 47(1):157–167

Karabasz A, Bzowska M, Szczepanowicz K (2020) Biomedical applications of multifunctional polymeric nanocarriers: a review of current literature. Int J Nanomedicine 15:8673–8696

Krishnan V, Mitragotri S (2020) Nanoparticles for topical drug delivery: potential for skin cancer treatment. Adv Drug Deliv Rev 153:87–108

Domínguez A, Suárez-Merino B, Goñi-De-Cerio F (2014) Nanoparticles and blood-brain barrier: the key to central nervous system diseases. J Nanosci Nanotechnol 14(1):766–779
Lu Y, Park K (2013) Polymeric micelles and alternative nanonized delivery vehicles for poorly soluble drugs. Int J Pharm 453(1):198–214
Mollazadeh S, Mackiewicz M, Yazdimamaghani M (2021) Recent advances in the redox-responsive drug delivery nanoplatforms: a chemical structure and physical property perspective. Mater Sci Eng C 118:111536
Li C, Wang J, Wang Y, Gao H, Wei G, Huang Y et al (2019) Recent progress in drug delivery. Acta Pharm Sin B 9(6):1145–1162
Li Z, Tan BH, Lin T, He C (2016) Recent advances in stereocomplexation of enantiomeric PLA-based copolymers and applications. Prog Polym Sci 62:22–72
Maghsoudnia N, Eftekhari RB, Sohi AN, Zamzami A, Dorkoosh FA (2020) Application of nano-based systems for drug delivery and targeting: a review. J Nanopart Res 22(8)
Cabral H, Kataoka K (2014) Progress of drug-loaded polymeric micelles into clinical studies. J Control Release 190:465–476
Qiu LY, Bae YH (2006) Polymer architecture and drug delivery. Pharm Res 23(1):1–30
Chu Z, Feng Y (2012) Empirical correlations between krafft temperature and tail length for amidosulfobetaine surfactants in the presence of inorganic salt. Langmuir 28(2):1175–1181
Zhang Q, Lin J, Wang L, Xu Z (2017) Theoretical modeling and simulations of self-assembly of copolymers in solution. Prog Polym Sci 75:1–30
Zhang X, Waymouth RM (2017) 1,2-dithiolane-derived dynamic, covalent materials: cooperative self-assembly and reversible cross-linking. J Am Chem Soc 139(10):3822–3833
Gu L, Faig A, Abdelhamid D, Uhrich K (2014) Sugar-based amphiphilic polymers for biomedical applications: from nanocarriers to therapeutics. Acc Chem Res 47(10):2867–2877
Hawker CJ, Bosman AW, Harth E (2001) New polymer synthesis by nitroxide mediated living radical polymerizations. Chem Rev 101(12):3661–3688
Perin F, Motta A, Maniglio D (2021) Amphiphilic copolymers in biomedical applications: synthesis routes and property control. Mater Sci Eng C 123:111952
Schmidt BVKJ (2018) Double hydrophilic block copolymer self-assembly in aqueous solution. Macromol Chem Phys 219(7):1700494
Garnier S, Laschewsky A (2006) New amphiphilic diblock copolymers: surfactant properties and solubilization in their micelles. Langmuir 22(9):4044–4053
Altintas O, Artar M, Ter Huurne G, Voets IK, Palmans ARA, Barner-Kowollik C et al (2015) Design and synthesis of triblock copolymers for creating complex secondary structures by orthogonal self-assembly. Macromolecules 48(24):8921–8932
Guo Y, Yu Y, Shi K, Zhang W (2021) Synthesis of ABA triblock copolymer nanoparticles by polymerization induced self-assembly and their application as an efficient emulsifier. Polym Chem 12(4):572–580
Gupta MK, Martin JR, Dollinger BR, Hattaway ME, Duvall CL (2017) Thermogelling, ABC triblock copolymer platform for resorbable hydrogels with tunable, degradation-mediated drug release. Adv Funct Mater 27(47):1704107
Thakur VK, Thakur MK, Gupta RK (2013) Graft copolymers from cellulose: synthesis, characterization and evaluation. Carbohydr Polym 97(1):18–25
Jiao W, Yang H, Wu Z, Liu J, Zhang W (2020) Self-assembled block polymer aggregates in selective solution: controllable morphology transitions and their applications in drug delivery. Expert Opin Drug Deliv 17(7):947–961
Lombardo D, Kiselev MA, Magazù S, Calandra P (2015) Amphiphiles self-assembly: basic concepts and future perspectives of supramolecular approaches. Adv Condens Matter Phys 2015
Bahadur A, Cabana-Montenegro S, Aswal VK, Lage EV, Sandez-Macho I, Concheiro A et al (2015) NaCl-triggered self-assembly of hydrophilic poloxamine block copolymers. Int J Pharm 494(1):453–462

Won YY, Brannan AK, Davis HT, Bates FS (2002) Cryogenic transmission electron microscopy (cryo-TEM) of micelles and vesicles formed in water by poly(ethylene oxide)-based block copolymers. J Phys Chem B 106(13):3354–3364
Church J, Willner MR, Renfro BR, Chen Y, Diaz D, Lee WH et al (2021) Impact of interfacial tension and critical micelle concentration on bilgewater oil separation. J Water Process Eng 39: 101684
Zhou W, Li C, Wang Z, Zhang W, Liu J (2016) Factors affecting the stability of drug-loaded polymeric micelles and strategies for improvement. J Nanopart Res 18(9):1–18
Shi Y, Luo HQ, Li NB (2011) Determination of the critical premicelle concentration, first critical micelle concentration and second critical micelle concentration of surfactants by resonance Rayleigh scattering method without any probe. Spectrochim Acta A Mol Biomol Spectrosc 78(5):1403–1407
Di Michele A, Brinchi L, Di Profio P, Germani R, Savelli G, Onori G (2011) Effect of head group size, temperature and counterion specificity on cationic micelles. J Colloid Interface Sci 358(1): 160–166
Zhao M, Zheng L (2011) Micelle formation by N-alkyl-N-methylpyrrolidinium bromide in aqueous solution. Phys Chem Chem Phys 13(4):1332–1337
Mandal A, Bisht R, Rupenthal ID, Mitra AK (2017) Polymeric micelles for ocular drug delivery: from structural frameworks to recent preclinical studies. J Control Release 248:96–116
Yang H, Miao Y, Chen L, Li Z, Yang R, Xu X et al (2020) Redox-responsive nanoparticles from disulfide bond-linked poly-(N-ε-carbobenzyloxy-L-lysine)-grafted hyaluronan copolymers as theranostic nanoparticles for tumor-targeted MRI and chemotherapy. Int J Biol Macromol 148: 483–492
Maibaum L, Dinner AR, Chandler D (2004) Micelle formation and the hydrophobic effect. J Phys Chem B 108(21):6778–6781
Li Y, Wang N, Huang X, Li F, Davis TP, Qiao R et al (2020) Polymer-assisted magnetic nanoparticle assemblies for biomedical applications. ACS Appl Bio Mater 3(1):121–142
Schmid R (2001) Recent advances in the description of the structure of water, the hydrophobic effect, and the like-dissolves-like rule. Monatshefte Chemie 132(11):1295–1326
Krieg E, Bastings MMC, Besenius P, Rybtchinski B (2016) Supramolecular polymers in aqueous media. Chem Rev 116(4):2414–2477
Heinzelmann G, Figueired W, Girardi M (2012) Micellar dynamics and water-water hydrogen-bonding from temperature-jump Monte Carlo simulations. Chem Phys Lett 550:83–87
Hu A, Cui Y, Wei X, Lu Z, Ngai T (2010) Hydrogen-bonding-induced complexation of polydimethylsiloxane-graft- poly(ethylene oxide) and poly(acrylic acid)-block-polyacrylonitrile micelles in water. Langmuir 26(18):14502–14508
Boles MA, Engel M, Talapin DV (2016) Self-assembly of colloidal nanocrystals: from intricate structures to functional materials. Chem Rev 116(18):11220–11289
Chen J, Spear SK, Huddleston JG, Rogers RD (2005) Polyethylene glycol and solutions of polyethylene glycol as green reaction media. Green Chem 7(2):64–82
Rusanov AI (1993) The mass action law theory of micellar solutions. Adv Colloid Interf Sci 45 (C):1–78
Russell TH, Edwards BJ, Khomami B (2014) Characterization of the Flory-Huggins interaction parameter of polymer thermodynamics. EPL 108(6):66003
Forrest ML, Zhao A, Won CY, Malick AW, Kwon GS (2006) Lipophilic prodrugs of Hsp90 inhibitor geldanamycin for nanoencapsulation in poly(ethylene glycol)-b-poly(ε-caprolactone) micelles. J Control Release 116:139–149
Lau D, Jian W, Yu Z, Hui D (2018) Nano-engineering of construction materials using molecular dynamics simulations: prospects and challenges. Compos Part B Eng 143:282–291
Klein ML, Shinoda W (2008) Large-scale molecular dynamics simulations of self-assembling systems. Science 321(5890):798–800
Seif M, Montazeri A (2022) Molecular dynamics simulation reveals the reliability of Brij-58 nanomicellar drug delivery systems for flurbiprofen. J Mol Liq 360:119496

Farafonov VS, Lebed AV, McHedlov-Petrossyan NO (2020) Computing pKa Shifts Using Traditional Molecular Dynamics: Example of Acid-Base Indicator Dyes in Organized Solutions. J Chem Theory Comput 16(9):5852–5865

Liu Q, Ji X, Wang S, Zou W, Li J, Lv D et al (2019) Effect of additives on surfactant micelle shape transformation: rheology and molecular dynamics studies. J Phys Chem C 123(5):2922–2932

Jafari M, Mehrnejad F, Rahimi F, Asghari SM (2018) The molecular basis of the sodium dodecyl sulfate effect on human ubiquitin structure: a molecular dynamics simulation study. Sci Rep 8(1):2150

Crespo EA, Schaeffer N, Coutinho JAP, Perez-Sanchez G (2020) Improved coarse-grain model to unravel the phase behavior of 1-alkyl-3-methylimidazolium-based ionic liquids through molecular dynamics simulations. J Colloid Interface Sci 574:324–336

Vicent-Luna JM, Romero-Enrique JM, Calero S, Anta JA (2017) Micelle formation in aqueous solutions of room temperature ionic liquids: a molecular dynamics study. J Phys Chem B 121(35):8348–8358

Liu K, Watanabe E, Kokubo H (2017) Exploring the stability of ligand binding modes to proteins by molecular dynamics simulations. J Comput Aided Mol Des 31(2):201–211

Upadhya R, Kosuri S, Tamasi M, Meyer TA, Atta S, Webb MA et al (2021) Automation and data-driven design of polymer therapeutics. Adv Drug Deliv Rev 171:1–28

Cern A, Golbraikh A, Sedykh A, Tropsha A, Barenholz Y, Goldblum A (2012) Quantitative structure - property relationship modeling of remote liposome loading of drugs. J Control Release 160(2):147–157

Wu W, Zhang C, Lin W, Chen Q, Guo X, Qian Y et al (2015) Quantitative structure-property relationship (QSPR) modeling of drug-loaded polymeric micelles via genetic function approximation. PLoS One 10(3):e0119575

Reeve JR, Thomas RK, Penfold J (2021) Surface activity of ethoxylate surfactants with different hydrophobic architectures: the effect of layer substructure on surface tension and adsorption. Langmuir 37(30):9269–9280

Jover A, Fraga F, Meijide F, Vázquez Tato J, Cautela J, Del Giudice A et al (2021) Revealing the complex self-assembly behaviour of sodium deoxycholate in aqueous solution. J Colloid Interface Sci 604:415–428

Sikder A, Vambhurkar G, Amulya E, Bagasariya D, Famta P, Shah S et al (2022) Advancements in redox-sensitive micelles as nanotheranostics: a new horizon in cancer management. J Control Release 349:1009–1030

Xi S, Wang L, Liu J, Chapman W (2019) Thermodynamics, microstructures, and solubilization of block copolymer micelles by density functional theory. Langmuir 35(14):5081–5092

Kadam Y, Singh K, Marangoni DG, Ma JH, Aswal VK, Bahadur P (2010) Thermodynamic of micelle formation of nonlinear block co-polymer Tetronic® T904 in aqueous salt solution. Colloids Surfaces A Physicochem Eng Asp 369(1–3):121–127

Tso SC, Mahler F, Höring J, Keller S, Brautigam CA (2020) Fast and robust quantification of detergent micellization thermodynamics from isothermal titration calorimetry. Anal Chem 92(1):1154–1161

Sarolia J, Kumar D, Shah SA, Bahadur P, Tiwari S (2022) Thermodynamics of pluronic 103 micellization in mannitol solution: analyses based on isothermal titration calorimetry. Colloids Surfaces A Physicochem Eng Asp 648:129240

Bos I, Timmerman M, Sprakel J (2021) FRET-based determination of the exchange dynamics of complex coacervate Core micelles. Macromolecules 54(1):398–411

Thoppil AA, Chennuri BK, Gardas RL (2020) Thermodynamics and micellization behavior of ethanolammonium carboxylate surface active ionic liquids in aqueous media. J Mol Liq 299: 112116

Barnes BC, Brennan JK, Byrd EFC, Izvekov S, Larentzos JP, Rice BM (2019) Toward a predictive hierarchical multiscale modeling approach for energetic materials. Challenges Adv Comput Chem Phys 28:229–282

Ramezani M, Shamsara J (2016) Application of DPD in the design of polymeric nano-micelles as drug carriers. J Mol Graph Model 66:1–8

Dzwinel W, Yuen DA (2000) Matching macroscopic properties of binary fluids to the interactions of dissipative particle dynamics. Int J Mod Phys C 11(1):1–25

Müller K, Fedosov DA, Gompper G (2015) Smoothed dissipative particle dynamics with angular momentum conservation. J Comput Phys 281:301–315

Li Z, Tang YH, Lei H, Caswell B, Karniadakis GE (2014) Energy-conserving dissipative particle dynamics with temperature-dependent properties. J Comput Phys 265:113–127

Dzwinel W, Yuen DA, Boryczko K (2002) Mesoscopic dynamics of colloids simulated with dissipative particle dynamics and fluid particle model. J Mol Model 8(1):33–43

Arienti M, Pan W, Li X, Karniadakis G (2011) Many-body dissipative particle dynamics simulation of liquid/vapor and liquid/solid interactions. J Chem Phys 134(20):204114

Wang Z, Zhou J, Wang J, Wang M, Lyu Q, Li C et al (2022) Interplay of distributions of multiple guest molecules in block copolymer micelles: a dissipative particle dynamics study. J Colloid Interface Sci 607:1142–1152

Types of Polymeric Micelles for Controlled Drug Release

4

Divya Mahajan, Divya Suares, R. Rachana, and Saritha Shetty

Abstract

Polymeric micelles are nano-sized drug delivery carrier systems with a core or shell structure composed of biocompatible and biodegradable amphiphilic block copolymers. Surface functional groups on polymeric micelles allow other molecules to attach to them. Notably, the architectural arrangement of these polymeric micelles offers the benefits of shielding water-sensitive molecules and solubilizing poorly water-soluble drugs via encapsulation. Additionally polymeric micelles serve as carriers having solid-like core that enables them to remain intact for longer hours and provide controlled release of drugs. These polymeric micelles primarily function by circumventing the host defense mechanism, by prolonging the drugs systemic circulation time, reducing the drugs toxicity, and facilitating targeted drug delivery, especially in case of tumors. Literature cites several types of polymeric micelles that have been developed to deliver actives for various illnesses. They can be categorized based on the systems in-built trigger mechanism, such as pH-sensitive, redox-sensitive, glucose-responsive, reacting oxygen species-responsive, hypoxia-responsive, enzyme-triggered, dual-responsive, multi-responsive and extrinsic stimuli-responsive polymeric micelles. Recent advances have proven the feasibility of generating nucleic acid-based, phytoconstituent-loaded, surface engineered, ligand-conjugated, protein-based, and biosensor-based polymeric micelles. The

D. Mahajan · D. Suares · S. Shetty (✉)
Shobhaben Pratapbhai Patel School of Pharmacy & Technology Management, SVKM's Narsee Monjee Institute of Management Studies (Deemed to be University), Mumbai, India
e-mail: Saritha.Shetty@nmims.edu

R. Rachana
Department of Biotechnology, JIIT Noida, Noida, UP, India

S. K. Singh et al. (eds.), *Polymeric Micelles: Principles, Perspectives and Practices*,
https://doi.org/10.1007/978-981-99-0361-0_4

applications of polymeric micelles have been explored in the treatment and diagnosis of conditions such as arthritis, cancer, diabetes, to name a few. Nevertheless, they have also been studied as potential theranostic agents.

Keywords

Micelles · Polymeric nanoformulations · Surface engineering · Biosensors · Protein-based polymeric micelles

4.1 Introduction

Amphiphilic copolymers are composed of various polymers containing a hydrophobic core as well as hydrophilic shell. These copolymers have the property of self-assembly and aggregation above the critical micelle concentration (CMC). The micelles formed usually are spherical in shape, but depending on the type of polymers used and other physiological conditions that influence their formation, they can be formed into lamellar, cylindrical, or flower-like shape, etc. The size of these polymeric micelles ranges from around 10–100 nm, which may be influenced by the methodology employed in forming micelles, the molecular weight of the copolymers, aggregation of these polymers, the number of hydrophobic and hydrophilic chains (Almeida et al. 2018; Mourya et al. 2011).

The polymeric micelles having a hydrophobic core can encapsulate hydrophobic drugs within them for targeted delivery (as depicted in Fig. 4.1). They can also carry the drugs, which are degraded in specific pH or aqueous solutions until they reach the site of action. They can remain stable for longer durations under sink conditions and thus are good candidates for controlled release of the drugs. Owing to their small size, they are able to extravasate through capillaries into the tumor cells for targeting the drug directly on the tumor cells (Kwon and Okano 1996). Nonetheless, the micelles can bypass the renal clearance and get trapped in the reticuloendothelial system, which consists of phagocytic cells. However, apart from phagocytic function, they are also involved in causing cytotoxicity against the tumor cells, which will enhance the anti-tumor activity (Baas et al. 1994; Miyata et al. 2011). Over the past three decades, several polymeric micellar systems that respond to different

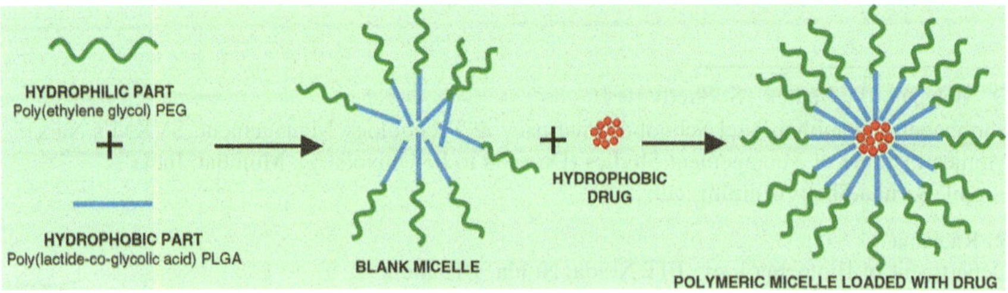

Fig. 4.1 Illustration of hydrophobic drug entrapped within the polymeric micellar system

kinds of stimuli have been developed and experimented in the preclinical stages (Hwang et al. 2020) and it has been found that the polymeric micelles not only provide drug delivery at a specific site but also help reduce toxicities and improve drug solubilization (Croy and Kwon 2006; Aliabadi and Lavasanifar 2006).

4.2 Types of Polymeric Micelles

The polymeric micelles have a tendency to be cleaved or disassemble in situ in the presence of various factors such as changes in pH, temperature, redox potential, enzymes, and so on. In the following chapter, the polymeric micelles have been characterized based on the trigger mechanisms and applications. Figure 4.2 illustrates the activation of polymeric micelles in presence of various stimuli, for anticancer activity.

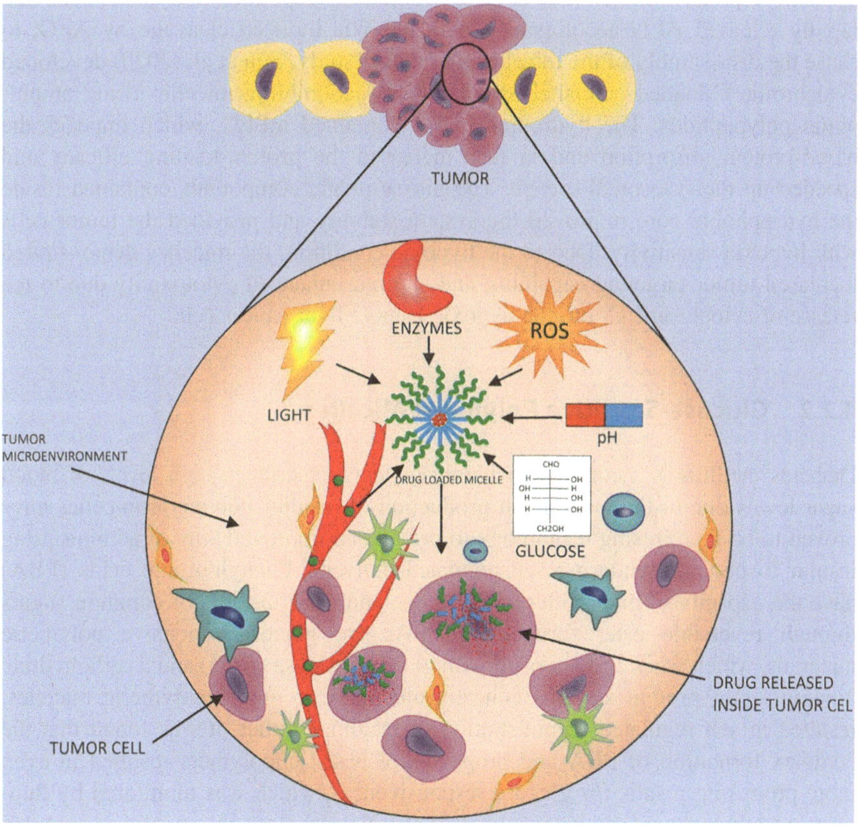

Fig. 4.2 Illustration of delivery of polymeric micelles based on various trigger mechanisms in tumor cells

4.2.1 Hypoxia-Responsive Polymeric Micelles

Hypoxia is a condition of inadequate supply of oxygen to the tissues. It is a characteristic feature for most tumor cells and hence a target for polymeric micelles to deliver the drugs in order to bring about tumor cell apoptosis. Feng et al. (2020) developed polymeric micelles loaded with doxorubicin (DOX) to target hypoxic cells and studied their effect on MFC-7 cells for breast adenocarcinoma. The polymers used were polyethylene glycol (PEG) and poly(glutamic acid(3-(2-nitro-imidazolyl)-propyl)) (P(LGlu-NI)). In the study, the hypoxia-responsive micelles contained 2-nitroimidazole in the micellar hydrophobic core. When in contact with the hypoxic conditions of the tumor cell, 2-nitroimidazole reduced to a hydrophilic amino imidazole. Due to this, the polymeric micellar system disintegrated and released DOX into the tumor microenvironment. On similar lines, Long et al. (2021) targeted hypoxic cells in the bone marrow for bone metastasis by using PEG and poly-L-lysine (PLL) based copolymer. The polymeric micellar system was loaded with alendronate (ALN), a bone target moiety and azobenzene (AZO), a hypoxia-responsive linker. The developed micelles responded to hypoxic cells and rapidly released ALN accompanied with hypoxia induced cleavage by AZO, to cause the disassembly of the micelles. In another study, Sun et al. (2020) developed cytochrome C-loaded core/shell-structured self-assembling micelles using amphiphilic polypeptides. The hydrophilic shell contained mPEG, which impedes the blood-protein absorption and in turn increased the protein-loading efficacy and speeded up the lysosomal escape. The nitroaromatic compounds contained inside the hydrophobic core improved the system stability and provided the tumor cells with hypoxia sensitivity. Due to the hypoxic condition, the micelles demonstrated increased tumor targeting capability and showed enhanced cytotoxicity due to the release of cytochrome C into the hypoxic HepG2 liver cancer cell.

4.2.2 Glucose-Sensitive Polymeric Micelles

Diabetes mellitus is a condition wherein patients experience high levels of blood sugar levels due to the insufficient production of insulin. Polymeric micelles have proved to be a promising technology to provide an improved way of administering insulin to patients rendering self-regulated delivery. Phenylboronic acids (PBA) have the capacity to bind with cis-1, 2-diols, and cis-1, 3-diols of common sugars through reversible ester formations. PBAs are glucose-responsive polymeric materials, which when incorporated with glucose oxidase (GOx) and a carbohydrate binding lectin protein such as concanavalin A (Con A) as polymeric micelles, resulted in self-regulated administration of insulin. Further, it was found that the complex formation of PBA and propane-diol based copolymers resulted in even more promising results for glucose responsiveness, which was monitored by fluorescein isothiocyanate labeled insulin delivery. Nevertheless, the developed complex was easily triggered at physiological pH levels (Gaballa and Theato 2019). Despite being quite effective, the system developed by Gaballa and co-workers experienced a few drawbacks. Since GOx is an aerobic dehydrogenation enzyme, it could bring about the oxidation of glucose into gluconic acid and lead to the formation of hydrogen peroxide as a by-product, which is harmful to the biological tissues. Thus, polymeric micelles constituting phenylboronic esters (PBE) on the side of amphiphilic block copolymers were synthesized. The presence of PBE enabled the detachment of glucose from the PBA-glucose complex, binds the free glucose molecule, and causes a prompt breakdown of the micelle to release the encapsulated insulin from the system (Liu et al. 2020). On similar lines, Zhang et al. (2014) developed polymeric micelles containing two monomers, wherein (2-phenylboronic esters-1,3-dioxane-5-ethyl) methyl acrylate (PBDEMA) was employed as the glucose responsive monomer and (2-pyrenylboronic esters-1,3-dioxane-5-ethyl) methyl acrylate (PPyBDEMA) as the strong hydrophobic monomer. The release of insulin from the polymeric micelles exhibited a slow release at normoglycemic conditions (1 mg/mL) and comparatively a faster release at hyperglycemic conditions (3 mg/mL). Due to the alternating physiological levels of glucose conditions, the insulin release was observed to be switched on and off, making the system a self-regulated insulin delivery system.

4.2.3 Reduction-Sensitive Polymeric Micelles

Reactive oxygen species (ROS) are actively used against cancer cells to damage the cells and cause their apoptosis. This characteristic of ROS species has been exploited for the delivery of chemotherapeutics during chemotherapy. Cancer cells are generally habitual to oxidative stress caused due to various therapies and in response, they show an overexpression of antioxidants such as glutathione (GSH). These cause a recovery from the damage by the ROS and bring about an ineffectual treatment. In order to enhance the cancer treatment, polymeric micelles composed of methoxy poly(ethylene glycol) (mPEG)-S-S-poly (ε-caprolactone) (PCL)-protoporphyrin (Por) were loaded with DOX, which demonstrated GSH-scavenging and ROS-generating activity. This treatment strategy combined with chemotherapy and photodynamic therapy showed remarkable anti-tumor activity with the tested in vitro activity of IC50 of 0.041 μg/mL (Xia et al. 2021). In another study, DOX was incorporated within polycaprolactone-b-polyethylene glycol methyl ether methacrylate (PCL-SS-PPEGMA) polymers containing disulfide bonds. These polymeric micelles responded to the reduction stimulation and released DOX owing to the cleavage of disulfide bonds due to presence of high quantities of GSH released by the tumor cells; as GSH is a thiol-containing tripeptide that helps in the reduction and breakdown of disulfide bonds (Yang et al. 2021). In a study by Zhang et al. (2020a) hyaluronic acid-G-cystamine dihydrochloride-poly-e-(benzyloxycarbonyl)-L-lysine, (HA-ss-PLLZ) were used for the co-delivery of anticancer agents, paclitaxel (PTX) and apatinib (APA) against multi-drug resistant (MDR) cases of cancer. It was observed that hyaluronic acid (HA) receptors like receptor for hyaluronan-mediated motility (RHAMM) and CD44, a cell surface adhesion receptor, were overexpressed in many of the cancer cells, making HA a preferred choice of macromolecule for the delivery of anticancer drugs. Due to the presence of GSH, the polymeric micelles erupted, leading to the release of the chemotherapeutic agents at elevated intracellular concentrations and causing damage to the MDR cancer cells. A study by Li et al. (2016) used a strategy based on the charges of the micelles to release the chemotherapeutic agent within the tumor cell. A self-assembly of copolymers made up of poly (lactide) and PAEMA/DMMA with disulfide bonds was originally negatively charged. However, an acidic environment converted them into positively charged polymeric micelles, which were easily taken up by the tumor cells. On entering the tumor cells, the polymeric micelles were reduced by GSH to release the chemotherapeutic agents enclosed within them.

4.2.4 pH-Sensitive Polymeric Micelles

Polymeric micelles can be designed to release actives by being disassembled in the presence of specific pH conditions around the diseased tissues and deliver the actives at the desired site. Change in pH in the environment of the polymeric micelles may cause ionization of the polymers leading to their breakdown. The polymers that can be used for this type of release could be anionic, cationic, or acid labile polymers (Gao et al. 2013). In a study carried out to overcome the limitations of MDR tumor cells, pH-sensitive DOX-loaded polymeric micelles were formulated. The polymers used to develop the polymeric micellar system were poly(L-histidine) (Mn: 5K)-b-PEG (Mn: 2K)-folate (poly His/PEG-folate) and poly(L-lactic acid) (Mn: 3K)-b-PEG (Mn: 2K)-folate (PLLA/PEG-folate). The polymers were attached with folates to ensure binding of the micelles to the expressed folate receptors for tumor targeting. This in turn instigated receptor-mediated endocytosis as the preliminary mechanism for overcoming the poly-glycoproteins, which are expressed in MDR cases of cancer (Goren et al. 2000). Ionization led to the disruption of the micelles and the levels of DOX from micelle system that got entrapped within the cancer cells were found to be 20 times more than the free DOX administered (Lee et al. 2005). In a study by Li et al. (2017) polymeric micelles were explored for the treatment of inflamed joints using prednisolone (PD) as a model drug. The polymers employed for the study formed acid labile hydrazone bonds. When the PD-loaded polymeric micelles reached the synovial fluid, its acidic nature caused the hydrolysis of the hydrazone bonds and subsequent release of PD. In vivo studies done on arthritic mice showed that the PD levels in the plasma from the micelles were much higher than that of the free PD administered. In another study done by Woraphatphadung et al. (2018), pH-sensitive polymers were used to form micelles to encapsulate curcumin, which were used for targeting colorectal cancer. N-naphthyl-N,O-succinyl chitosan (NSCS) and N-octyl-N,O-succinyl chitosan (OSCS) were the polymers used. Depending on the pH variability of the various simulated fluids, it was found that the polymers released the maximum amount of curcumin in the simulated fluids of the colon. The curcumin-loaded polymeric micelles that were composed of NSCS showed high activity against HT29-colorectal cancer cells.

4.2.5 Enzyme-Responsive Polymeric Micelles

Our body releases several enzymes physiologically and during pathological conditions. The expression of some specific enzymes can be exploited as targets to deliver drugs through polymeric micelles. In prostate cancer, the enzyme matrixmetaloproteinase-2 (MMP-2) is overexpressed and can be used to cleave polymeric micelles. Cabazitaxel-loaded DUPA-coupled polymeric micelles composed of cholesterol as the hydrophobic core and PEG were developed. The release of encapsulated drug from the ligand-coupled polymeric micellar system was dependent on the cleavage of MMP2-responsive peptide. Owing to this, micelle demonstrated increased cellular uptake in prostate cancer cells as compared to free drug. Significantly, the micelle displayed greater inhibition of tumor growth in mice bearing prostate cancer xenografts when compared to unmodified micelle and free drug (Barve et al. 2020). In another study, DOX-loaded polymeric micelles comprising of poly(ethylene glycol)-blocked-poly(L-lysine) (PEG-b-PLL) as the polymer backbone and biotin as the targeting ligand molecule were designed. The micelles contained an enzyme-sensitive linkage that was prone to cleavage by the enzyme protease within the tumor cell, thus weakening the bond between the PLL segment and DOX and releasing the drug for its anti-tumor activity (Chen et al. 2015).

4.2.6 Dual-Responsive Polymeric Micelles

Literature cites few polymeric micelle systems, wherein two stimuli were explored to release the drug, thus providing dual-responsive polymeric micelles. The various stimuli included, but not restricted to, as combination of change in pH, temperature, presence of glucose, enzyme, and so on. Wei and researchers designed hydrazine functionalized Adriamycin (ADR)-loaded polymeric micelles, which were responsive to both pH and reduction. The micelles composed of block copolymers, PEO-b-P (MAA-g-Hyd), bind with ADR by forming a hydrazone structure. The micelle is further stabilized by crosslinking with dithiodiethanoic acid, which adds a disulfide bond to the micelle. The disulfide bond cleaves in presence of GSH in a cancer cell and the cleavage of hydrazone takes place in slight acidic conditions. The researchers found that low pH conditions led to an increase in the release of the drug; specifically, the release of the drug was 22.4%, 34.3%, and 53.6% in 48 h when pH was 7.4, 5.0, and 4.0, respectively. During the 120 h period, the release reached a total of 24.0%, 37.6%, and 63.6%, respectively. The dual responsiveness study was carried out by using logic gates to see if the system gave an "AND" response, which signifies a high output in both cases, i.e., high pH responsiveness and high output in reduction responsiveness (Wei et al. 2011). On similar lines, in order to overcome the poor water solubility and instability issues, the drug camptothecin (CPT) was designed into dual-responsive polymeric micelles. CPT was chemically bonded to mPEG through redox linkers, while polycaprolactone and mPEG units were connected through AZO linkages. Additionally, phenylboronic acid was added to the combination. The micelle once inside the cell cleaved to let out the encapsulated drug at the target site, while the two polymeric linkages were cleaved by GSH or azoreductase enzyme, making the polymeric micelle sensitive to both enzyme and redox degradation (Zhang et al. 2017). A similar approach of using the pH and reduction responsive polymeric micelles was used to damage DNA in mitochondria during cancer therapy. In these micelles, triphenylphosphonium (TPP) grafted poly(ethylene glycol) (PEG)-poly(D,L-lactide)(PLA); (TPP-PEG-SS-PLA) were used and have been coated with chondroitin sulfate (CS). The CS is initially negatively charged and in the acidic environment of lysosomes/endosomes, the CS gets removed and the polymer is exposed. Successively, in the cytoplasm, the micelles reach the mitochondrial outer membrane by TPP-mediated targeting, thereby causing a reduction in the membrane potential. This in turn causes an opening of the permeability transition pore. This initiates cell apoptosis due to the overproduction of ROS in the mitochondria. The released drug directly diffuses into the mitochondria resulting in mitochondrial DNA damage (Zhang et al. 2020b). Liu and co-workers developed a micelle system for the release of insulin, which was designed to provide a dual response to glucose as well as hydrogen peroxide. Poly (ethylene glycol)-block-poly(amino phenylboronic ester) (PEG-b-PAPBE) were the polymers explored for the study. Phenylboronic ester in the polymer breaks down in the presence of glucose and is hydrolyzed by H_2O_2. This leads to the release of the encapsulated insulin from the micelle. Another strategy that may be employed to elicit similar results would be by adding glucose oxidase (GoX) inside the micelle. This would oxidize the glucose leading to the release of hydrogen peroxide, which in turn could cause the hydrolysis of the ester bonds (Liu et al. 2020).

Dual-response based polymeric micelles for the treatment of RA were also explored, wherein pH- and enzyme-triggered polymeric micelles were developed. The inflamed tissues and joints are generally characterized by low pH conditions and overexpression of MMPs. In a study, dexamethasone-loaded polymeric micelles were designed, wherein both the triggers were utilized to cleave the polymeric micelles, resulting in the localized release of the drug (He et al. 2021). Another interesting approach for dual response includes the use of high intensity focused ultrasound (HIFU) in combination with redox responsiveness. PEG-S-S-PLA block copolymer was used to develop the polymeric micelles. While in combination with chemotherapy, the polymeric micelles were disrupted due to the cleavage of disulfide bonds and released the encapsulated drug. This was attributed to the irradiation through HIFU, an external stimulus, or cleavage in situ due to the presence of GSH within the cells. The study results reported that the irradiation by HIFU was site-specific and breaking of the S-S bonds responded rapidly to the radiations as compared to the redox agent present within the cells. This technique has been tested by gel permeation chromatography and the results were found to be favorable (Li et al. 2010), (Table 4.1).

Table 4.1 Summary of different types of polymeric micelles, their stimuli, polymers used, actives encapsulated, agents used, and conditions that can be treated

Sr. no.	Type of polymeric micelle	Stimuli used for disassembly of polymer	Polymers used	Drugs encapsulated/ released	Agents added to micelles to enhance selectivity	Condition
1.	Hypoxia responsive	Hypoxic tumor cell	PEG + P(LGlu-NI), PLL	DoX, Aldronate	AZO, as a linker	Breast cancer, bone metastasis
2.	Glucose sensitive	Glucose	PBA, Con A, PBDEMA, PPyBDEMA	Insulin	GoX, PBE	Diabetes mellitus
3.	Reduction sensitive	GSH, reducing agents	PCL-SSPPEGMA, HA-ss-PLLZ	DOX, PTX, APA	–	MDR cancer
4.	pH sensitive	Varying pH conditions	polyHis/PEG-folate, PLLA/ PEG-folate, NSCS	PD, Curcumin, DOX	Folate, hydrazone bonds	RA, colorectal cancer, MDR cancer
5.	Enzyme responsive	MMP-2, protease	PEG, cholesterol, DUPA, PEG-b-PLL	Cabazitaxel, DOX	Biotin, as targeting ligand	Prostate cancer
6.	Dual responsive	pH + reduction, enzymes + reduction, glucose + H_2O_2	PEO-b-P (MAA-g-Hyd), mPEG, PBA, PEG-b-PAPBE	ADR, CPT, insulin	Chondroitin sulfate	Cancer, diabetes, RA

4.2.7 Multi-Responsive Polymeric Micelles

Due to the promising outcomes of dual-responsive polymeric micelles, researchers have also tried investigating the influence of two or more stimuli in the delivery of actives via the polymeric micelles. A multi-responsive polymer sensitive to light, pH, and temperature was synthesized using a pyrene-functionalized poly (dimethylaminoethyl methacrylate) polymer. The pyrene contains two segments in which the quaternized segment formed the light responsive shell, while the unquaternized segment formed the temperature-/pH-responsive core. Under the influence of UV irradiation, the polymeric micelles disrupted and released the encapsulated material. When the temperature of the surrounding solution was increased, the micelles shrunk. At an acidic pH of 3, the micelles were swollen or dissociated, while at an alkaline pH of 10, the micelles shrunk. The system displayed a controlled release of Nile Red from the micelle under various stimuli, thus proposing multiple stimuli-responsive approach as a promising platform for nanocarrier delivery of actives (Dong et al. 2013).

4.2.8 Polymeric Micelles Used in Imaging and Detection

Apart from delivering actives for the management or treatment of various disease conditions, polymeric micelles have also been investigated for theranostic purposes. In a study, tetraphenylethene (TPE) moieties with aggregation-induced emission (AIE) were used as functional hydrophobic chains to induce copolymer self-assembly. This led to the formation of polymeric micelles having strong fluorescence, which were used for imaging. The advantage of the developed TPE-containing DOX-loaded polymeric micelles was that the micelles were formed in the aqueous medium at very low critical micelle concentrations. In addition to this, the presence of TPE did not hinder the release of DOX from the micelle, which was triggered by esterase enzyme in tumor cell. Thus, the results of the study exhibited drug release for anti-tumor activity, followed by imaging due to fluorescence quenching (Yan et al. 2020). In another study, TPE molecules were incorporated within the polymers, styrene, and poly(N-methyl-pyridinium iodide). The developed polymeric micelles demonstrated strong fluorescence, in which the glucose liberated hydrogen peroxide in the presence of glucose oxidase, followed by oxidation of I^-, which further led to the formation of I_2. This exhibited strong fluorescence that was studied to detect the presence of D-glucose in aqueous medium (Shen et al. 2012).

4.3 Types of Polymeric Micelles Depending on Their Application

Recent developments have demonstrated the viability of producing polymeric micelles based on nucleic acids, phytoconstituents, surface engineering, ligand conjugation, proteins, and biosensors. The application of these micelles is discussed in the section that follows.

4.3.1 Nucleic Acid-Based Polymeric Micelles

Nucleic acids (NAs) have the capability to alter gene expression and hence tackle life threatening disorders. Due to the sensitivity of NAs to various physiological barriers, they lose some therapeutic efficacy when they encounter these barriers, but this can be overcome by the encapsulation of NAs within the polymeric micelles. These carriers can provide release of the contents at a specific site and give a targeted delivery (Jarak et al. 2021).

Delivery of small interfering RNA (siRNA) is challenging, given that it gets degraded by endosomal compartments and polyion exchange in the cellular environment. In order to overcome this, polymeric micelles can be designed, wherein SiRNAs can be shielded within the polymeric micelles. Furthermore, they can be stabilized with disulfide bonds and substances like cholesterol may be incorporated for their intracellular cleavage. Additionally, it has been found that the incorporation of polyaspartamides into the block polymers during the formulation of the micelles has enhanced the safe delivery of NAs to the targeted site. In order to reduce the electrostatic interactions of siRNA with the cellular environment, further stabilization can be provided to the micelles, by incorporating polymers like PEG-b-poly(L-lysine) copolymers constituting the iminothiolane groups, which help in the crosslinking of the polymers through disulfide bonds. These strategies have helped in the transfection of the genes by the siRNA using polymeric micelles as the carrier system (Cabral and Kataoka 2014).

4.3.2 Phytoconstituent-Loaded Polymeric Micelles

Curcumin, a phytoconstituent, has been widely used for the treatment of several disorders such as pancreatic or colon cancer, multiple myeloma, psoriasis, myelodysplastic syndromes, and so on (Goel et al. 2008; Liu et al. 2013). It was recommended by Ravindran et al. (2009) that the encapsulation of curcumin within a polymeric micelle made up of pluronic/polycaprolactone (pluronic/PCL) block copolymers increased the solubility of curcumin in water. These micelles were seen to be taken up by the colorectal adenocarcinoma cells and showed in vitro toxicity. A marketed preparation of the nanomicelles which has displayed high oral bioavailability, SinaCurcumin®, has been marketed by a Tehran based company called Exir Nano Sina Company and was developed by Nanotechnology Research Centre of Mashhad University of Medical Science (Ravindran et al. 2009; Rahimi et al. 2016). Woraphatphadung et al. (2018) developed a chitosan-based pH controlled polymeric micelle, wherein the loaded curcumin was majorly released in the simulated colonic fluid and was seen to give enhanced anticancer activity against colorectal cancer cells.

In another study, the hydrophobic core of polymeric micelles was used as the porter for hydrophobic drugs to treat diabetes (Ahmad et al. 2014). Silymarin encapsulated micelles were tested on streptozotocin induced diabetic rats and the results showed a high degree of reduction in the cellular levels of fasting glucose in the first week and a great amount of suppression of blood glucose levels in the rats in the second week (El-Far et al. 2016).

Morin, a bioflavonoid, has shown significant results in lowering lipogenesis, gluconeogenesis, inflammation as well as oxidative stress. It also has some insulin-mimetic actions (Paoli et al. 2013). Although this drug is reported to have a low oral bioavailability, when it was loaded in polymeric micelle, it showed 3.6 times better cellular penetration and 2.4 times higher permeability and also improved bioavailability in the systemic circulation (Choi et al. 2015; Padhi et al. 2021).

4.3.3 Surface-Engineered Polymeric Micelles

In order to impart multifunctionality to the polymeric micelles and to increase their cellular uptake, the surface of polymeric micelles can be engineered by adding various components to it and decorating it with various targeting ligands like phenylboronic acids, folic acids, biotin, antibodies, etc. (Agwa and Sabra 2022).

Folic acid conjugated poly(styrene-co-maleic anhydride) polymeric micelles loaded with an extremely potent synthetic curcumin difluorinated analog, which is hydrophobic in nature and photolabile, were designed. The incorporation of actives into a polymeric micelle provided photostability and exhibited anticancer activity against ovarian cancer cells. The polymeric micelles displayed targeted delivery due to the overexpressed folate receptors. In the tumor cell, the drug was released from the micelles and the cells were seen to undergo apoptosis due to the increase in the tumor suppressor phosphatase and tensin homolog, along with simultaneous inhibition of nuclear factor kappa-B (Luong et al. 2017) (Fig. 4.3).

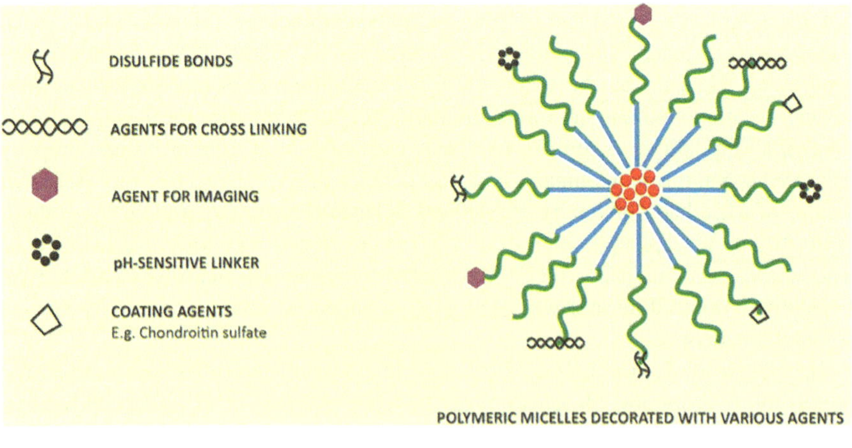

Fig. 4.3 Illustration of polymeric micelles decorated with various agents

4.3.4 Ligand-Conjugated Polymeric Micelles

Many ligands such as antibody fragments, transferrins, epidermal growth factors, α_2-glycoprotein, and folate can be used to decorate the polymeric micelles to give more sensitivity for targeting various sites (Kedar et al. 2010).

NK012 is a polymeric micellar formulation made up of PEG and polyglutamate (PGlu) block copolymers, conjugated with 7-ethyl-10-hydroxy-campothecin (SN-38), an analog of camptothecin, which can act as a DNA topoisomerase-1 inhibitor. According to the study carried out by Nakajima et al. (2008), it was found that the IC50 values of NK012 were 5.8 times higher than the administration of free SN-38. It was also found that when 5-fluorouracil was conjugated with NK012, the micelles showed high anticancer activity in human colon cancer cell xenografts (Oerlemans et al. 2010).

In a study carried out by Zhao and Yung (2008), poly(D,L-lactide-coglycolide)–poly(ethylene glycol)–folate (PLGA–PEG–FOL) copolymer were used to form a micelle which was loaded with DOX. To find out the targeting capacity of the folate to cancer cells, three cell lines were used (i.e., KB, MATB III, C6) and the action of the micelles was compared with normal fibroblast cells. On carrying out an analysis of the cell cycle, it was found that the percentage of apoptosis of the normal fibroblasts was considerably low in comparison to the cancer cells, which were targeted by the DOX-loaded folate conjugated micelles.

In another study, an antibody, Cetuximab (C225), was conjugated on the surface of a polymeric micelle composed of methoxy poly(ethylene glycol)-*b*-poly (lactide) (mPEG-*b*-PLA) and was loaded with Chlorin e6 (Ce6). These micelles were taken up by the epidermal growth factor receptors (EGFR) which are overexpressed in A431 cells. It was observed that the IC50 value of Ce6-mediated photodynamic therapy decreased from 0.42 to 0.173 μM. The study results demonstrated that active EGFR targeting of photosensitizer-loaded micelles may be beneficial in minimizing the dose-dependent toxicity of photosensitizers and be a viable strategy for drug delivery (Chang et al. 2018).

4.3.5 Protein-Based Polymeric Micelles

Polymeric micelles have proved to be virtuous carrier system for the delivery of various proteins (Cohen et al. 2012). Since proteins have greater selectivity at relatively lower concentrations, protein-based drugs are highly used for treating disorders such as insulin for diabetes and erythropoietin for renal anemia (Taluja et al. 2007).

In a study by Gao et al. (2012), the polymeric micelles were designed by self-assembling human serum albumin, a model protein, and a degradable block copolymer methoxy poly(ethylene glycol)-poly(β-amino ester) (PEG-PAE) with piperidine and imidazole rings. The cationic group was incorporated in the hydrophilic PEG to form PEG-b-PAE and an imidazole ring was added to the complex to improve the stability of micelles at fluctuating pH conditions. The in vitro cytotoxicity studies were carried out through MTT assay on MDA-MB-435 cell lines. The micelle system showed no obvious toxicity at a polymer concentration of less than 200 μg/mL. In order to evaluate the capability of pH-tunable positively charged polymeric micelle as a vehicle for protein delivery in vivo in acidic conditions, a disease rat model of cerebral ischemia was used, wherein Cy5.5-labled albumin-encapsulated polymeric micelle was administered intravenously to the rats. The study results displayed a steady rise in the fluorescence signals of the brain ischemic area, owing to the pH-tuning in acidic environment. This demonstrates the use of Cy5.5-labled protein-loaded polymeric micelles as a diagnostic tool for in vivo optical imaging.

Wang et al. (2010) used a combination of polymer and a phage protein for a highly sensitive targeting. In the study, MCF-7 cell-specific micelles were self-assembled at very low cmc value by using a combination of amphiphilic polyethylene glycol-phosphatidylethanolamine (PEG-PE) conjugate, MCF-7-specific landscape phage fusion coat protein, and paclitaxel. The developed polymeric micelles showed high cytotoxicity against MCF-7 cells, but failed to show activity against non-target C166 cells. The study results concluded that tumor cell-specific phage proteins can be sensibly used as targeting ligands for polymeric micelle-based pharmaceutical preparations.

In order to overcome the limitations of protein delivery, such as stability issues, short half-life, and adverse immune responses, Tao et al. (2020) developed protein-loaded polymeric micelles using ionic complexation and pH-cleavable covalent bonding. The micelle system contained myoglobin protein and carboxydimethyl maleic anhydride (CDM), a pH-sensitive moiety. The CDM moiety was found to form bonds with the amide group which were stable at pH 7.4 but cleaved at the pathophysiological pH of 6.5. Addition of poly(ethylene glycol)-poly(L-lysine) block copolymers provided the opportunity to load proteins of varying sizes and isoelectric points within the micelle.

4.3.6 Biosensor-Based Polymeric Micelles

Polymers self-assembled as micelles are a promising option for fabricating biosensors, given that these polymeric micelles respond to various external stimuli like pH, temperature, reacting oxygen species (ROS), and so on (Shukla et al. 2016).

Fluorophore pyrene, lipophilic N,N-dimethyl-N-dodecylamine, and 2-dodecylpyridine are the components that self-assemble into polymeric micelles. PHEA-PEG$_{5000}$-C16, a polyaspartamide polymer, forms the hydrophobic core, whose viscosity and polarity can be detected using fluorescent probes. These micelles act as an "off-on-off" fluorescence sensors for a pH window. This micellar biosensor is in the state of "on" under the physiological pH of 6–8 and by the addition of charged co-surfactants, the pH can be shifted minutely in both directions of the axis and can be measured (Diaz-Fernandez et al. 2010).

A molecularly imprinted polymeric micelle (MIPMs) was used to develop a voltammetric glucose sensor via direct electrodeposition technique, by using glucose as a template. The micelle composed of a photo-crosslinkable amphiphilic copolymer. The photo-crosslinking property of the micelle provided a robust film of molecularly imprinted micelle (MIP) in situ on the electrode and showed higher detection of the receptor sites due to their high surface areas. These glucose-MIP sensors showed higher stability and reversibility and can be explored in the future, due to their good performance, ease of control, and ease of application (Yang et al. 2011).

4.4 Conclusion

Polymeric micelles are a unique and promising system for delivery of drugs especially to target diseases like diabetes, cancer, etc. Since these polymeric micelles can respond to multiple stimuli (Nishiyama and Kataoka 2006), they can be targeted at various regions of the body and can be used to deliver the drugs that cannot be given in their free form. Furthermore, they provide several advantages that make them a preferred choice to deliver the drugs, such as attenuating toxicities, providing controlled release, enhancing drug delivery, and in some cases increasing the efficacy in comparison to free drug administration. Due to the promising results of these polymeric micellar systems, they make a good candidate for the delivery of drugs, nucleic acids, proteins, etc. and can also be used for theranostic/diagnostic purposes.

References

Agwa MM, Sabra SA (2022) Surface engineering of polymeric micelles for imparting multifunctionality. In: Polymeric micelles for drug delivery. Woodhead Publishing, Philadelphia, pp 285–314

Ahmad Z, Shah A, Siddiq M, Kraatz HB (2014) Polymeric micelles as drug delivery vehicles. RSC Adv 4(33):17028–17038

Aliabadi HM, Lavasanifar A (2006) Polymeric micelles for drug delivery. Expert Opin Drug Deliv 3(1):139–162

Almeida M, Magalhães M, Veiga F, Figueiras A (2018) Poloxamers, poloxamines and polymeric micelles: definition, structure and therapeutic applications in cancer. J Polym Res 25(1):1–14

Baas J, Senninger N, Elser H (1994) The reticuloendothelial system. An overview of function, pathology and recent methods of measurement. Z Gastroenterol 32(2):117–123

Barve A, Jain A, Liu H, Zhao Z, Cheng K (2020) Enzyme-responsive polymeric micelles of cabazitaxel for prostate cancer targeted therapy. Acta Biomater 113:501–511

Cabral H, Kataoka K (2014) Progress of drug-loaded polymeric micelles into clinical studies. J Control Release 190:465–476

Chang MH, Pai CL, Chen YC, Yu HP, Hsu CY, Lai PS (2018) Enhanced antitumor effects of epidermal growth factor receptor targetable cetuximab-conjugated polymeric micelles for photodynamic therapy. Nanomaterials 8(2):121

Chen WH, Luo GF, Lei Q, Jia HZ, Hong S, Wang QR, Zhuo RX, Zhang XZ (2015) MMP-2 responsive polymeric micelles for cancer-targeted intracellular drug delivery. Chem Commun 51(3):465–468

Choi YA, Yoon YH, Choi K, Kwon M, Goo SH, Cha JS, Choi MK, Lee HS, Song IS (2015) Enhanced oral bioavailability of morin administered in mixed micelle formulation with PluronicF127 and Tween80 in rats. Biol Pharm Bull 38(2):208–217

Cohen S, Coué G, Beno D, Korenstein R, Engbersen JF (2012) Bioreducible poly (amidoamine) s as carriers for intracellular protein delivery to intestinal cells. Biomaterials 33(2):614–623

Croy SR, Kwon GS (2006) Polymeric micelles for drug delivery. Curr Pharm Des 12(36): 4669–4684

Diaz-Fernandez YA, Mottini E, Pasotti L, Craparo EF, Giammona G, Cavallaro G, Pallavicini P (2010) Multicomponent polymeric micelles based on polyaspartamide as tunable fluorescent pH-window biosensors. Biosens Bioelectron 26(1):29–35

Dong J, Wang Y, Zhang J, Zhan X, Zhu S, Yang H, Wang G (2013) Multiple stimuli-responsive polymeric micelles for controlled release. Soft Matter 9(2):370–373

El-Far YM, Zakaria MM, Gabr MM, El Gayar AM, El-Sherbiny IM, Eissa LA (2016) A newly developed silymarin nanoformulation as a potential antidiabetic agent in experimental diabetes. Nanomedicine 11(19):2581–2602

Feng H, Chu D, Yang F, Li Z, Fan B, Jin L, Li J (2020) Hypoxia-responsive polymeric micelles for enhancing cancer treatment. Front Chem 8:742

Gaballa H, Theato P (2019) Glucose-responsive polymeric micelles via boronic acid–diol complexation for insulin delivery at neutral pH. Biomacromolecules 20(2):871–881

Gao GH, Park MJ, Li Y, Im GH, Kim JH, Kim HN, Lee JW, Jeon P, Bang OY, Lee JH, Lee DS (2012) The use of pH-sensitive positively charged polymeric micelles for protein delivery. Biomaterials 33(35):9157–9164

Gao GH, Li Y, Lee DS (2013) Environmental pH-sensitive polymeric micelles for cancer diagnosis and targeted therapy. J Control Release 169(3):180–184

Goel A, Jhurani S, Aggarwal BB (2008) Multi-targeted therapy by curcumin: how spicy is it? Mol Nutr Food Res 52(9):1010–1030

Goren D, Horowitz AT, Tzemach D, Tarshish M, Zalipsky S, Gabizon A (2000) Nuclear delivery of doxorubicin via folate-targeted liposomes with bypass of multidrug-resistance efflux pump. Clin Cancer Res 6(5):1949–1957

He L, Qin X, Fan D, Feng C, Wang Q, Fang J (2021) Dual-stimuli responsive polymeric micelles for the effective treatment of rheumatoid arthritis. ACS Appl Mater Interfaces 13(18): 21076–21086

Hwang D, Ramsey JD, Kabanov AV (2020) Polymeric micelles for the delivery of poorly soluble drugs: from nanoformulation to clinical approval. Adv Drug Deliv Rev 156:80–118

Jarak I, Pereira-Silva M, Santos AC, Veiga F, Cabral H, Figueiras A (2021) Multifunctional polymeric micelle-based nucleic acid delivery: current advances and future perspectives. Appl Mater Today 25:101217

Kedar U, Phutane P, Shidhaye S, Kadam V (2010) Advances in polymeric micelles for drug delivery and tumor targeting. Nanomedicine 6(6):714–729

Kwon GS, Okano T (1996) Polymeric micelles as new drug carriers. Adv Drug Deliv Rev 21(2): 107–116

Lee ES, Na K, Bae YH (2005) Doxorubicin loaded pH-sensitive polymeric micelles for reversal of resistant MCF-7 tumor. J Control Release 103(2):405–418

Li Y, Tong R, Xia H, Zhang H, Xuan J (2010) High intensity focused ultrasound and redox dual responsive polymer micelles. Chem Commun 46(41):7739–7741

Li SX, Liu L, Zhang LJ, Wu B, Wang CX, Zhou W, Zhuo RX, Huang SW (2016) Synergetic enhancement of antitumor efficacy with charge-reversal and reduction-sensitive polymer micelles. Polym Chem 7(32):5113–5122

Li C, Li H, Wang Q, Zhou M, Li M, Gong T, Zhang Z, Sun X (2017) pH-sensitive polymeric micelles for targeted delivery to inflamed joints. J Control Release 246:133–141

Liu J, Chen S, Lv L, Song L, Guo S, Huang S (2013) Recent progress in studying curcumin and its nano-preparations for cancer therapy. Curr Pharm Des 19(11):1974–1993

Liu X, Li C, Lv J, Huang F, An Y, Shi L, Ma R (2020) Glucose and H2O2 dual-responsive polymeric micelles for the self-regulated release of insulin. ACS Appl Bio Mater 3(3): 1598–1606

Long M, Liu X, Huang X, Lu M, Wu X, Weng L, Chen Q, Wang X, Zhu L, Chen Z (2021) Alendronate-functionalized hypoxia-responsive polymeric micelles for targeted therapy of bone metastatic prostate cancer. J Control Release 334:303–317

Luong D, Kesharwani P, Alsaab HO, Sau S, Padhye S, Sarkar FH, Iyer AK (2017) Folic acid conjugated polymeric micelles loaded with a curcumin difluorinated analog for targeting cervical and ovarian cancers. Colloids Surf B Biointerfaces 157:490–502

Miyata K, Christie RJ, Kataoka K (2011) Polymeric micelles for nano-scale drug delivery. React Funct Polym 71(3):227–234

Mourya VK, Inamdar N, Nawale RB, Kulthe SS (2011) Polymeric micelles: general considerations and their applications. Indian J Pharm Educ Res 45(2):128–138

Nakajima TE, Yasunaga M, Kano Y, Koizumi F, Kato K, Hamaguchi T, Yamada Y, Shirao K, Shimada Y, Matsumura Y (2008) Synergistic antitumor activity of the novel SN-38-incorporating polymeric micelles, NK012, combined with 5-fluorouracil in a mouse model of colorectal cancer, as compared with that of irinotecan plus 5-fluorouracil. Int J Cancer 122(9): 2148–2153

Nishiyama N, Kataoka K (2006) Current state, achievements, and future prospects of polymeric micelles as nanocarriers for drug and gene delivery. Pharmacol Ther 112(3):630–648

Oerlemans C, Bult W, Bos M, Storm G, Nijsen JFW, Hennink WE (2010) Polymeric micelles in anticancer therapy: targeting, imaging and triggered release. Pharm Res 27(12):2569–2589

Padhi S, Dash M, Behera A (2021) Nanophytochemicals for the treatment of type II diabetes mellitus: a review. Environ Chem Lett 19(6):4349–4373

Paoli P, Cirri P, Caselli A, Ranaldi F, Bruschi G, Santi A, Camici G (2013) The insulin-mimetic effect of Morin: a promising molecule in diabetes treatment. Biochim Biophys Acta 1830(4): 3102–3111

Rahimi HR, Nedaeinia R, Shamloo AS, Nikdoust S, Oskuee RK (2016) Novel delivery system for natural products: nano-curcumin formulations. Avicenna J Phytomed 6(4):383

Ravindran J, Prasad S, Aggarwal BB (2009) Curcumin and cancer cells: how many ways can curry kill tumor cells selectively? AAPS J 11(3):495–510

Shen X, Shi Y, Peng B, Li K, Xiang J, Zhang G, Liu Z, Chen Y, Zhang D (2012) Fluorescent polymeric micelles with tetraphenylethylene moieties and their application for the selective detection of glucose. Macromol Biosci 12(11):1583–1590

Shukla SK, Govender PP, Tiwari A (2016) Polymeric micellar structures for biosensor technology. In: Advances in biomembranes and lipid self-assembly, vol 24. Academic, New York, pp 143–161

Sun XS, Jang MS, Fu Y, Lee JH, Lee DS, Li Y, Yang HY (2020) Intracellular delivery of cytochrome C using hypoxia-responsive polypeptide micelles for efficient cancer therapy. Mater Sci Eng C 114:111069

Taluja A, Youn YS, Bae YH (2007) Novel approaches in microparticulate PLGA delivery systems encapsulating proteins. J Mater Chem 17(38):4002–4014

Tao A, Huang GL, Igarashi K, Hong T, Liao S, Stellacci F, Matsumoto Y, Yamasoba T, Kataoka K, Cabral H (2020) Polymeric micelles loading proteins through concurrent ion complexation and pH-cleavable covalent bonding for in vivo delivery. Macromol Biosci 20(1):1900161

Wang T, Petrenko VA, Torchilin VP (2010) Paclitaxel-loaded polymeric micelles modified with MCF-7 cell-specific phage protein: enhanced binding to target cancer cells and increased cytotoxicity. Mol Pharm 7(4):1007–1014

Wei C, Guo J, Wang C (2011) Dual stimuli-responsive polymeric micelles exhibiting "AND" logic gate for controlled release of Adriamycin. Macromol Rapid Commun 32(5):451–455

Woraphatphadung T, Sajomsang W, Rojanarata T, Ngawhirunpat T, Tonglairoum P, Opanasopit P (2018) Development of chitosan-based pH-sensitive polymeric micelles containing curcumin for colon-targeted drug delivery. AAPS PharmSciTech 19(3):991–1000

Xia H, Liang Y, Chen K, Guo C, Wang M, Cao J, Han S, Ma Q, Sun Y, He B (2021) Reduction-sensitive polymeric micelles as amplifying oxidative stress vehicles for enhanced antitumor therapy. Colloids Surf B Biointerfaces 203:111733

Yan K, Zhang S, Zhang K, Miao Y, Qiu Y, Zhang P, Jia X, Zhao X (2020) Enzyme-responsive polymeric micelles with fluorescence fabricated through aggregation-induced copolymer self-assembly for anticancer drug delivery. Polym Chem 11(48):7704–7713

Yang Y, Yi C, Luo J, Liu R, Liu J, Jiang J, Liu X (2011) Glucose sensors based on electrodeposition of molecularly imprinted polymeric micelles: a novel strategy for MIP sensors. Biosens Bioelectron 26(5):2607–2612

Yang C, Yin L, Yuan C, Liu W, Guo J, Shuttleworth PS, Yue H, Lin W (2021) DPD simulations and experimental study on reduction-sensitive polymeric micelles self-assembled from PCL-SS-PPEGMA for doxorubicin controlled release. Colloids Surf B Biointerfaces 204:111797

Zhang G, Zhang X, Shen H, Yang J, Yang J (2014) Smarter glucose-sensitivity of polymeric micelles formed from phenylborate ester-co-pyrenylboronic ester for insulin delivery at physiological pH. RSC Adv 4(91):49964–49973

Zhang L, Wang Y, Zhang X, Wei X, Xiong X, Zhou S (2017) Enzyme and redox dual-triggered intracellular release from actively targeted polymeric micelles. ACS Appl Mater Interfaces 9(4): 3388–3399

Zhang X, Ren X, Tang J, Wang J, Zhang X, He P, Yao C, Bian W, Sun L (2020a) Hyaluronic acid reduction-sensitive polymeric micelles achieving co-delivery of tumor-targeting paclitaxel/apatinib effectively reverse cancer multidrug resistance. Drug Deliv 27(1):825–835

Zhang X, Wang Y, Wei G, Zhao J, Yang G, Zhou S (2020b) Stepwise dual targeting and dual responsive polymer micelles for mitochondrion therapy. J Control Release 322:157–169

Zhao H, Yung LYL (2008) Selectivity of folate conjugated polymer micelles against different tumor cells. Int J Pharm 349(1–2):256–226

Drug Solubilization and Drug Release from Polymeric Micelles

5

Ganesh Vambhurkar, Naitik Jain, Dadi A. Srinivasarao, Paras Famta, Shashi Bala Singh, and Saurabh Srivastava

Abstract

Amphiphilic polymeric block copolymers find frequent applications due to their ability to deliver hydrophobic moieties to the physiological system. Their ability to solubilize water-insoluble drugs in the core has improved the pharmacokinetics and therapeutic potential of such drugs. The hydrophobic core of the polymeric micelles provides protection to degradation-prone drugs. The thermodynamically stable polymeric micelles can be engineered to liberate the loaded cargo in various fashions. Drug release patterns such as biphasic, stimuli-responsive, and physiological/pathological environment-responsive have been reported to improve the therapeutic potential in multiple diseases such as neurological disorders and cancer. The advantages of polymeric micelles over conventional drug delivery systems have been explained. The concepts behind engineering polymeric micelles with a particular drug-releasing phenomenon will be deliberated throughout the chapter. Additionally, the need for functionalization and the development of functionalized micelles in various diseases have been discussed in this chapter.

Keywords

Polymeric micelles · Solubility · Drug release · CMC

G. Vambhurkar · N. Jain · D. A. Srinivasarao · P. Famta · S. Srivastava (✉)
Pharmaceutical Innovation and Translational Research Lab (PITRL)., Department of Pharmaceutics, National Institute of Pharmaceutical Education and Research (NIPER) Hyderabad, Hyderabad, Telangana, India
e-mail: srinivasarao.ananda@niperhyd.ac.in; saurabh@niperhyd.ac.in

S. B. Singh
Department of Biological Sciences, National Institute of Pharmaceutical Education and Research (NIPER) Hyderabad, Hyderabad, Telangana, India

S. K. Singh et al. (eds.), *Polymeric Micelles: Principles, Perspectives and Practices*,
https://doi.org/10.1007/978-981-99-0361-0_5

5.1 Introduction

Polymeric micelles are colloidal drug delivery systems consisting of supra molecular nano-constructs of size ~10 to 100 nm. Self-assembly of amphiphilic molecules, mainly di- or tri-block copolymers, forms micelles (Trivedi and Uday 2012). These amphiphilic molecules exist as separate entities at lower concentrations. Whereas at higher concentrations, these can self-assemble and form micelles. The concentration where amphiphilic molecules start forming micelles is called critical micellar concentration (CMC). At CMC, the interface and bulk phase are saturated with monomers; hence, further increment in their concentration causes aggregation of monomers resulting in micelle formation, which in turn is evidenced by liberating the system's free energy (Patist et al. 2001). The formed micellar structures comprise two parts: the core (the inner structure of micelles) and the shell (the outer surface of micelles). In most cases, the hydrophilic portion of block copolymers forms the shell/outer surface of micelles, and the hydrophobic portion includes micelles' core/inner structure. In contrast, the converse is the case with reverse micelles. The most commonly employed hydrophilic polymers for micelle formulation are: poly(ethylene oxide)/poly(ethylene glycol), poly(vinyl pyrrolidone), and poly(*N*-isopropyl acryl amide). The polymers that form hydrophobic core structure include poly(L-lactide), poly(lactide-*co*-glycolic acid), poly(ε-caprolactone), poly(propylene oxide), poly(L-aspartic acid), poly(L-histidine) (Aliabadi and Lavasanifar 2006). The critical parameters that affect micelle formation are (i) the composition of the amphiphilic structure, (ii) the molecular weight of hydrophobic and/or hydrophilic moieties, (iii) the extent of hydrophilicity/phobicity, (iv) wetting property, (v) hydrophilic–hydrophobic balance of the copolymer, (vi) temperature, (vii) solvent system employed, (viii) concentration of amphiphilic molecules, (ix) aggregation number of amphiphiles, (x) process of micellization, (xi) solvation ability of amphiphilic moieties, etc. (Trivedi and Uday 2012; Patist et al. 2001; Aliabadi and Lavasanifar 2006; Förster and Plantenberg 2002; Hwang et al. 2020; Lombardo et al. 2015). These parameters, in turn, govern the size, shape, drug entrapment efficiency, and drug release kinetics of polymeric micelles (Aliabadi and Lavasanifar 2006; Förster and Plantenberg 2002; Zhang et al. 2017).

Polymeric micelles have evolved as potential carriers owing to their small size, ease of preparation, ability to enhance the solubility of hydrophobic drugs, and facilitate the transport of drugs across cellular barriers. The hydrophobic nature of the micellar core can facilitate the solubilization of hydrophobic drug moieties by interacting via weak Van der walls and ionic or hydrophobic interactions. As a result, micelles enhance the oral bioavailability of hydrophobic drugs (Xu et al. 2013). Further, the hydrophilic shell of micelles also plays a vital role in preventing opsonization to reduce immune clearance (Jhaveri and Torchilin 2014). Due to their small size and optimal surface characteristics, nanomicelles can endocytose easily and get transported across cellular barriers, potentially improving therapeutic agents' bioavailability in various tissues (Gaurav et al. 2013).

Further, composition, the pattern of self-assembly of block copolymers, and their physicochemical properties govern the release kinetics (i.e., immediate release/

sustained release/controlled or stimuli-responsive release) (Ghezzi et al. 2021). In conventional micelle-based formulations, drug release takes place via diffusion. Such diffusion-based drug release may occur from intact micelles or after disassociation/disassembly/distortion of micelle structure. Literature reveals that the drug release kinetics from polymeric micelles can be varied by altering the interaction between therapeutic agents and block copolymers used to fabricate micelles. Since premature drug release before reaching the intended site in the body leads to compromised efficiency of micelle-based delivery systems, appropriate strategies have been adopted to achieve drug release, preferably in the area of interest. Such strategies include cross-linking polymeric micelles and using stimuli-responsive polymers (Ghezzi et al. 2021; Zhou et al. 2018a; Kim et al. 2021; Barve et al. 2020; Shi et al. 2017).

5.2 Drug Release from Micelles

Drug release profile plays a vital role in governing the pharmacokinetics and pharmacodynamics of administered therapeutic agents. Therefore, appropriate formulation parameters must be adopted during the development of micelle-based formulations to achieve the desired release rate. Since the drug release is majorly directed by the composition of the micelles and the pattern of arrangement of amphiphilic block copolymers, the polymeric materials employed for the fabrication of micelles play a pivotal role in governing drug release from micelles. Further, drug entrapment and release kinetics can also be governed by the type of micelles (normal/reverse) employed for drug delivery applications. In general, conventional or normal micelles are composed of hydrophilic shell and hydrophobic core and hence, widely employed to entrap and deliver hydrophobic drugs.

Conversely, reverse micelles are composed of a hydrophobic shell and hydrophilic core and are employed to deliver hydrophilic drugs. Reverse micelles showed promising results in protecting sensitive hydrophilic moieties such as proteins, enzymes, antibodies, and nucleic acids (Qiu et al. 2007). Since the biological environment contains aqueous fluids, normal micelles can show predictable drug release profiles compared to reverse micelles (Vrignaud et al. 2011). The structure of normal and reverse micelles have been depicted in the Fig. 5.1.

Fig. 5.1 Types of micelles

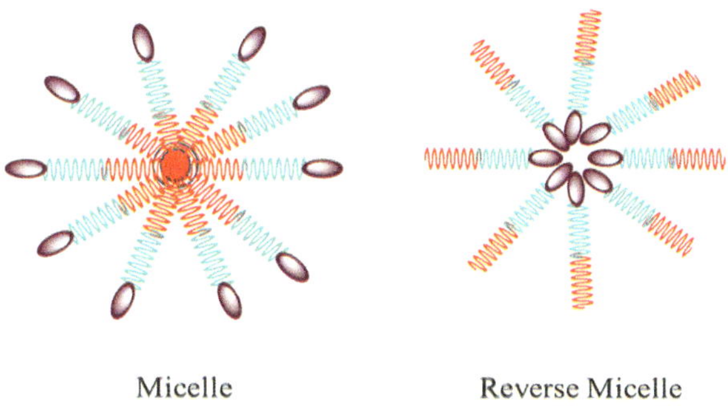

The major drawback associated with polymeric micelles is premature drug release before they reach the site of interest. To overcome this challenge, scientists explored the stabilization process, i.e., the stabilization of polymeric core by forming reversible or weak bonding between the therapeutic agent and micellar structure (Xu et al. 2013; Talelli et al. 2015; Lu et al. 2018). Although this approach offered satisfactory results, the emergence of advanced strategies such as stimuli-responsive drug delivery gained more attention. Stimuli-responsive micelles can help to achieve spatio-temporal controlled drug release at the tissue of interest. The stimuli-responsive drug delivery strategies explored thus far include enzyme-responsive micelles, light-responsive micelles, magnetic field-responsive micelles, pH-responsive micelles, redox-responsive micelles, thermoresponsive micelles, ultrasound responsive micelles, or multi responsive micelles (Ghezzi et al. 2021; Zhou et al. 2018a; Kim et al. 2021; Barve et al. 2020). Various studies demonstrated improved efficacy via stimuli-responsive micellar systems for treating various diseases. In a study, Barve et al. demonstrated enhanced efficacy of enzyme-responsive polymeric micelles for tumor management. In this study, authors formulated matrix metalloproteinase-2 (MMP-2) enzyme-responsive polymeric micelles for targeted delivery of cabazitaxel to prostate cancer cells. The micelles were fabricated using two amphiphilic block copolymers (PEG, an enzyme-responsive peptide, cholesterol; and PEG, cholesterol, a targeting ligand). Since MMP-2 expression is higher in prostate cancer cells, the micelles released a higher amount of drug in the tumor microenvironment and showed improved therapeutic efficacy in the prostate cancer xenograft mice model compared to free cabazitaxel and unmodified micelles (Barve et al. 2020).

Further, light-responsive polymeric micelles showed promising results in cancer therapy by controlling drug release at the specific site, as demonstrated by Kim et al., where light-responsive micelles conjugated with diazonaphthoquinone (DNQ) were fabricated that could change their polarity, i.e., hydrophobic micelles to hydrophilic micelles under UV light. Such light responsiveness resulted in enhanced release of encapsulated therapeutic agent, docetaxel, and as a result, improved cytotoxicity toward breast cancer cells (Kim et al. 2021). Another stimuli-responsive system, pH-responsive polymeric micelles were developed by Zhou et al. for delivery of doxorubicin to the tumor. Lipid modified polymer, namely 1,2-distearoyl-*sn*-glycero-3-phosphoethanolamine-*N*-[methoxy(polyethylene glycol)] conjugated poly(β-amino esters) was employed for pH sensitive release and demonstrated enhanced cytotoxicity of micelles toward tumor cells such as B16F10, HepG2, and HeLa cell lines (Zhou et al. 2018a). Taken together, stimuli-responsive micelle-based delivery systems show promise for spatio-temporal controlled release of entrapped therapeutic agents. The drug loading and drug release from micelle based- delivery systems have been depicted in the Fig. 5.2.

The micelles are stable above CMC and the structure disassembles when the concentration of the amphiphilic molecules is diluted below the CMC (Owen et al. 2012). Along with CMC, the kinetic stability index of micelles also plays an important role in the disassembly of micelles postdilution which is described as the tendency of micelles to disassemble over time when the concentration of block

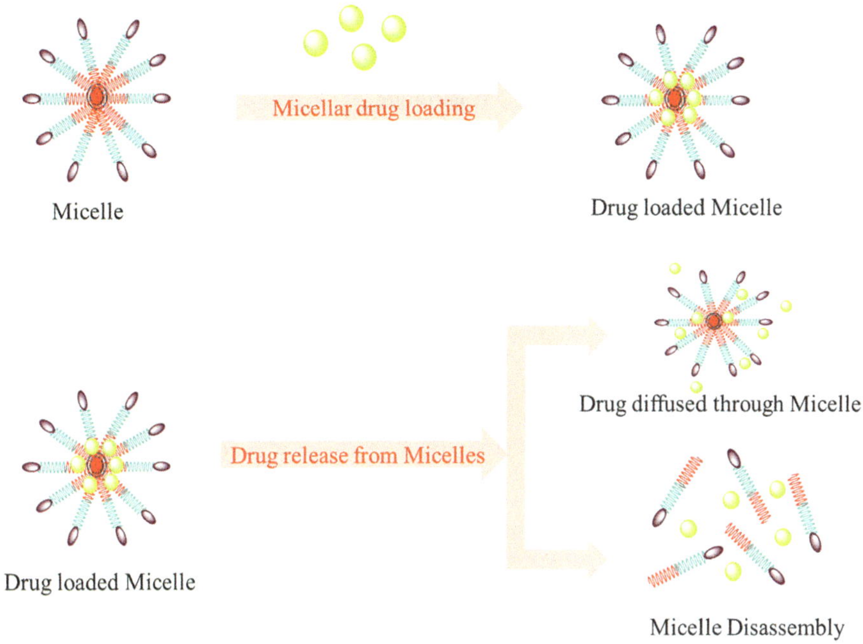

Fig. 5.2 Drug release from micellar structure

copolymers is reduced below CMC (Owen et al. 2012). Polymeric micelles demonstrate the advantage of low CMC and higher kinetic stability, enabling them to form stable structures with the ability to deliver drugs to the desired target organ (Wakaskar 2017). In comparison low molecular weight surfactants lose their structures in the range of microseconds, disabling the successful delivery of the drug. Drug release from polymeric micelles follows two mechanisms, either diffusion of the drug from the core of micelles or drug release due to the disassembly of the micellar structure (Fig. 5.3) (Ghezzi et al. 2021). To achieve site-specific drug release various stimuli-responsive micelles have been developed. Such stimuli-responsive micelles may aid in developing tolerability to toxic and potent drugs such as chemotherapeutics (Zhou et al. 2018b). The tumor microenvironment demonstrates various atypical conditions such as high levels of glutathione, acidic pH, and hypoxia (Whiteside 2008). Recent literature demonstrate that redox-, pH-, and hypoxia-sensitive micelles have demonstrated better anticancer activity and lower side effects (Sikder et al. 2022; Son et al. 2021; Feng et al. 2020). Stimuli-responsive micelles are disassembled in the presence of specific stimuli which leads to the release of loaded drugs.

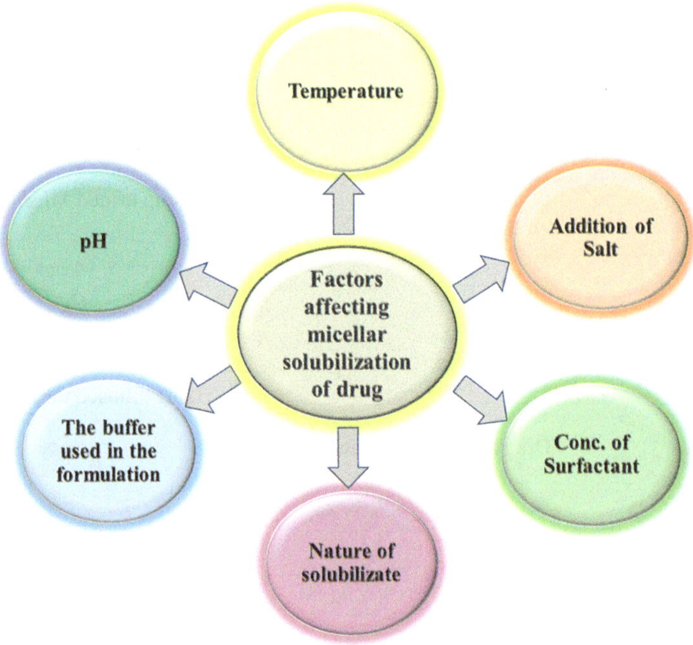

Fig. 5.3 Factors affecting micellar solubilization of drug substances

5.3 Role of Micelles in Protecting Drugs from the Biological Environment

The biological system consists of complex cellular and extracellular tissue compartments, aqueous fluids with varying ionic concentration and/or pH, and other miscellaneous units, each performing a particular physiological function. Such a complex physiological environment may adversely affect the solubility, stability, pharmacokinetics, and pharmacodynamics of administered therapeutic agents, ultimately leading to compromised therapeutic efficacy. These drawbacks can be minimized by employing a suitable drug delivery system, wherein drugs are either entrapped, encapsulated, conjugated, or adsorbed on a suitable material (Allen and Cullis 2004; Vargason et al. 2021). Micelles can protect entrapped therapeutic agents from the harsh biological environment (such as acid and/or alkali-mediated degradation or denaturation, enzyme-mediated degradation, etc.) and offer site-specific drug delivery (Aliabadi and Lavasanifar 2006). Since micelles are composed of amphipathic/amphiphilic moieties (that contain hydrophilic and hydrophobic domains), they tend to aggregate by entropy effect, which results in the incorporation of drugs into their structure. The resultant hydrophilic exterior surface stabilizes the structure of the micelle by forming weak interactions (such as hydrogen bonds) with the surrounding microenvironment/aqueous fluids.

Further, such arrangement of micelles helps to (i) protect entrapped therapeutic agents from harsh microenvironments such as gastric pH, (ii) reduce the extent of metabolism of rapidly metabolizing drugs, and (iii) minimize immune clearance of entrapped drugs (Hwang et al. 2020). Therefore, micelles can be explored as drug delivery vehicles for delivering sensitive therapeutic agents across harsh biological microenvironments. In a study, Almeida et al. demonstrated the potential of amphiphilic chitosan micelle-based drug delivery system in preventing camptothecin degradation. Literature reveals that camptothecin exists in two forms based on surrounding pH conditions: the active lactone form (at acidic pH) and the inactive carboxylated form (at alkaline pH). Since alkaline pH solutions/fluids convert camptothecin to inactive carboxylated form, it is desirable to maintain the stability of camptothecin in vivo at alkaline pH conditions using suitable drug delivery strategies. In an attempt to resolve this issue, a novel copolymer consisting of *O*-methyl-*O′*-succinylpolyethylene glycol 5000 (mPEG)-chitosan-oleic acid and the resultant copolymer conjugate (i.e., chitosan conjugated with hydrophobic and hydrophilic moieties) was used for fabrication of camptothecin loaded micelles. The stability of free camptothecin and micelle-entrapped camptothecin was studied in simulated gastric fluid (SGF, pH 1.2) and simulated intestinal fluid (SIF, pH 6.8). Since camptothecin is unstable in SIF (alkaline pH cause degradation/carboxylation of camptothecin), authors compared the stability of free camptothecin and micelle-entrapped camptothecin in SIF. It was observed that free camptothecin was degraded in SIF, whereas micelle-entrapped camptothecin was stable. Therefore, authors inferred that the fabricated chitosan micelles protected the structure of camptothecin (active lactone form) by preventing its hydrolysis to inactive carboxylated form when incubated in simulated intestinal fluid (Almeida et al. 2020).

Furthermore, micelles can be employed not only for improving the stability and targetability of chemical moieties but also for protein or oligonucleotide-based therapeutic moieties (such as plasmid DNA, siRNA, and mRNA) as well. Since many nucleic acid-based entities require supraphysiological ionic strengths to maintain their integrity and are susceptible to degradation by nucleases, it is desirable to formulate them in an appropriate delivery system, such as micelles, for efficient in vivo delivery. In a study, Agarwal et al. employed cationic poly(ethylene glycol)-polylysine block copolymer as a coating material for different DNA origami structures. The authors coated DNA origami structures via electrostatic interactions and named them DNA origami polyplex micelles (DOPMs). The stability of resultant DOPM and intact DNA origami structures was studied by incubating them with a buffer containing DNase I or nucleases in fetal bovine serum-supplemented RPMI media. The agarose gel electrophoresis study revealed that the polyplex structures were stable and protected against DNAse I, whereas control DNA origami structures were completely degraded. This study demonstrated that the stability of nucleic acid-based structures was improved while using micelles as a delivery system (Agarwal et al. 2017).

In addition to their protective role, micelle-based delivery systems can minimize the metabolism of drugs in the physiological environment. Cheng et al. improved the circulatory half-life of a rapidly metabolizing drug, docetaxel, by employing a micelle-based drug delivery system. In this study, authors prepared docetaxel nanocrystals by high-pressure homogenization. Subsequently, the resultant nanocrystals were employed to fabricate nanocrystal-loaded micelles using methoxy polyethylene glycol-poly(D,L-lactide) block copolymer [mPEG-PLA]. When administered to rats, the resulting amphiphilic copolymer-based docetaxel-loaded micelles prolonged the retention time of docetaxel in the blood (as evidenced by enhanced mean residence time and half-life) when compared to free docetaxel and docetaxel nanocrystals. These results inferred that micelle-based drug delivery systems could be employed for reducing the rate and extent of metabolism of rapidly metabolizing drugs. As a consequence, micelles can potentially improve the circulation time and bioavailability of drugs (Cheng et al. 2021).

Further, numerous studies demonstrated that micelles, due to the presence of a hydrophilic shell, can escape immune recognition by the reticuloendothelial system (RES), as a result, prolong in vivo circulation time of entrapped therapeutic agents. In a study, Shen et al. demonstrated the ability of cholesterol-conjugated polyoxyethylene sorbitol oleate (CPSO) micelles to evade phagocytosis by escaping from macrophages. In this study, rhodamine B (RhB) and paclitaxel (PTX) loaded CPSO micelles were fabricated using a dialysis-ultrasonic method, and the resultant optimized formulation was evaluated for intracellular uptake in human alveolar basal epithelial cells (A549 cells) and phagocytosis escape in alveolar macrophages (NR8383 cells). The confocal laser scanning microscopy study revealed that RhB-loaded CPSO micelles get endocytosed by A549 cells (as evidenced by enhanced fluorescence intensity) and escape phagocytosis by NR8383 cells (as evidenced by lower fluorescence intensity). This study revealed micelles could be designed to escape phagocytosis by macrophages (Shen et al. 2020).

5.4 Factors Affecting the Solubilization of Drugs in Micelles

Since micelles are the structures of copolymers or surfactant, they helps in the solubilization of hydrophobic drugs in the aqueous environment (Kulthe et al. 2012). According to the literature, authors came to know that nowadays, whatever drug substances were invented or synthesized, most of them are hydrophobic. Therefore, micelles play an essential role in achieving higher bioavailability of hydrophobic drugs (Xu et al. 2013). Some dominant factors which affect the solubilization of hydrophobic drug substances by the micellar structures are as follows: surfactant concentration above the CMC, the addition of salt, temperature, pH, nature of solubilizate, the polarity of the solubilizate, molecular structure of solubilizate, use of buffers in the formulation of micelles, use of surfactant mixture instead of single surfactant (Tehrani-Bagha and Holmberg 2013).

5.4.1 Use of Concentration of Surfactant above the CMC Level

Solubilization of hydrophobic drugs by the micelles is dependent on surfactant concentration. The concentration of surface-active agents below the CMC level is unable to solubilize the hydrophobic drug because below CMC level. Beneath the CMC level, surfactants were present in the molecular form as a single moiety. These individual structures of the surface-active agents are not capable of solubilizing the hydrophobic drugs. Therefore, for the solubilization of water-hating drug substances, the concentration of the surfactant should be above the CMC level; at/above the CMC level, surfactant significantly affects the solubility of the hydrophobic drug substances. As the concentration of the surfactants increases above the CMC level, solubilization of the hydrophobic drug substance increases. An increase in solubility with an increase in surfactant concentration above the CMC level was observed up to some extent. After a specific concentration, the viscosity of the solution starts to elevate, which is considered an indication of rod shape micelles formation. So, beyond this point, the linear relation between the increased surfactant concentration and hydrophobic drug solubilization does not exist. So, we can say that up to some extent increase in the concentration of surface-active agents will help in the improvement of drug solubilization (Tehrani-Bagha et al. 2007). Anura S. indulkar et al. observed the impact of nature and surfactant concentration on the amorphous and crystalline solubility of atazanavir supersaturated solution. Findings from the study suggested that the crystalline solubility of the atazanavir increased in sodium dodecyl sulfate (SDS) compared to the absence of SDS in the supersaturated solution of the atazanavir. From observed results, researchers concluded that the concentration of surfactant, i.e., SDS, and the addition of other additives would positively affect the solubility of the drug substance (Indulkar et al. 2017).

5.4.2 Addition of Salt or Electrolyte

Innumerable studies confirm that adding salts or electrolytes positively affects the micellar solubilization of hydrophobic drugs. Micelles possess positively charged head groups on the surface and negatively charged tails in the inner core. Due to the presence of charged groups on the micelles, there is a possibility of ionic interactions between electrolytes and the charged moieties (that were present on the micellar structure).

In the case of micelles prepared from ionic surfactant, the addition of electrolyte causes ionic reactions between the charged groups present on the surface of the micelles. The reaction between the surface charge of micelles and electrolytes results in the neutralization of the micellar surface charge. This shielding or neutralization of surface charge leads to a more hydrophobic surfactant. Hydrophobic surfactants can form micelles at the lowest concentration compared to the other surfactant, which will help manufacture more stable micellar formation. Generally, low CMC value surfactants will create a more stable micellar formulation.

After a conclusive literature search, authors observed that the addition of NaCl salt in the polymeric solution of dodecyl trimethylammonium chloride (DTAC) led to an increase in the solubility of Sudan IV (hydrophobic dye), which is two times higher than the polymeric solution in the absence of NaCl (Ikeda et al. 1980).

Adding salt to the surfactant solution increases the solubilization power and aggregation number of the surfactant. In contrast, it decreases the CMC concentration, which fulfills the requirement of a standard surfactant for stable micelle formation. An increase in the concentration of NaCl in the surfactant solution increases the solubilization power up to a specific concentration. Once the concentration of NaCl crosses 1.5 M, then the shape of micelles transfers from spherical to rod or worm shape, which may cause an increase in the viscosity of the solution. Beyond this limit, adding more salt did not work, and the solubilization of the hydrophobic drug did not increase. Therefore an appropriate amount of electrolytes or salts must be added (Ozeki and Ikeda 1980).

5.4.3 Temperature

In typical solutions, an increment in temperature positively increases the solubility of the drug substance. Similarly, in the case of micelles also rise in temperature leads to improved micellar solubilization of hydrophobic drug substances. Different surfactants like non-ionic surfactants (Penta(ethylene glycol) monoundecyl ether), anionic surfactants (SDS), and cationic surfactants (dodecyl trimethylammonium bromide (DTAB)) are the temperature sensitive surfactants which are mainly used for the micelles formulation to solubilize the hydrophobic drugs. Generally, non-ionic surfactants prominently act as thermoresponsive surfactants in micellar formulations. Researchers conducted an experiment on the solubilization of hydrophobic dye to identify thermoresponsive properties of different polymers like anionic, cationic, and non-ionic polymers. After completing the study, researchers found that non-ionic surfactant shows prominent temperature-dependent solubility of the hydrophobic dye compared to cationic and anionic types of surfactants (Datyner 1978).

Due to an increase in temperature of the non-ionic surfactant solution, the polyoxyethylene chain of the surfactant lost the hydration water, which led to a rise in critical packing parameters of the surfactant, which is considered the leading cause for the translation of spherical-shaped micelles into elongated micelles (Gharanjig et al. 2011). Increment in the micellar solubilization of hydrophobic drugs is mainly due to the elongated micelles.

5.4.4 Effect of pH

The relationship between pH and micellar solubilization has not been explored much. It is believed that there is no relation between the micellar solubilization of the drug and the pH of the medium, but this is not true most of the time as we increase the pH of the polymeric solution of DTAB, SDS, and penta(ethylene glycol) monoundecyl ether above 8, an increase in solubilization of hydrophobic dyes, i.e., Sudan I was observed. It only happens with cationic surfactants, whereas in the case of non-ionic and anionic surfactants, no significant difference was observed in the micellar solubilization of surfactant. An increase in solubilization of this dye is observed due to the deprotonation of the phenolic hydroxyl group. Deprotonation causes the ionization of functional groups in the dye, which further makes ionic bonding with cations on the cationic surfactant. Due to electrostatic forces, bond formation between dye and surfactant will happen, which is the predominant region for the increment of solubilization of hydrophobic dye in micellar solution with an increase in pH of the solution.

In some cases, this high pH of the micellar solution causes the ionization of dye which behaves as an anionic surfactant. So due to the formation of an anionic surfactant formation of mixed micelles takes place, which is responsible for the reduction of the CMC value and the enhancement in the solubilization power of the surfactant. This deprotonation will not be possible in the case of non-ionic and anionic surfactants. Hence the increase in pH value did not positively affect the solubilization power of the surfactant (Tehrani-Bagha et al. 2013).

5.4.5 Position of Solubilizate in Micellar Structure

In micelles, we can see the existence of the polarity gradient, i.e., the outer portion of the micellar structure is polar. In contrast, the inner part consists of lipophilic tails of the surfactant. According to the like-dissolve-like phenomenon, polar substances retain on the outer portion of the micelles, i.e., near the head portion of the surfactant or in the palisade region of the micelles. Similarly, non-polar or lipophilic drugs reside in the inner core of the micelles. It is not similar in all cases; sometimes, the drugs having functional groups like OH or NH_2 may interact with the head portion of the surfactant. Due to these interactions, the drug substance resides in the palisade region, i.e., just beneath the outer part of the micelles (Gharanjig et al. 2011).

5.4.6 Use of Mixed Micelles

The surfactant having a low CMC level is always preferred. It will help in the formulation of more stable micelles as compared to the other surfactants. In the formation of mixed micelles, more than one surfactant is used, which will give the synergetic effect or improve the performance of the surfactant, which will usually be more than the summation of the individual surfactant performance in the micellar formulation. A combination of more than one surfactant will help reduce the CMC level of the surfactant and increase the solubilization power of the surfactant. The mixed micelles concept helps to reduce the concentration of cationic surfactants, most of which are toxic in higher amounts. In some cases, a mixture of ionic surfactants will negatively affect the solubilization power of micelles due to changes in the environment of the micelles (Muto et al. 1988).

5.4.7 Structure of Surfactant

Solubilization of any hydrophobic or hydrophilic drugs prominently depends on the solubilization power of the surface-active agents used in the formulation of the micelles. Solubilization of any drug or substance in the micellar formulation mutually depends on the structural properties of surface-active agents used in the formulation and the drug substance that has to be solubilized. The functional groups present on the surface-active agents, as well as functional groups present on the drug substance, decide the orientation of the drug substance in the micellar structure. Molecular interaction between functional groups of drugs and surfactants determines the solubilization power of the surfactant for the specific drug substance.

According to some case studies, it was observed that the solubilization power of the cationic surfactant is directly proportional to the number of carbon atoms present in the alkyl chain. But in some instances, it was observed that the replacement of the CH_3 group by CF_3 from the DTAB surfactant led to a decline in the solubility of the hydrophobic dye in an aqueous solution of DTAB. Fluorination of the DTAB leads to the formation of the unfavorable inner core of the micellar structure and decreases the aggregation number, which hampers the solubilization power of the surfactant (Tehrani-Bagha et al. 2013).

5.5 Advantages over Conventional Drug Delivery

Conventional drug delivery strategies majorly aim to deliver therapeutic agents into the body using drug solutions, suspensions, or emulsion-based delivery systems (that have particle/globule dimensions in a few micrometers range). The major drawback associated with such drug delivery systems is (i) compromised stability of the drug/dosage form resulting in less shelf life; (ii) suitability for systemic administrations (suspensions and emulsions are not suitable for systemic administration); (iii) susceptibility of drugs to the harsh physiological environment; (iv) low bioavailability; (v) non-specific drug distribution in body tissues; and (vi) (Allen and Cullis 2004; Vargason et al. 2021; Wen et al. 2015) no targeting ability. Such disadvantages of conventional drug delivery formulations can be addressed using micelles. Nanomicelles, owing to their small size and high surface area to volume ratio, can potentially offer advantages such as (i) high penetrability across cellular barriers; (ii) providing high stability of entrapped therapeutic agents by preventing their degradation in harsh biological environments; (iii) enhance the solubility of lipophilic drugs; (iv) extend the circulatory half-life of entrapped therapeutic agents by minimizing their recognition by the immune system; (v) enable targeted drug delivery to a specific tissue site; (vi) amenable for conjugation of targeting moieties; and (vii) offer controlled or sustained drug delivery, stimuli (pH, temperature, ultrasound, light, magnetism, redox potential) responsive delivery, etc. (Aliabadi and Lavasanifar 2006; Hwang et al. 2020; Srinivasarao et al. 2019). Nanomicelles are popularly known for enhancing the solubility of lipophilic drugs such as paclitaxel. In a study, Zhang et al. illustrated enhancement in paclitaxel solubility using micelles fabricated using Solutol HS15 and tocopherol polyethylene glycol succinate (TPGS). The resultant micelles not only improved the solubility of paclitaxel but also enhanced cell uptake efficiency and cytotoxicity in drug-resistant MCF-7/Adr cells compared to taxol-based formulations.

Further, these micelles enhanced the bioavailability of paclitaxel after oral administration and inhibited tumor growth in a rodent model (Zhang et al. 2017). In addition, micelles can also improve the retention of drugs in body tissues such as the eye, resulting in enhanced bioavailability of administered drugs (Ghezzi et al. 2022). Ghezzi et al. demonstrated that cyclosporin-loaded non-ionic amphiphilic micelles composed of tocopherol polyethylene glycol 1000 succinate (TPGS) and Solutol® HS15 offered enhanced permeation across superficial ocular layers and improved micelle retention in cornea and scleral tissues. Such enhanced permeation and improved retention of cyclosporin micelles in ocular tissues showed a reservoir effect (for cyclosporin) and sustained presentation of the drug in other ocular tissues. Therefore, micelle-based delivery systems can help to improve patient compliance *by* offering sustained drug delivery (Ghezzi et al. 2022).

Further, micelle-based drug delivery carriers have shown promising applications in targeted drug delivery, as illustrated by Zhang et al. by fabricating docetaxel and anti-Nucleostemin siRNA entrapped polyethylene oxide (PEO)-polycaprolactone (PCL) micelles decorated with DCL {*N*-[*N*-[(*S*)-1,3-dicarboxypropyl] carbamoyl]-(*S*)-lysine} and TAT ligands. DCL ligand can target (active targeting) the micelles to prostate cancer (PCa) cells, whereas TAT ligand helps enhance micelles' penetration across cell membranes of PCa cells. The resultant micelles were evaluated for their in vivo efficacy in the castration-resistant prostate cancer (CRPC) mice tumor model. The study results revealed that the surface-modified micelle-based dual drug delivery system suppressed CRPC tumor proliferation by inhibiting the G1/S and G2/M mitotic cycle via synergistic interaction. Consequently, micelle-based drug delivery systems effectively stopped tumor growth (Zhang et al. 2021). Together, micelles showed enhanced efficacy in improving the pharmacokinetics and pharmacodynamics of entrapped therapeutic agents for better alleviating pathological conditions.

5.6 Role of Functionalized Micelles in Various Diseases

Micelles are one of the prominent delivery systems because of their ability to encapsulate and improve the pharmacokinetic profiles of hydrophobic drugs and their stability in circulation (Xu et al. 2013). Despite several advantages offered by polymeric micelles, they face issues like toxicity and immunogenicity, poor selectivity, less efficient drug delivery, and low membrane disrupting capabilities, which can be addressed by functionalization with appropriate ligands to improve the targeting capabilities (Figueiras et al. 2022). Targeted therapy is needed to improve patient outcomes by enhancing therapeutic effectiveness and reducing side effects, and functionalization is key to targeted therapy. Functionalization of micelles is a widely explored domain in various diseases wherein various options are available for functionalization but are not limited to peptides, antibodies, polymers, glycoproteins, folate, transferrin, and aptamers (Jhaveri and Torchilin 2014; Sutton et al. 2007).

5.6.1 Cancer

Cancer is a disease characterized by the uncontrolled proliferation of cells. Despite numerous studies and research, none of the available therapies can eliminate cancer or prevent its relapse. Functionalizing micelles with ligands that bind to overexpressed receptors on cancer cells is a popular approach. Several receptors, like folate receptors, are overexpressed on various cancer cells, including ovarian, breast, brain, and lung (Fernández et al. 2018). It is well established that exploiting targeting ligands endow micelles with superior capabilities in terms of improved cellular uptake, reduced off-target effects, and prolonged circulation time (Jhaveri and Torchilin 2014). Zhang et al. utilized folate modification on the surface of micelles containing superparamagnetic iron oxide nanoparticles (SPIONS) and sorafenib to target hepatocellular carcinoma, where these micelles specifically reached cancer cells overexpressing folate receptors. Folate-functionalized micelles displayed greater cellular internalization by HepG2 cells compared to naked micelles. Moreover, functionalized micelles had a higher apoptosis rate of about 17.01%, whereas naked micelles had only an 11.04% apoptotic rate. Moreover, micelles could be visualized by MRI because of the presence of SPIONS (contrast agent) and could assist in image-guided chemotherapy (Zhang et al. 2013).

Metastasis is a major reason that contributes to treatment complications and a major contributor to mortality (Guan 2015). Developing micelles that deliver the drugs at the metastatic sites could improve cancer management. A more significant fibronectin expression characterizes metastatic sites compared to primary tumors (Wang and Hielscher 2017). Given the above strategy, Gong et al. developed PEGylated polymeric micelles encapsulating doxorubicin and vinorelbine, functionalized micelles with cysteine-arginine-glutamic acid-lysine-alanine (CREKA) to target fibronectin-expressing metastatic tumors. These engineered micelles had superior targeting capabilities and efficiently inhibited fibronectin expression compared to non-functionalized micelles. Modified micelles prolonged the circulatory half-life of doxorubicin and vinorelbine, where 51.48% of doxorubicin and 58.69% of vinorelbine were detected in the blood. Still, only 18.36% of doxorubicin and 45.30% of vinorelbine were detected after 2 hours of administration of non-functionalized micelles. CREKA-modified micelles efficiency suppressed metastatic tumors compared to non-modified micelles (Gong et al. 2020). In addition to functionalization, this combination of drugs effectively suppressed the metastasis.

Multi-drug resistant cancer is another setback limiting the treatment of cancers. Several siRNA has been shown to minimize or overcome tumor resistance by downregulating P-gp (Navarro et al. 2012; Liu et al. 2019). Combining siRNA with chemotherapy displays synergistic effects in addition to reducing drug resistance (Sun et al. 2018). Yalamarty et al. developed 2C5 antibody functionalized dendrimer-based mixed micelles (MDM) encapsulating small interfering RNA (siRNA) and doxorubicin. Western blot analysis confirmed the downregulation of P-gp by siRNA where 2C5-functionalized MDM compared to non-functionalized MDM owing to targeted delivery. Additionally, greater cellular uptake was observed in MDA-MB-231 and SKOV-3TR cells compared to non-functionalized micelles (Yalamarty et al. 2022).

5.6.2 Neurological Disorders

Polymeric micelles have widely been explored for brain diseases due to their ability to traverse across BBB by enhancing membrane fluidity and steric stabilization due to PEGylation (Waris et al. 2022). The ease of functionalization makes them an attractive carrier for drugs in various CNS-related diseases, including epilepsy, psychosis, Alzheimer's disease, Parkinson's disease, glioblastoma, and cerebral ischemic injury (Kaur et al. 2022). Lactoferrin receptors are overexpressed on the surface of the brain endothelial cells in Alzheimer's and Parkinson's disease and could be effectively utilized for brain targeting (Le Gao et al. 2010). Agwa and co-workers used lactoferrin functionalized conjugated linoleic acid (CLA) micelles to deliver CLA in Alzheimer's. In vivo biodistribution studies revealed a more significant accumulation of functionalized micelles in brain tissue than in other organs owing to the presence of Lactoferrin receptors on the surface of brain endothelial cells. Furthermore, it reduced brain oxidative stress, inflammation, apoptosis and acetylcholine esterase activity, and improved cognitive capabilities (Agwa et al. 2020). CLA-based micelles could be a potential carrier system to deliver drugs across BBB to treat neurodegenerative disorders like Alzheimer's. A major obstacle to treating brain disorders is the passage of drugs across the BBB. Strategies such as nose-to-brain delivery have been shown to circumvent BBB (Giunchedi et al. 2020). Interestingly, functionalization with TAT peptide enabled nose-to-brain targeting of mPEG-PDLLA micelles (Ahmad et al. 2020).

5.6.3 Infectious Diseases

Combating intracranial infections is challenging due to the complexity and selective permeability of the blood–brain barrier, which poses a hurdle for brain targeting (Ziai and Lewin 2006). Management of CNS infections is more challenging than peripheral infections due to the limited stability of drugs in blood circulation and poor permeability across BBB (Kasinathan et al. 2015). Glucose transporter-1 (GLUT1) is abundantly expressed on the surface of BBB, transporting D-glucose and other hexoses across BBB (Koepsell 2020). Shao and co-workers developed a micellar delivery system for improving the permeability of Itraconazole (first line antifungal drug) across the BBB. Micelles comprised of PEG-*b*-poly(L-lysine)-*b*-poly(L-phenylalanine) (PEG-pLys-pPhe). Additionally, a disulfide linkage was introduced at the pLys end to impart stimuli responsiveness (Shao et al. 2015). Then endowing it with brain targeting capabilities by functionalization with dehydroascorbic acid (DHA), having a high affinity toward GLUT1 on BBB. The functionalized micelles remarkably prolonged the circulatory half-life in blood and improved the delivery of Itraconazole in the brain, effectively managing intracranial infection (Shao et al. 2015).

Amphotericin B is another important agent for managing systemic fungal infections; however, it exerts infusion-related and chronic toxicities (Cavassin et al. 2021). Wang et al. encapsulated Amphotericin B in PEG-PC micelles functionalized with urea (PEG-PUC) and PEG-PC micelles functionalized with phenylboronic acid (PEG-PBC) for reducing toxicity. To test the formulation's effectiveness, it was compared with the marketed product Fungizone®. PEG-PBC micelles showed sustained release, whereas PEG-PUC micelles displayed burst release. Both micelles showed comparable antifungal activity to free Amphotericin B or Fungizone®. PEG-PBC micelles caused very little or no hemolysis, whereas PEG-PBC micelles caused slightly lower hemolysis compared to free Amphotericin B and Fungizone®. Interestingly, when PEG-PUC and PEG-PBC were mixed in equimolar ratios to form micelles, these micelles had similar antifungal activity to free Amphotericin B and Fungizone® and remarkably decreased the nephrotoxicity (Wang et al. 2016).

5.6.4 Pulmonary Diseases

Tuberculosis is one of the deadly diseases whose treatment is complicated, and the emergence of resistance to therapy further limits the treatment options. Rifampicin, one of the most effective drugs for treating tuberculosis, possesses high hydrophobicity, a barrier to drug delivery (Mazlan et al. 2021). Tripodo et al. developed inulin-based micelles functionalized with vitamin E (INVITE) to deliver Rifampicin in treating mycobacterium tuberculosis infection. Compared to Rifampicin, INVITE prolonged the drug release for up to seven days. Furthermore, INVITE displayed antimicrobial activity comparable to free Rifampicin (Tripodo et al. 2019). INVITE could be a potential platform for delivering hydrophobic drugs against Mycobacterium.

5.6.5 Imaging and Diagnosis

Various imaging techniques like magnetic resonance imaging (MRI), nuclear imaging, and computed tomography scan (CT) play a crucial role in diagnosing and evaluating the response to the therapies, where the contrast agent is administered that highlights the area of interest (Oerlemans et al. 2010). However, imaging agents lack specificity, which limits their distribution at desired sites. To address the above issue, Yoo et al. fabricated peptide amphiphile micelles using DSPE-PEG2000-DTPA (Gd), functionalized with peptide CREKA to target fibrin; hence imaging fibrin-containing atherosclerotic plaques.. Clot-binding assays revealed that CREKA-based micelles targeted clots eight times higher than non-functionalized micelles. Additionally, in vivo MRI and optical imaging studies of the aortas and hearts confirmed fibrin targeting conferred by the CREKA peptide (Yoo et al. 2016).

Inflammation is an innate host defense mechanism that protects against harmful stimuli or foreign substances (Chen et al. 2018). Various cell adhesion molecules (VCAM-1, ICAM-1, E, and P selectin) mediate this complex process (Panés et al. 1999). Pagoto and co-workers, to improve the visualization of inflammation, developed DSPE-PEG2000 micelles loaded with MRI contrast agent Gd-DOTAMA $(C18)_2$; the micelles were functionalized with a cyclic peptide specific to VCAM-1 receptor (a biomarker of endothelium activated by inflammation) (Pagoto et al. 2016). This delivery system improved the MRI visualization of inflammation. Various studies that employed functionalized micelles for targeted therapeutic applications have been presented in the Table 5.1.

5.7 Summary

Micelles have successfully enabled the solubilization and disposition of hydrophobic drugs. The loading of drugs in the core of micelles protects the degradation-prone drugs from various degrading stimuli. Polymeric micelles demonstrate advantages such as low CMC and high kinetic stability enabling efficient drug delivery to various physiological compartment. Low CMC value of the amphiphilic block copolymers is important for stability of micellar structure post-dilution in the physiological fluids. The main mechanisms behind drug release from micelles are diffusion and disassembly of micelles in the physiological system. Micelles can be efficiently functionalized with polymers and ligands to engineer the dispositioning of micelles in the physiological system. The development of stimuli-responsive micelles has demonstrated excellent results in the neoplastic and inflammatory diseases due to the site-specific fashion of drug release.

Table 5.1 Summarizing some of the functionalized micelles employed in various diseases

Disease/ category	Polymers	Ligand	Drug	Application	References
Triple-negative breast cancer	PEG-P(CL-DTC) and ATN-PEG-P(CL-DTC)	ATN peptide	Paclitaxel	Induces immunogenic cell death More effectively inhibits metastatic 4 T1 tumors	(Qiu et al. 2022)
Breast cancer	Pluronic 127	Folic acid	Fisetin	6-fold greater intraperitoneal bioavailability than free fisetin solution	(Pawar et al. 2018)
Hepatocellular carcinoma	PLGA-grafted dextran	A54 peptide	Doxorubicin SPIONS	Specifically targeted hepatoma BEL-7402 cells Less toxic compared to commercial adriamycin injection	(Situ et al. 2016)
Prostate cancer	PEO-PCL	Spermine ligand, DCL ligand, and TAT peptide	Docetaxel Anti-nucleostemin siRNA	Improved cell penetration and targeting	(Zhang et al. 2021)
Glioma	PEG-PLGA	T7 peptide	Carmustine	Enhanced tumor accumulation	(Bi et al. 2016)
Eye infection	Pluronic F127 and phospholipid	Maleimide	Voriconazole	Improved antifungal effect and reduced toxicity	(Wu et al. 2022)
Antiviral therapy	PLLA-b-PEG	Methyl-*b*-neuraminic acid	Amantadine	Extended-release profile and enhanced antiviral activity	(Ahn et al. 2010)
Antibacterial	PEO-*b*-PCL	Phosphonium	–	Selective disruption of bacterial cell membrane	(Hisey et al. 2017)

Acknowledgments The authors would like to acknowledge the research funding support by Department of Pharmaceuticals (DoP), Ministry of Chemicals and Fertilizers, Govt. of India to National Institute of Pharmaceutical Education and Research (NIPER) Hyderabad, INDIA.

Declaration of Interest The authors declare that there is no conflict of interest.

References

Trivedi R, Uday K (2012) Nanomicellar formulations for sustained drug delivery: strategies and underlying principles. Nanomedicine (Lond) 100:130–134

Patist SG, Oh R, Leung DO (2001) Shah, kinetics of micellization: its significance to technological processes. Colloids Surfaces A Physicochem Eng Asp 176:3–16. https://doi.org/10.1016/S0927-7757(00)00610-5

Aliabadi HM, Lavasanifar A (2006) Polymeric micelles for drug delivery. Expert Opin Drug Deliv 3:139–162. https://doi.org/10.1517/17425247.3.1.139

Förster S, Plantenberg T (2002) From self-organizing polymers to nanohybrid and biomaterials. Angew Chemie Int Ed 41:688–714. https://doi.org/10.1002/1521-3773(20020301)41:5<688::aid-anie688>3.0.co;2-3

Hwang D, Ramsey JD, Kabanov AV (2020) Polymeric micelles for the delivery of poorly soluble drugs: from nanoformulation to clinical approval. Adv Drug Deliv Rev 156:80–118. https://doi.org/10.1016/j.addr.2020.09.009

Lombardo D, Kiselev MA, Magazù S, Calandra P (2015) Amphiphiles self-assembly: basic concepts and future perspectives of supramolecular approaches. Adv Condens Matter Phys 2015:1–22. https://doi.org/10.1155/2015/151683

Zhang T, Luo J, Fu Y, Li H, Ding R, Gong T, Zhang Z (2017) Novel oral administrated paclitaxel micelles with enhanced bioavailability and antitumor efficacy for resistant breast cancer. Colloids Surfaces B Biointerfaces 150:89–97. https://doi.org/10.1016/j.colsurfb.2016.11.024

Xu W, Ling P, Zhang T (2013) Polymeric micelles, a promising drug delivery system to enhance bioavailability of poorly water-soluble drugs. J Drug Deliv 2013:1–15. https://doi.org/10.1155/2013/340315

Jhaveri AM, Torchilin VP (2014) Multifunctional polymeric micelles for delivery of drugs and siRNA. Front Pharmacol 5:77. https://doi.org/10.3389/fphar.2014.00077

Gaurav S, Elena VB, Alexander VK (2013) Different internalization pathways of polymeric micelles and unimers and their effects on vesicular transport. Bioconjug Chem 19:51–57

Ghezzi M, Pescina S, Padula C, Santi P, Del Favero E, Cantù L, Nicoli S (2021) Polymeric micelles in drug delivery: An insight of the techniques for their characterization and assessment in biorelevant conditions. J Control Release 332:312–336. https://doi.org/10.1016/j.jconrel.2021.02.031

Zhou XX, Jin L, Qi RQ, Ma T (2018a) PH-responsive polymeric micelles self-assembled from amphiphilic copolymer modified with lipid used as doxorubicin delivery carriers. R Soc Open Sci 5:171654. https://doi.org/10.1098/rsos.171654

Kim KN, Oh KS, Shim J, Schlaepfer IR, Karam SD, Lee JJ (2021) Light-responsive polymeric micellar nanoparticles with enhanced formulation stability. Polymers (Basel) 13:1–11. https://doi.org/10.3390/polym13030377

Barve A, Jain H, Liu Z, Zhao K (2020) Cheng, enzyme-responsive polymeric micelles of cabazitaxel for prostate cancer targeted therapy. Acta Biomater 113:501–511. https://doi.org/10.1016/j.actbio.2020.06.019

Shi Y, Lammers T, Storm G, Hennink WE (2017) Physico-chemical strategies to enhance stability and drug retention of polymeric micelles for tumor-targeted drug delivery. Macromol Biosci 17. https://doi.org/10.1002/mabi.201600160

Qiu L, Zhang J, Yan M, Jin Y, Zhu K (2007) Reverse self-assemblies based on amphiphilic polyphosphazenes for encapsulation of water-soluble molecules. Nanotechnology 18:475602. https://doi.org/10.1088/0957-4484/18/47/475602

Vrignaud S, Anton N, Gayet P, Benoit JP, Saulnier P (2011) Reverse micelle-loaded lipid nanocarriers: a novel drug delivery system for the sustained release of doxorubicin hydrochloride. Eur J Pharm Biopharm 79:197–204. https://doi.org/10.1016/j.ejpb.2011.02.015

Talelli M, Barz M, Rijcken CJF, Kiessling F, Hennink WE, Lammers T (2015) Core-crosslinked polymeric micelles: principles, preparation, biomedical applications and clinical translation. Nano Today 10:93–117. https://doi.org/10.1016/j.nantod.2015.01.005

Lu Y, Zhang E, Yang J, Cao Z (2018) Strategies to improve micelle stability for drug delivery. Nano Res 11:4985–4998. https://doi.org/10.1007/s12274-018-2152-3

Owen SC, Chan DPY, Shoichet MS (2012) Polymeric micelle stability. Nano Today 7:53–65. https://doi.org/10.1016/J.NANTOD.2012.01.002

Wakaskar RR (2017) Polymeric micelles and their properties. J Nanomed Nanotechnol 8:433. https://doi.org/10.4172/2157-7439.1000433

Zhou Q, Zhang L, Yang TH, Wu H (2018b) Stimuli-responsive polymeric micelles for drug delivery and cancer therapy. Int J Nanomedicine 13:2921. https://doi.org/10.2147/IJN.S158696

Whiteside TL (2008) The tumor microenvironment and its role in promoting tumor growth. Oncogene 27:5904. https://doi.org/10.1038/ONC.2008.271

Sikder G, Vambhurkar E, Amulya D, Bagasariya P, Famta S, Shah DK, Khatri SB, Singh VR, Sinha S (2022) Srivastava, advancements in redox-sensitive micelles as nanotheranostics: a new horizon in cancer management. J Control Release 349:1009–1030. https://doi.org/10.1016/J.JCONREL.2022.08.008

Son Y, Lee J, Baek M, Park D, Han SK, Min D, Lee BS (2021) Kim, PH-responsive amphiphilic polyether micelles with superior stability for smart drug delivery. Biomacromolecules 22:2043–2056. https://doi.org/10.1021/ACS.BIOMAC.1C00163/SUPPL_FILE/BM1C00163_SI_001.PDF

Feng H, Chu D, Yang F, Li Z, Fan B, Jin L, Li J (2020) Hypoxia-responsive polymeric micelles for enhancing cancer treatment. Front Chem 8:742. https://doi.org/10.3389/FCHEM.2020.00742/BIBTEX

Allen TM, Cullis PR (2004) Drug delivery systems: entering the mainstream. Science 303:1818–1822. http://science.sciencemag.org/content/sci/303/5665/1818.full.pdf

Vargason AM, Anselmo AC, Mitragotri S (2021) The evolution of commercial drug delivery technologies. Nat Biomed Eng 5:951–967. https://doi.org/10.1038/s41551-021-00698-w

Almeida A, Araújo M, Novoa-Carballal R, Andrade F, Gonçalves H, Reis RL, Lúcio M, Schwartz S, Sarmento B (2020) Novel amphiphilic chitosan micelles as carriers for hydrophobic anticancer drugs. Mater Sci Eng C 112:110920. https://doi.org/10.1016/j.msec.2020.110920

Agarwal NP, Matthies M, Gür FN, Osada K, Schmidt TL (2017) Block copolymer micellization as a protection strategy for DNA origami. Angew Chemie 129:5552–5556. https://doi.org/10.1002/ange.201608873

Cheng M, Liu Q, Gan T, Fang Y, Yue P, Sun Y, Jin Y, Feng J, Tu L (2021) Nanocrystal-loaded micelles for the enhanced in vivo circulation of docetaxel. Molecules 26:4481. https://doi.org/10.3390/molecules26154481

Shen D, Shen Y, Chen Q, Huang B, Mi Y, Shan Y, Yu G, Webster TJ (2020) Macrophage escape by cholesterol-polyoxyethylene sorbitol oleate micelles for pulmonary delivery. Nanomedicine 15:489–509. https://doi.org/10.2217/nnm-2019-0376

Kulthe SS, Choudhari YM, Inamdar NN, Mourya V (2012) Polymeric micelles: authoritative aspects for drug delivery. Des Monomers Polym 15:465–521. https://doi.org/10.1080/1385772X.2012.688328

Tehrani-Bagha AR, Holmberg K (2013) Solubilization of hydrophobic dyes in surfactant solutions. Materials (Basel). 6:580–608. https://doi.org/10.3390/ma6020580

Tehrani-Bagha AR, Bahrami H, Movassagh B, Arami M, Amirshahi SH, Menger FM (2007) Dynamic adsorption of gemini and conventional cationic surfactants onto polyacrylonitrile.

Colloids Surfaces A Physicochem Eng Asp 307:121–127. https://doi.org/10.1016/j.colsurfa.2007.05.011

Indulkar AS, Mo H, Gao Y, Raina SA, Zhang GGZ, Taylor LS (2017) Impact of micellar surfactant on supersaturation and insight into solubilization mechanisms in supersaturated solutions of atazanavir. Pharm Res 34:1276–1295. https://doi.org/10.1007/s11095-017-2144-0

Ikeda S, Ozeki S, Tsunoda MA (1980) Micelle molecular weight of dodecyldimethylammonium chloride in aqueous solutions, and the transition of micelle shape in concentrated NaCl solutions. J Colloid Interface Sci 73:27–37. https://doi.org/10.1016/0021-9797(80)90117-4

Ozeki S, Ikeda S (1980) The viscosity behavior of aqueous NaCl solutions of dodecyldimethylammonium chloride and the flexibility of its rod-like micelle. J Colloid Interface Sci 77:219–231. https://doi.org/10.1016/0021-9797(80)90434-8

Datyner (1978) The solubilization of disperse dyes by dispersing agents at 127 °C. J Soc Dye Colour 94:256–260. https://doi.org/10.1111/j.1478-4408.1978.tb03417.x

Gharanjig K, Sadeghi-Kiakhani M, Tehrani-Bagha AR, Khosravi A, Menger FM (2011) Solubility of two disperse dyes derived from n-alkyl and n-carboxylic acid naphthalimides in the presence of gemini cationic surfactants. J Surfactant Deterg 14:381–389. https://doi.org/10.1007/s11743-011-1253-8

Tehrani-Bagha AR, Singh RG, Holmberg K (2013) Solubilization of two organic dyes by anionic, cationic and nonionic surfactants. Colloids Surfaces A Physicochem Eng Asp 417:133–139. https://doi.org/10.1016/j.colsurfa.2012.10.006

Muto Y, Asada M, Takasawa A, Esumi K, Meguro K (1988) The efficiency of solubilization in mixed micelles of nonionic and anionic surfactants. J Colloid Interface Sci 124:632–638. https://doi.org/10.1016/0021-9797(88)90200-7

Wen H, Jung H, Li X (2015) Drug delivery approaches in addressing clinical pharmacology-related issues: opportunities and challenges. AAPS J 17:1327–1340. https://doi.org/10.1208/s12248-015-9814-9

Srinivasarao DA, Lohiya G, Katti DS (2019) Fundamentals, challenges, and nanomedicine-based solutions for ocular diseases. Wiley Interdiscip Rev Nanomedicine Nanobiotechnology 11:e1548. https://doi.org/10.1002/wnan.1548

Ghezzi M, Ferraboschi I, Delledonne A, Pescina S, Padula C, Santi P, Sissa C, Terenziani F, Nicoli S (2022) Cyclosporine-loaded micelles for ocular delivery: investigating the penetration mechanisms. J Control Release 349:744–755. https://doi.org/10.1016/j.jconrel.2022.07.019

Zhang Y, Wang Y, Meng L, Huang Q, Zhu Y, Cui W, Cheng Y, Liu R (2021) Targeted micelles with chemotherapeutics and gene drugs to inhibit the G1/S and G2/M mitotic cycle of prostate cancer. J Nanobiotechnol 19:17. https://doi.org/10.1186/s12951-020-00756-6

Figueiras A, Domingues C, Jarak I, Santos AI, Parra A, Pais A, Alvarez-Lorenzo C, Concheiro A, Kabanov A, Cabral H, Veiga F (2022) New advances in biomedical application of polymeric micelles. Pharmaceutics 14:1700. https://doi.org/10.3390/pharmaceutics14081700

Sutton D, Nasongkla N, Blanco E, Gao J (2007) Functionalized micellar Systems for Cancer Targeted Drug Delivery. Pharm Res 24:1029–1046. https://doi.org/10.1007/s11095-006-9223-y

Fernández M, Javaid F, Chudasama V (2018) Advances in targeting the folate receptor in the treatment/imaging of cancers. Chem Sci 9:790–810. https://doi.org/10.1039/C7SC04004K

Zhang L, Gong F, Zhang F, Ma J, Zhang P, Shen J (2013) Targeted therapy for human hepatic carcinoma cells using folate-functionalized polymeric micelles loaded with superparamagnetic iron oxide and sorafenib in vitro. Int J Nanomedicine 8:1517–1524. https://doi.org/10.2147/IJN.S43263

Guan X (2015) Cancer metastases: challenges and opportunities. Acta Pharm Sin B 5:402–418. https://doi.org/10.1016/j.apsb.2015.07.005

Wang JP, Hielscher A (2017) Fibronectin: how its aberrant expression in tumors may improve therapeutic targeting. J Cancer 8:674–682. https://doi.org/10.7150/jca.16901

Gong Z, Chen M, Ren Q, Yue X, Dai Z (2020) Fibronectin-targeted dual-acting micelles for combination therapy of metastatic breast cancer. Signal Transduct Target Ther 5:12. https://doi.org/10.1038/s41392-019-0104-3

Navarro G, Sawant RR, Biswas S, Essex S, Tros de Ilarduya C, Torchilin VP (2012) P-glycoprotein silencing with siRNA delivered by DOPE-modified PEI overcomes doxorubicin resistance in breast cancer cells. Nanomedicine 7:65–78. https://doi.org/10.2217/nnm.11.93

Liu W, Lo Y-L, Hsu C, Wu Y-T, Liao Z-X, Wu W-J, Chen Y-J, Kao C, Chiu C-C, Wang L-F (2019) CS-PEI/Beclin-siRNA downregulate multidrug resistance proteins and increase paclitaxel therapeutic efficacy against NSCLC. Mol Ther Nucleic Acids 17:477–490. https://doi.org/10.1016/j.omtn.2019.06.017

Sun W, Chen X, Xie C, Wang Y, Lin L, Zhu K, Shuai X (2018) Co-delivery of doxorubicin and anti-BCL-2 siRNA by pH-responsive polymeric vector to overcome drug resistance in in vitro and in vivo HepG2 hepatoma model. Biomacromolecules 19:2248–2256. https://doi.org/10.1021/ACS.BIOMAC.8B00272/SUPPL_FILE/BM8B00272_SI_001.PDF

Yalamarty SSK, Filipczak N, Li X, Pathrikar TV, Cotter C, Torchilin VP (2022) Co-delivery of siRNA and chemotherapeutic drug using 2C5 antibody-targeted dendrimer-based mixed micelles for multidrug resistant cancers. Pharmaceutics 14:1470. https://doi.org/10.3390/pharmaceutics14071470

Waris A, Ali AU, Khan M, Asim D, Zamel K, Fatima A, Raziq MA, Khan N, Akbar A, Baset MAS (2022) Abourehab, applications of various types of nanomaterials for the treatment of neurological disorders. Nanomater 12:2140. https://doi.org/10.3390/NANO12132140

Kaur J, Gulati M, Kapoor B, Jha NK, Gupta PK, Gupta G, Chellappan DK, Devkota HP, Prasher P, Ansari MS, Aba Alkhayl FF, Arshad MF, Morris A, Choonara YE, Adams J, Dua K, Singh SK (2022) Advances in designing of polymeric micelles for biomedical application in brain related diseases. Chem Biol Interact 361:109960. https://doi.org/10.1016/J.CBI.2022.109960

Le Gao H, Pang ZQ, Fan L, Hu KL, Wu BX, Jiang XG (2010) Effect of lactoferrin- and transferrin-conjugated polymersomes in brain targeting: in vitro and in vivo evaluations. Acta Pharmacol Sin 31:237. https://doi.org/10.1038/APS.2009.199

Agwa MM, Abdelmonsif DA, Khattab SN, Sabra S (2020) Self- assembled lactoferrin-conjugated linoleic acid micelles as an orally active targeted nanoplatform for Alzheimer's disease. Int J Biol Macromol 162:246–261. https://doi.org/10.1016/j.ijbiomac.2020.06.058

Giunchedi P, Gavini E, Bonferoni MC (2020) Nose-to-brain delivery. Pharmaceutics 12:138. https://doi.org/10.3390/pharmaceutics12020138

Ahmad E, Lv Y, Zhu Q, Qi J, Dong X, Zhao W, Chen Z, Wu W, Lu Y (2020) TAT modification facilitates nose-to-brain transport of intact mPEG-PDLLA micelles: evidence from aggregation-caused quenching probes. Appl Mater Today 19:100556. https://doi.org/10.1016/j.apmt.2020.100556

Ziai WC, Lewin JJ (2006) Advances in the Management of Central Nervous System Infections in the ICU. Crit Care Clin 22:661–694. https://doi.org/10.1016/j.ccc.2006.11.009

Kasinathan N, Jagani HV, Alex AT, Volety SM, Rao JV (2015) Drug delivery strategies for drug delivery to the central nervous system by systemic route strategies for drug delivery to the central nervous system by systemic route. Drug Deliv 22:243–257. https://doi.org/10.3109/10717544.2013.878858

Koepsell H (2020) Glucose transporters in brain in health and disease. Pflugers Arch 472:1299–1343. https://doi.org/10.1007/s00424-020-02441-x

Shao K, Zhang Y, Ding N, Huang S, Wu J, Li J, Yang C, Leng Q, Ye L, Lou J, Zhu L, Jiang C (2015) Functionalized nanoscale micelles with brain targeting ability and intercellular microenvironment biosensitivity for anti-intracranial infection applications. Adv Healthc Mater 4:291–300. https://doi.org/10.1002/adhm.201400214

Cavassin FB, Baú-Carneiro JL, Vilas-Boas RR, Queiroz-Telles F (2021) Sixty years of amphotericin B: An overview of the Main antifungal agent used to treat invasive fungal infections. Infect Dis Ther 10:115–147. https://doi.org/10.1007/s40121-020-00382-7

Wang Y, Ke X, Voo ZX, Yap SSL, Yang C, Gao S, Liu S, Venkataraman S, Obuobi SAO, Khara JS, Yang YY, Ee PLR (2016) Biodegradable functional polycarbonate micelles for controlled release of amphotericin B. Acta Biomater 46:211–220. https://doi.org/10.1016/j.actbio.2016.09.036

Mazlan MKN, Tazizi MHDM, Ahmad R, Noh MAA, Bakhtiar A, Wahab HA, Gazzali AM (2021) Antituberculosis targeted drug delivery as a potential future treatment approach. Antibiotics 10:1–19. https://doi.org/10.3390/antibiotics10080908

Tripodo G, Perteghella S, Grisoli P, Trapani A, Torre ML, Mandracchia D (2019) Drug delivery of rifampicin by natural micelles based on inulin: physicochemical properties, antibacterial activity and human macrophages uptake. Eur J Pharm Biopharm 136:250–258. https://doi.org/10.1016/j.ejpb.2019.01.022

Oerlemans C, Bult W, Bos M, Storm G, Nijsen JFW, Hennink WE (2010) Polymeric micelles in anticancer therapy: targeting, imaging and triggered release. Pharm Res 27:2569–2589. https://doi.org/10.1007/s11095-010-0233-4

Yoo SP, Pineda F, Barrett JC, Poon C, Tirrell M, Chung EJ (2016) Gadolinium-functionalized peptide amphiphile micelles for multimodal imaging of atherosclerotic lesions. ACS Omega 1:996–1003. https://doi.org/10.1021/acsomega.6b00210

Chen L, Deng H, Cui H, Fang J, Zuo Z, Deng J, Li Y, Wang X, Zhao L (2018) Inflammatory responses and inflammation-associated diseases in organs. Oncotarget 9:7204. https://doi.org/10.18632/ONCOTARGET.23208

Panés J, Perry M, Granger DN (1999) Leukocyte-endothelial cell adhesion: avenues for therapeutic intervention. Br J Pharmacol 126:537. https://doi.org/10.1038/SJ.BJP.0702328

Pagoto R, Stefania F, Garello F, Arena G, Digilio S, Aime E (2016) Terreno, paramagnetic phospholipid-based micelles targeting VCAM-1 receptors for MRI visualization of inflammation. Bioconjug Chem 27:1921–1930. https://doi.org/10.1021/acs.bioconjchem.6b00308

Qiu X, Qu Y, Guo B, Zheng H, Meng F, Zhong Z (2022) Micellar paclitaxel boosts ICD and chemo-immunotherapy of metastatic triple negative breast cancer. J Control Release 341:498–510. https://doi.org/10.1016/j.jconrel.2021.12.002

Pawar A, Singh S, Rajalakshmi S, Shaikh K, Bothiraja C (2018) Development of fisetin-loaded folate functionalized pluronic micelles for breast cancer targeting. Artif Cells Nanomed Biotechnol 46:347–361. https://doi.org/10.1080/21691401.2018.1423991

Situ J-Q, Wang X-J, Zhu X-L, Xu X-L, Kang X-Q, Hu J-B, Lu C-Y, Ying X-Y, Yu R-S, You J, Du Y-Z (2016) Multifunctional SPIO/DOX-loaded A54 homing peptide functionalized dextran-g-PLGA micelles for tumor therapy and MR imaging. Sci Rep 6:35910. https://doi.org/10.1038/srep35910

Bi Y, Liu L, Lu Y, Sun T, Shen C, Chen X, Chen Q, An S, He X, Ruan C, Wu Y, Zhang Y, Guo Q, Zheng Z, Liu Y, Lou M, Zhao S, Jiang C (2016) T7 peptide-functionalized PEG-PLGA micelles loaded with Carmustine for targeting therapy of glioma. ACS Appl Mater Interfaces 8:27465–27473. https://doi.org/10.1021/acsami.6b05572

Wu B, Sai S, Li K, Sun X, Han J, Tian B (2022) Maleimide-functionalized phospholipid/Pluronic F127 mixed micelles for efficient ophthalmic delivery of voriconazole against Candida albicans. Colloids Surfaces B Biointerfaces. 209:112180. https://doi.org/10.1016/J.COLSURFB.2021.112180

Ahn YS, Baik HJ, Lee BR, Lee ES, Oh KT, Lee DH, Youn YS (2010) Preparation of multifunctional polymeric micelles for antiviral treatment. Macromol Res 18:747–752. https://doi.org/10.1007/s13233-010-0802-8

Hisey B, Ragogna PJ, Gillies ER (2017) Phosphonium-functionalized polymer micelles with intrinsic antibacterial activity. Biomacromolecules 18:914–923. https://doi.org/10.1021/acs.biomac.6b01785

Physical and Analytical Techniques Used for the Characterization of Polymeric Micelles

6

Amulya Etikala, Shweta Nene, Shashi Bala Singh, and Saurabh Srivastava

Abstract

Nanomedicines are notably the intriguing strategies for addressing diverse challenges allied with drug delivery. Polymeric micelles, a self-assembled nanocarrier of amphiphilic polymers, help in overcoming drug-related issues like poor solubility and permeability. Diminished size, enhanced solubilization properties, ease of fabrication and sterilization make polymeric micelles more fascinating than other nanocarriers. The first and most imperative step after micelle formation is their characterization which gives detailed information about its size, morphology, critical micellar concentration (CMC), chemical structure, and stability. Characterization is considered the backbone for developing stable, safe, and effective drug delivery systems (DDS). Instead, the benefits of one or more characterization techniques can be further clubbed in order to get better understanding of micelles behavior. This chapter will present a brief description of various physical and analytical characterization techniques for evaluating polymeric micelles.

Keywords

Polymeric micelles · Nanocarrier · Transmission electron microscopy · Scanning electron microscopy · Critical micellar concentration

A. Etikala · S. Nene · S. Srivastava (✉)
Pharmaceutical Innovation and Translational Research Lab (PITRL), Department of Pharmaceutics, National Institute of Pharmaceutical Education and Research (NIPER) Hyderabad, Hyderabad, Telangana, India
e-mail: saurabh@niperhyd.ac.in

S. B. Singh
Department of Biological Sciences, National Institute of Pharmaceutical Education and Research (NIPER) Hyderabad, Hyderabad, Telangana, India

S. K. Singh et al. (eds.), *Polymeric Micelles: Principles, Perspectives and Practices*,
https://doi.org/10.1007/978-981-99-0361-0_6

6.1 Introduction

A key necessity of the healthcare industry has been the search for innovative medications in the management of diseases without compromising on safety and efficacy. Over the past few decades, nanocarriers have gained stupendous interest due to their extensive applications in the field of medicine (Su and Kang 2020).

Polymeric micelles are one such nanocarrier-based drug delivery systems (DDS) that have piqued the interest of researchers for the treatment and diagnosis of a variety of disorders, owing to significant benefits over conventional therapies (Joseph et al. 2017). Polymeric micelles offer drug targeting to disease tissue (Deng et al. 2012), in vivo biodegradability (Bölgen 2018) and biocompatibility (Atanase 2021), increased accumulation (Nair et al. 2018), and reduced adverse effects (Rana et al. 2017). Kataoka's group pioneered the application of polymeric micelles as DDS in the early 1990s by developing doxorubicin (DOX) conjugated block copolymer micelles (Kedar et al. 2010). The architecture of polymeric micelles generally includes the inner core and the outer shell (Tiwari and Rohiwal 2018). The lipophilic portion of the block copolymer, i.e. core, incorporates the poorly water-soluble drug. In contrast, the outermost covering of the copolymer's hydrophilic block safeguards the drug from an aqueous milieu. Polymeric micelles can be synthesized via physical, chemical, or a combination of physical and chemical procedures (Yadav et al. 2018). Physical approaches are often more accessible, while chemical methods assure higher systemic stability.

The development of safe and efficient theranostic nanomedicine requires robust and accurate characterization of the interactions between nanoengineered materials and biosystems (Tsuji et al. 2006). Characterization of a micelle's physical and chemical characteristics is critical for assuring reproducibility as well as for exploring how the physical and chemical features of micelles influence their biological activities (Mahmoudi 2021). Characterization techniques present insights on critical micellar concentration (CMC), size, shape, viscosity, stability, and chemical structure. In this chapter, we have highlighted the sophisticated physical and analytical techniques used in the characterization of polymeric micelles.

6.2 Physical and Analytical Characterization of Polymeric Micelles

Characterization is crucial for developing and implementing micelles for various applications. Several characterization techniques are developed in response to the micelle's distinct and unique physicochemical features. Micelles are analyzed to investigate physical and chemical aspects such as morphology, composition, size, shape, surface area, optical properties, and surface composition. Micelles characterization should not be confined to a single approach since numerous measurements

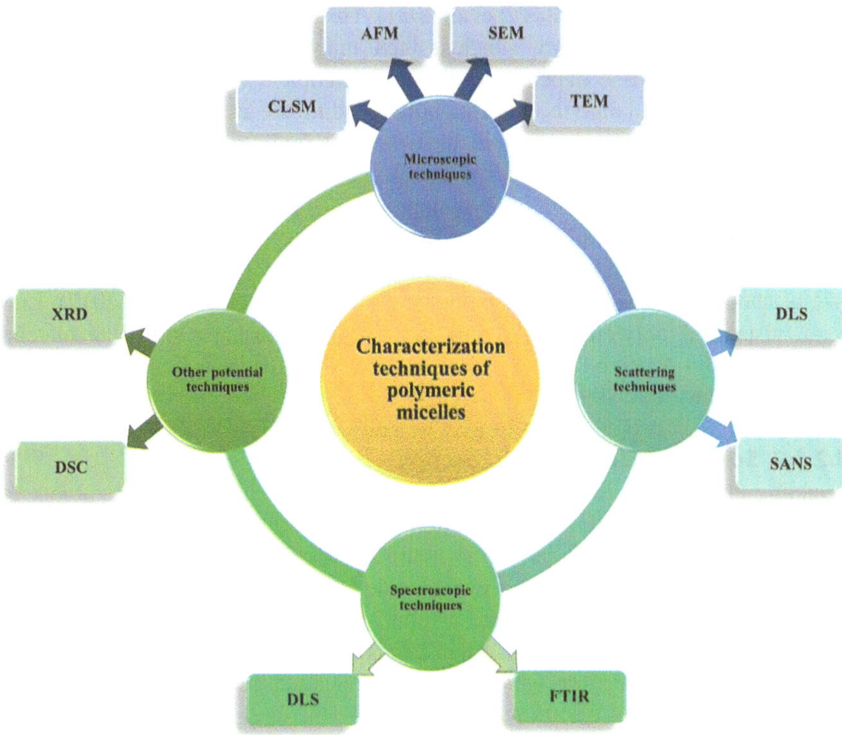

Fig. 6.1 Various characterization techniques of polymeric micelles

are generally required to capture all relevant features of micelles. This section is further categorized into microscopic, spectroscopic, scattering, and other potential techniques. Figure 6.1 depicts the various characterization techniques of polymeric micelles.

6.2.1 Microscopic Techniques

6.2.1.1 Atomic Force Microscopy

Atomic force microscopy (AFM) is considered a high-resolution microscopic technique widely used to evaluate the morphology and size of the micelles (Spyratou et al. 2009). Scanning modes used in AFM include static mode, dynamic/tapping mode which gives information regarding nanocarrier morphology and interactions between particles and biomolecules. Even redox and temperature-related variations can be characterized utilizing AFM (Kim et al. 2021). AFM is a feasible technique for estimating a micelle's size as it can distinguish particles of varied dimensions and work at room temperature (Ghezzi et al. 2021). This technique involves the micelles deposition on a solid support, which is then analyzed using a sharp probe. The movement of the probe depends on the fixed sample's topography. The sample should be firmly fixed onto the solid support to endure the lateral forces exhibited by the scanning tip to ensure accurate measurement findings (Ghezzi et al. 2021). Therefore, it is necessary to employ the tapping mode, a method recognized to reduce sample distortion brought on by interactions between the surface and the tip (Kotta et al. 2022). Curcumin-loaded mono methoxy poly(ethylene glycol)-poly (ε-caprolactone) (MPEG-PCL) micelles characterization by AFM demonstrated a size of about 89 nm with spherical morphology (Kheiri Manjili et al. 2017). Liaw et al. visualized poly(ethylene oxide)-poly(β-benzyl L-aspartate)-pyrene micelles using AFM and observed the spherical morphology of micelles and size of about 50-80 nm (Liaw et al. 1998).

6.2.1.2 Scanning Electron Microscopy

Scanning electron microscopy (SEM) is a microscopic method that explores the nano dimension of materials using electron beams (Venkateshaiah et al. 2020). This technique uses a milligram sample to assess the particle size, shape, and texture. Direct visualization provided by SEM provides details on the sample's crystalline structure, chemical composition, and external morphology (Yaneva and Georgieva 2017). The basic concept of SEM is that a tungsten filament or a field emission gun produces a beam of electrons which are then accelerated through a high voltage and passed through a set of electromagnetic lens and apertures to yield a thin beam of electrons (Marassi and Nobili 2009). The so-formed beams scan the specimen's surface resulting in the formation of electrons from the specimen, which are then captured by a detector. Birhan and his co-workers prepared a diselenide bond possessing Bi(mPEG-PLGA)-Se2 micelles using poly(lactide-*co*-glycolide)-*b*-poly (ethylene glycol) methyl ether (mPEG-PLGA) and 3,3′-diselanediyldipropanoic acid (DSeDPA). SEM was employed to evaluate its morphology which depicted the spherical nature of micelles with a size of around 100 nm. Compared to the dynamic light scattering (DLS) result, a decrease in the hydrodynamic diameter of micelles was observed in SEM, mainly due to water evaporation during SEM sample preparation (Birhan et al. 2019). Cholesterol end-capped (CO) micelles analyzed using SEM depicted a size below 200 nm, whereas enhanced size (200–300 nm) was observed after quercetin loading (Gao et al. 2020).

6.2.1.3 Transmission Electron Microscopy

In transmission electron microscopy (TEM), the electron beams pass through the ultrathin sample (<100 nm) and form an image. It is an excellent tool for analyzing the structural and chemical features of the materials at the nanoscale (Rostamabadi et al. 2020). TEM provides detailed information about the material behavior and properties by combining the principles of diffraction, microanalysis, and imaging.

TEM resolution is considerably high compared to a light microscope (Malatesta 2021). J. Glisoni fabricated glucosylated pluronic F127 (F127-O-Glu) micelles and characterization by TEM depicted the spherical morphology of micelles (Glisoni and Sosnik 2014). TEM analysis of Artesunate micelles portrayed that the micelles are round in appearance (Long et al. 2021). Poly(L-lactide) micelles depicted alteration in the shapes of micelles from diamond shape to cylindrical upon modifying the polymeric block length (Yu et al. 2017). Poly(styrene-*block*-acrylic acid-*block*-ethylene oxide) micelles were found to be circular with a core diameter of 25 nm (Alam et al. 2014).

6.2.1.4 Confocal Laser Scanning Microscopy

Confocal laser scanning microscopy (CLSM) offers significantly higher spatial resolution and contrast than conventional wide-field optical microscopy (Justyna 2017). By blocking maximum amount of light from the specimen, a CLSM provides clear pictures of specimens that would otherwise seem blurry with conventional microscopes. CLSM creates a three-dimensional (3D) picture of the specimen by assembling several tiny slices of it. Additional benefits include the capacity to capture serial optical sections from thick specimens, a reduced background signature and control over field depth. One of the most significant developments in fluorescence imaging in recent years is CLSM which is regarded as a crucial tool in the biological study (Hovis and Heuer 2010). Characterization of DOX-loaded micelles by CLSM showed higher fluorescence in HeLa cells compared to the free drug (Wang et al. 2018). OVCAR-3 cells were incubated with micelles and observed under CLSM. Results portrayed that the concentration of micelles in the cells is temperature dependent, i.e. at 37 °C, the number of micelles in OVCAR-3 cells enhanced, whereas at 4 °C, no uptake was observed in the cells (Kim et al. 2013).

6.2.2 Spectroscopic Techniques

6.2.2.1 Fourier-Transform Infrared Spectroscopy

Fourier-transform infrared spectroscopy (FTIR) involves the usage of infrared radiations, which helps in analyzing various polymeric, organic and inorganic materials (Titus et al. 2019). It is also known as vibrational spectroscopy, as it measures changes in vibrational energies (Singh et al. 2014). When the natural fundamental vibrational frequency becomes equal to the frequency of infrared radiations, molecules absorb the radiations and give absorption spectra. In general, each and every molecule possess distinct fingerprint, making FTIR a crucial tool for chemical identification (Shabanian et al. 2020). Interaction between two or more materials can be identified by measuring the alterations in their characteristic absorption bands. FTIR shows a vital role in characterizing the polymeric micelles, which involves the identification and detection of unknown moieties, additives, and contaminants (Dhakal et al. 2019). S. Patil et al. prepared Galangin-loaded galactosylated pluronic F68 (GF68-Gal) micelles and characterized them using FTIR. Galangin displayed the absorption bands at 3176 cm^{-1}, 1669 cm^{-1}, and 2901 cm^{-1}. Same bands were observed in GF68-Gal micelles, demonstrating the drug's stability within the micelle (Patil et al. 2019). Silibinin polymeric micelles prepared using pluronic F-68 were characterized using FTIR. Pure drug showed peaks at 3435.35 cm^{-1}, 2927.62 cm^{-1}, 1641.42 cm^{-1}, 1512.16 cm^{-1}, 1472.62 cm^{-1}, and 1268.18 cm^{-1}. Silibinin micelles did not show any significant changes in the peaks, indicating compatibility between silibinin and pluronic F-68 (Shankar and Agrawal 2015).

6.2.2.2 Nuclear Magnetic Resonance

Nuclear magnetic resonance (NMR) involves the assessment of absorption of electromagnetic radiation in the radio frequency region of around 4 to 400 MHz. The integrity of polymers used in polymeric micelles can be elucidated by NMR, which is generally considered a qualitative and quantitative technique (Espina et al. 2009). Proton NMR is the most frequently used NMR which emphasizes its application in evaluating polymer structure, exclusively for active targeting in which polymers are decorated with targeting moieties (Zia et al. 2019). Wang et al. used NMR to assess the intercalation of DOX into the core of polyethylene glycol (PEG)-phosphatidylethanolamine (PE) micelles. Results depicted the insertion of anthracycline ring of DOX between PE phospholipids, while amino sugar was positioned in the micelle's outer shell (Wang et al. 2010). Puig-Rigall et al. characterized polymeric micelles of Miltefosine (MF) by NMR to evaluate drug positioning in the micellar structure. Results depict that MF and D-α-tocopheryl PEG succinate (TPGS) formed mixed micelle, where the lipophilic fragments of surfactants are in contact with each other and the formation of the hydrophilic shell is by the zwitterionic head of the drug and PEG blocks of TPGS (Puig-Rigall et al. 2020).

6.2.3 Scattering Techniques

6.2.3.1 Dynamic Light Scattering

DLS (photon correlation spectroscopy), an analytical technique employed for analyzing the size and surface charge of the nanocarriers is widely used in the field of biotechnology, pharmacy, and biochemistry (Das et al. 2019). The size range between a few nanometers to a few micrometers can be examined via DLS (Ullmann et al. 2019). This method relies on the principle that in a gas or liquid, particles undergo random motion, i.e. Brownian movement (Caputo et al. 2019). The Stokes–Einstein equation explains the diffusion of particles. DLS possesses a variety of advantages such as no sample preparation, requires little amount of sample, offers complete sample recovery following measurement and stability of nanocarriers as a function of time can be studied (Falke and Betzel 2019). Micelle size is a crucial factor that is correlated to CMC and solubilizing effectiveness. Biodistribution and pharmacokinetic features of micelles are regulated by means of size and surface charge. Hence, prediction of size and charge by DLS is essential to evaluate the behavior of micelles when they are in contact with biological fluids or tissues (Honary and Zahir 2013). Zhang et al. fabricated methoxy PEG-poly(D,L-lactide) (MePEG-PDLLA) micelles and evaluated the average size using Zetasizer 3000 HS, Malvern Instruments. Results demonstrated that micelles possess size in the range of 25–30 nm and the size was found to be increased upon enhancing the concentration of poly(D,L-lactide) block (Zhang et al. 2007). Huang and his co-workers prepared pH-sensitive micelles using (poly(β-amino esters)-*g*-cholesterol)-*b*-PEG-b-(poly (β-amino esters)-*g*-cholesterol). Findings from the study suggested that upon decreasing the pH from 7.4 to 5.0, increase in the size as well as zeta potential was observed from 205.4 nm to 285.7 nm and +12.7 mV to +47.0 mV, respectively (Huang et al. 2017). DOX-loaded polymeric micelles depicted a size of about 144.8 ± 6.3 nm, 0.248 ± 0.027 polydispersity index (PDI), and zeta potential of about 18.68 ± 2.12 mV (Aji Alex et al. 2016).

6.2.3.2 Small Angle Neutron Scattering (SANS)

Small angle neutron scattering (SANS) is a potent and unique technique for studying structures and inhomogeneities with dimensions ranging from 1 to 1000 nm, which are significantly bigger than interatomic distances in condensed matter (Honecker et al. 2022). SANS is considered a vital neutron scattering technology for analyzing bulk polymers. The internal structure of polymeric micelles can be measured via SANS. Rigall and group have formulated MF-loaded polyethylene oxide-based polymeric micelles and evaluated their structural characteristics with the help of SANS, DLS, and NMR spectroscopy. The SANS study revealed that MF leads to the formation of self-aggregates in stable monodisperse micelles. The concentration of MF (0.2%) above its CMC showed a size around 3–3.5 nm from the data obtained from SANS scattering patterns at the temperatures of 20 °C and 37 °C. There was a slight decrease in the aggregation number and micelles shell hydration was observed at 50 °C. The SANS study for the MF-loaded TPGS micelles revealed the formation of a stable structure of micelles with temperature having a hydrated shell. The SANS study for Tetronic-MF micellar systems confirmed the formation of mixed micelles (Puig-Rigall et al. 2020).

6.2.4 Other Potential Techniques

6.2.4.1 Differential Scanning Calorimetry

Differential scanning calorimetry (DSC) is a simple, non-perturbing, high-efficient thermal analysis technique developed in the early 1960s (Chiu and Prenner 2011). In this technique, temperature and time-dependent changes in the physical properties of samples will be measured. In this technique, the energy needed to maintain zero temperature difference between the sample and reference is assessed. The core principle behind this technique is that when the sample goes through phase transitions, more or less heat is required to flow to it than the reference to retain both at a similar temperature. DSC is generally classified as heat-flux and power-compensated DSC based on the mode of operation (P. G, T.T. M, B. R. 2010). DSC is usually employed in the pharmaceutical field due to its wide range of applications like the thermal characterization of nanocarriers, alterations in glass-transition temperature of materials used in nanocarrier fabrication, and self-assembly of supramolecular nanocarriers. Using DSC, Gaber and his team characterized beclomethasone dipropionate (BDP) loaded polymeric micelles. BDP displayed a peak at 212.7 °C, whereas BDP-loaded micelles depicted a peak at 58 °C without any BDP peak, demonstrating that the drug was dispersed within the micellar structure (Gaber et al. 2006). Deferoxamine mesylate (DFO) loaded polymeric micelles analyzed using DSC did not show an endothermic peak at 190 °C, whereas DFO alone depicted a peak at 190 °C. This indicates that the drug is dissolved in the formulation and the thermotropic nature of micelles was not altered (Salimi et al. 2019). Sunitinib malate (SM) and SM-MPEG-PCL micelle's physicochemical characteristics were analyzed using DSC. The pure drug showed an endothermic peak at 200.86 °C, which is absent in the spectrum of micelles. This indicates that the drug is entrapped within the micellar formulation (Streets et al. 2020).

Sunitinib-Loaded MPEG-PCL Micelles for the Treatment of Age-Related Macular Degeneration Effect of shell-crosslinking of micelles on endocytosis and exocytosis: acceleration of exocytosis by crosslinking Effect of shell-crosslinking of micelles on endocytosis and exocytosis: acceleration of exocytosis by crosslinking.

6.2.4.2 X-Ray Diffraction

X-ray diffraction (XRD) plays a crucial role in determining the sample's identity based on its crystallinity (Bunaciu et al. 2015). Elemental proportions can be analyzed even if the sample is in mixture form. XRD data provides information about the degree of crystallinity, structural state, and deviation of the specific element from its ideal composition. Only a portion of the X-ray beam gets transmitted due to its interaction with the atomic planes and the remaining rays get absorbed, scattered, refracted, and diffracted by the sample. X-rays get diffracted differently by each element based on the type of atoms and their configuration. The technique of XRD relies on the constructive interference of monochromatic X-rays with a sample which is in crystalline form. A cathode ray tube emits the X-rays, which are then filtered to give monochromatic radiation, collimated and targeted at the sample. When the conditions are in accordance with Bragg's law, the interaction between incident rays and the sample results in constructive interference (Bunaciu et al. 2015). Liang and his team characterized paclitaxel (PTX)-loaded micelles and XRD data revealed that PTX existed as an amorphous or molecular state inside the micelles (Liang et al. 2018). Micellar curcumin-spray dried powder (MC-SDP) was prepared using curcumin, sucrose, casein, and TPGS. XRD analysis of MC-SDP did not exhibit any sharp peaks, which were generally observed in curcumin, sucrose, and TPGS. Loss of the crystallinity of curcumin is due to the interaction between curcumin and the lipophilic groups of casein (Wijiani et al. 2020).

6.2.4.3 Critical Micellar Concentration

The point at which surfactant molecules aggregate together in a liquid to form micelles is referred to as the CMC (Mabrouk et al. 2021). The thermodynamic stability of micelles is generally defined by CMC (Kapse et al. 2019). The thermodynamic stability is stronger when CMC values are lower (Anoune et al. 2002). Critical association concentration (CAC) is occasionally used instead of CMC to distinguish polymeric micelles from surfactants. Fluorescence spectroscopy, DLS, and surface tension measurement are frequently employed for CMC determination (Majumder et al. 2020). Birhan and his team prepared Bi(mPEG-PLGA)-Se_2 micelles utilizing mPEG-PLGA and DSeDPA. Bi(mPEG-PLGA)-Se_2 micelles depicted a CMC value of about 1.39×10^{-5} g/mL, which is 2.6-fold lower than the CMC of mPEG-PLGA (3.55×10^{-5} g/mL). This infers enhanced stability of micelles which is mainly due to the presence of lipophilic diselenide linkage in the micelle's core (Birhan et al. 2019). Weina et al. prepared vorinostat-loaded polymeric micelles to manage breast cancer. CMC value of drug-loaded micelles and control carrier was found to be 25 μg/mL and 31 μg/mL, respectively. The lower value of CMC of drug-loaded micelles depicts the enhanced stability of micelles in physiological fluid (Ma et al. 2020). Table 6.1 depicts outcomes obtained from various characterization techniques of micelles.

6.3 Conclusion

Polymeric micelles are recognized as potential nanocarriers for the targeted delivery of a wide range of therapeutics because of their remarkable biocompatibility, increased systemic circulation, low toxicity and ability to enhance solubility. Apart from formulation and development, characterization plays a critical role in optimization, maintaining stability, bench-to-bed side translation, thereby defining its quality and purity. Physical and analytical characterization techniques are essential in the effective qualitative and quantitative evaluation of polymeric micelles. Each characterization technique has its unique role in confirming the formation, morphological features, in vitro as well as in vivo fate of polymeric micelles. With growing interest of micelles in research and development, it is crucial that the researchers recognize the unpredicted challenges allied with reproducibility in formation of micelles along with their characterization. It is important for researchers to comprehend how characterization techniques might aid in eliminating synthesis errors.

Table 6.1 Key findings from various characterization techniques of polymeric micelles

Micelle system	Outcomes	References
Artemisinin loaded PCL–PEG–PCL micelles	Characterization of micelles by AFM depicted the size of about 70 nm with homogenous spherical morphology	(Manjili et al. 2018)
Gambogic acid delivery using *N*-octyl-*N*-arginine-chitosan micelles	Fan et al. characterized the micelles using AFM and confirmed that the micelles are sphere-shaped with a diameter of 160 nm	(Yu et al. 2014)
Pluronic F127-tocopherol micelles	DLS depicted the size, PDI, and zeta potential of about 51.87 ± 6.39 nm, 0.184 ± 0.015, and −8.43 ± 2.27 mV, respectively, suggesting that the smaller size of micelles would facilitate tumor-specific accumulation by improving the permeability and retention effect	(Liu et al. 2017)
Lauric acid loaded polymeric micelles	Findings from DLS illustrated that the micelles possess size in the range of 27–89 nm	(Tran et al. 2016)
TPGS/Solutol polymeric micelles loaded with PTX	TEM image depicted the smooth spherical morphology of micelles	(Li et al. 2019)
Cyclic arginine-glycine-aspartic acid decorated micellar mertansine prodrug (cRGD-MMP)	Findings from CLSM portrayed the efficient cellular uptake of cRGD-MMP by the MDA-MB-231 cells	(Zhong et al. 2017)
MPEG-PCL micelles loaded with diclofenac	Characterization of MPEG-PCL by DSC depicted two endothermic melting peaks at 49 °C and 57.3 °C, whereas diclofenac exhibited sharp endothermic melting point peak at 178 °C. The physical mixture of MPEG-PCL and diclofenac have shown all the characteristic peaks of polymer as well as drug which indicates the crystalline nature of diclofenac. DSC of diclofenac-loaded MPEG-PCL micelles illustrated the disappearance of drug's melting peak and the polymer peaks shifted to lower temperature. Overall findings from DSC demonstrated the decline in crystalline nature of diclofenac in micelles and existence interaction between drug and polymer	(Li et al. 2012)
Redox responsive core-crosslinked micelles loaded with methotrexate	Gulfam et al. used bis-alkyne-ethyl disulfide cross-linker to incorporate redox sensitive bridge in the core of micelles. ^{1}H NMR and ^{13}C NMR were used to evaluate the structure of cross-linker. Crosslinking was confirmed by the presence of ^{1}H NMR and ^{13}C NMR signals at 4.39 ppm and 62.48 ppm, respectively	(Gulfam et al. 2017)

Acknowledgments The authors would like to acknowledge the research funding support by Department of Pharmaceuticals (DoP), Ministry of Chemicals and Fertilizers, Govt. of India to National Institute of Pharmaceutical Education and Research (NIPER) Hyderabad, INDIA.

Declaration of Interest The authors declare that there is no conflict of interest.

References

Su S, Kang PM (2020) Recent advances in nanocarrier-assisted therapeutics delivery systems. Pharmaceutics 12(9):1–27

Joseph M, Trinh HM, Mitra AK (2017) Peptide and protein-based therapeutic agents. In: Emerging nanotechnologies for diagnostics, drug delivery and medical devices, pp 145–167

Deng C, Jiang Y, Cheng R, Meng F, Zhong Z (2012) Biodegradable polymeric micelles for targeted and controlled anticancer drug delivery: promises, progress and prospects. Nano Today 7(5): 467–480

Bölgen N (2018) Biodegradable polymeric micelles for drug delivery applications. In: Stimuli responsive polymeric nanocarriers for drug delivery applications, vol 1. Types and Triggers, pp 635–651

Atanase LI (2021) Micellar drug delivery systems based on natural biopolymers. Polymers (Basel) 13(3):1–33

Nair HA, Rajawat GS, Nagarsenker MS. Stimuli-responsive micelles: a nanoplatform for therapeutic and diagnostic applications. Drug Target Stimuli Sensit Drug Deliver Syst. 2018. 303–342

Rana S, Bhattacharjee J, Barick KC, Verma G, Hassan PA, Yakhmi JV (2017) Interfacial engineering of nanoparticles for cancer therapeutics. Nanostruct Cancer Therapy:177–209

Kedar U, Phutane P, Shidhaye S, Kadam V (2010) Advances in polymeric micelles for drug delivery and tumor targeting. Nanomed Nanotechnol Biol Med 6(6):714–729

Tiwari AP, Rohiwal SS (2018) Synthesis and bioconjugation of hybrid nanostructures for biomedical applications. Hybrid Nanostruct Cancer Theranostics:1–10

Yadav HKS, Almokdad AA, Shaluf SIM, Debe MS (2018) Polymer-Based Nanomaterials for Drug-Delivery Carriers. In: Nanocarriers for Drug Delivery: Nanoscience and Nanotechnology in Drug Delivery. Elsevier, Amsterdam, pp 531–556

Tsuji JS, Maynard AD, Howard PC, James JT, Lam C, Warheit DB et al (2006) Research strategies for safety evaluation of nanomaterials, part IV: risk assessment of nanoparticles. Toxicol Sci 89(1):42–50

Mahmoudi M (2021) The need for robust characterization of nanomaterials for nanomedicine applications. Nat Commun 12(1):5246

Spyratou E, Mourelatou EA, Makropoulou M, Demetzos C (2009) Atomic force microscopy: a tool to study the structure, dynamics and stability of liposomal drug delivery systems. Expert Opin Drug Deliv 6(3):305–317

Kim YM, Lee J, Jeon DJ, Oh SE, Yeo JS (2021) Advanced atomic force microscopy-based techniques for nanoscale characterization of switching devices for emerging neuromorphic applications. Appl Microsc 51(1):7

Ghezzi M, Pescina S, Padula C, Santi P, Del Favero E, Cantù L et al (2021) Polymeric micelles in drug delivery: an insight of the techniques for their characterization and assessment in biorelevant conditions. J Control Release 332:312–336

Kotta S, Aldawsari HM, Badr-Eldin SM, Nair AB, Kamal YT (2022) Progress in polymeric micelles for drug delivery applications. Pharmaceutics. 14(8):1636

Kheiri Manjili H, Ghasemi P, Malvandi H, Mousavi MS, Attari E, Danafar H (2017) Pharmacokinetics and in vivo delivery of curcumin by copolymeric mPEG-PCL micelles. Eur J Pharm Biopharm 116:17–30

Liaw J, Aoyagi T, Kataoka K, Sakurai Y, Okano T (1998) Visualization of PEO-PBLA-pyrene polymeric micelles by atomic force microscopy. Pharm Res 15(11):1721–1726

Venkateshaiah A, Padil VVT, Nagalakshmaiah M, Waclawek S, Černík M, Varma RS (2020) Microscopic techniques for the analysis of micro and nanostructures of biopolymers and their derivatives. Polymers (Basel) 12(3):512

Yaneva Z, Georgieva N (2017) Physicochemical and morphological characterization of pharmaceutical nanocarriers and mathematical modeling of drug encapsulation/release mass transfer processes. In: Nanoscale Fabrication, Optimization, Scale-up and Biological Aspects of Pharmaceutical Nanotechnology. William Andrew, Norwich, NY, pp 173–218

Marassi R, Nobili F (2009) Measurement methods|structural and chemical properties: transmission electron microscopy. In: Encyclopedia of Electrochemical Power Sources. Newnes, Oxford and Boston, pp 769–789

Birhan YS, Hailemeskel BZ, Mekonnen TW, Hanurry EY, Darge HF, Andrgie AT et al (2019) Fabrication of redox-responsive bi(mPEG-PLGA)-Se2 micelles for doxorubicin delivery. Int J Pharm 567:118486

Gao M, Yang Y, Bergfel A, Huang L, Zheng L, Bowden TM (2020) Self-assembly of cholesterol end-capped polymer micelles for controlled drug delivery. J Nanobiotechnol 18(1):13

Rostamabadi H, Falsafi SR, Jafari SM (2020) Transmission electron microscopy (TEM) of nanoencapsulated food ingredients. In: Characterization of nanoencapsulated food ingredients. Elsevier, Amsterdam, pp 53–82

Malatesta M (2021) Transmission electron microscopy as a powerful tool to investigate the interaction of nanoparticles with subcellular structures. Int J Mol Sci 22(23)

Glisoni RJ, Sosnik A (2014) Novel poly(ethylene oxide)-b-poly(propylene oxide) copolymer-glucose conjugate by the microwave-assisted ring opening of a sugar lactone. Macromol Biosci 14(11):1639–1651

Long M, Xu J, Fang W, Mao J, Zhang J, Liu S et al (2021) Enhanced delivery of artesunate by stimuli-responsive polymeric micelles for lung tumor therapy. J Drug Deliv Sci Technol 66: 102812

Yu W, Inam M, Jones JR, Dove AP, O'Reilly RK (2017) Understanding the CDSA of poly(lactide) containing triblock copolymers. Polym Chem 8(36):5504–5512

Alam MM, Zhao W, Zhai S, Yusa SI, Noguchi H, Nakashima K (2014) Fabrication of copper (II) oxide hollow nanosphere using ABC block copolymer templates and its application as anode materials in lithium ion batteries. Chem Lett 43(9):1426–1428

Justyna W (2017) Confocal microscopy: principles and modern practices. Curr Protoc Cytom 176(5):139–148

Hovis DB, Heuer AH (2010) The use of laser scanning confocal microscopy (LSCM) in materials science. J Microsc 240(3):173–180

Wang SW, Lin YK, Fang JY, Lee RS (2018) Photo-responsive polymeric micelles and prodrugs: synthesis and characterization. RSC Adv 8(51):29321–29337

Kim Y, Pourgholami MH, Morris DL, Lu H, Stenzel MH (2013) Effect of shell-crosslinking of micelles on endocytosis and exocytosis: acceleration of exocytosis by crosslinking. Biomater Sci 1(3):265–275

Titus D, James Jebaseelan Samuel E, Roopan SM (2019) Nanoparticle characterization techniques. In: Green synthesis, characterization and applications of nanoparticles. Elsevier, Amsterdam, pp 303–319

Shabanian M, Hajibeygi M, Raeisi A (2020) FTIR characterization of layered double hydroxides and modified layered double hydroxides. In: Layered double hydroxide polymer nanocomposites. Woodhead Publishing, Sawston, pp 77–101

Singh KS, Majik MS, Tilvi S (2014) Vibrational spectroscopy for structural characterization of bioactive compounds. Compr Anal Chem 65:115–148

Dhakal S, Schmidt WF, Kim M, Tang X, Peng Y, Chao K (2019) Detection of additives and chemical contaminants in turmeric powder using FT-IR spectroscopy. Foods 8(5):143

Patil S, Ujalambkar V, Rathore A, Rojatkar S, Pokharkar V (2019) Galangin loaded galactosylated pluronic F68 polymeric micelles for liver targeting. Biomed Pharmacother 112:108691
Shankar G, Agrawal YK (2015) Formulation and evaluation of polymeric micelles for a poorly absorbed drug. Res J Pharm Biol Chem Sci 6(3):1314–1321
Espina R, Yu L, Wang J, Tong Z, Vashishtha S, Talaat R et al (2009) Nuclear magnetic resonance spectroscopy as a quantitative tool to determine the concentrations of biologically produced metabolites: implications in metabolites in safety testing. Chem Res Toxicol 22(2):299–310
Zia K, Siddiqui T, Ali S, Farooq I, Zafar MS, Khurshid Z (2019) Nuclear magnetic resonance spectroscopy for medical and dental applications: a comprehensive review. Eur J Dent 13(1): 124–128
Wang Y, Wang R, Lu X, Lu W, Zhang C, Liang W (2010) Pegylated phospholipids-based self-assembly with water-soluble drugs. Pharm Res 27(2):361–370
Puig-Rigall J, Fernández-Rubio C, González-Benito J, Houston JE, Radulescu A, Nguewa P et al (2020) Structural characterization by scattering and spectroscopic methods and biological evaluation of polymeric micelles of poloxamines and TPGS as nanocarriers for miltefosine delivery. Int J Pharm 578:119057
Das S, Mukherjee A, Sengupta G, Singh VK (2019) Overview of nanomaterials synthesis methods, characterization techniques and effect on seed germination. In: Nano-Materials as Photocatalysts for Degradation of Environmental Pollutants: Challenges and Possibilities. Elsevier, Amsterdam, pp 371–401
Ullmann C, Babick F, Stintz M (2019) Microfiltration of submicron-sized and nano-sized suspensions for particle size determination by dynamic light scattering. Nano 9(6):829
Caputo F, Clogston J, Calzolai L, Rösslein M, Prina-Mello A (2019) Measuring particle size distribution of nanoparticle enabled medicinal products, the joint view of EUNCL and NCI-NCL. A step by step approach combining orthogonal measurements with increasing complexity. J Control Release 299:31–43
Falke S, Betzel C (2019) Dynamic Light Scattering (DLS) Principles, Perspectives, Applications to Biological Samples. Springer International Publishing, New York. https://doi.org/10.1007/978-3-030-28247-9_6
Honary S, Zahir F (2013) Effect of zeta potential on the properties of nano-drug delivery systems - a review (part 2). Trop J Pharm Res 12(2):265–273
Zhang J, Jiang W, Zhao X, Wang Y (2007) Preparation and characterization of polymeric micelles from poly(D, L-lactide) and Methoxypolyethylene glycol block copolymers as potential drug carriers. Tsinghua Sci Technol 12(4):493–496
Huang X, Liao W, Zhang G, Kang S, Zhang CY (2017) pH-sensitive micelles self-assembled from polymer brush (PAE-g-cholesterol)-b-PEG-b-(PAE-g-cholesterol) for anticancer drug delivery and controlled release. Int J Nanomedicine 12:2215–2226
Aji Alex MR, Veeranarayanan S, Poulose AC, Nehate C, Kumar DS, Koul V (2016) Click modified amphiphilic graft copolymeric micelles of poly(styrene-alt-maleic anhydride) for combinatorial delivery of doxorubicin and plk-1 siRNA in cancer therapy. J Mater Chem B 4(45):7303–7313
Honecker D, Bersweiler M, Erokhin S, Berkov D, Chesnel K, Venero DA et al (2022) Using small-angle scattering to guide functional magnetic nanoparticle design. Nanoscale Adv 4(4): 1026–1059
Chiu M, Prenner E (2011) Differential scanning calorimetry: an invaluable tool for a detailed thermodynamic characterization of macromolecules and their interactions. J Pharm Bioallied Sci 3(1):39–59
Gill P, Moghadam TT, Ranjbar B (2010) Differential scanning calorimetry techniques: applications in biology and nanoscience. J Biomol Tech 21(4):167–193
Gaber NN, Darwis Y, Peh KK, Tan YTF (2006) Characterization of polymeric micelles for pulmonary delivery of beclomethasone dipropionate. J Nanosci Nanotechnol 6(9–10): 3095–3101

Salimi A, Makhmal Zadeh BS, Kazemi M (2019) Preparation and optimization of polymeric micelles as an oral drug delivery system for deferoxamine mesylate: in vitro and ex vivo studies. Res Pharm Sci 14(4):293–307

Streets J, Bhatt P, Bhatia D, Sutariya V (2020) Sunitinib-loaded MPEG-PCL micelles for the treatment of age-related macular degeneration. Sci Pharm 88(3):1–12

Bunaciu AA, Udriştioiu E, gabriela, Aboul-Enein HY. (2015) X-ray diffraction: instrumentation and applications. Crit Rev Anal Chem 45(4):289–299

Liang N, Sun S, Gong X, Li Q, Yan P, Cui F (2018) Polymeric micelles based on modified glycol chitosan for paclitaxel delivery: preparation, characterization and evaluation. Int J Mol Sci 19(6)

Wijiani N, Isadiartuti D, Agus Syamsur Rijal M, Yusuf H (2020) Characterization and dissolution study of micellar curcumin-spray dried powder for oral delivery. Int J Nanomedicine 15:1787–1796

Mabrouk MM, Hamed NA, Mansour FR (2021) Spectroscopic methods for determination of critical micelle concentrations of surfactants; a comprehensive review. Appl Spectrosc Rev 58:1-29

Kapse A, Anup N, Patel V, Saraogi GK, Mishra DK, Tekade RK (2019) Polymeric micelles: a ray of hope among new drug delivery systems. Drug Delivery Syst:235–289

Anoune N, Nouiri M, Berrah Y, Gauvrit JY, Lanteri P (2002) Critical micelle concentrations of different classes of surfactants: a quantitative structure property relationship study. J Surfactant Deterg 5(1):45–53

Majumder N, Das NG, Das SK (2020) Polymeric micelles for anticancer drug delivery. Ther Deliv 11(10):613–635

Ma W, Sun J, Xu J, Luo Z, Diao D, Zhang Z et al (2020) Sensitizing triple negative breast cancer to tamoxifen chemotherapy via a redox-responsive Vorinostat-containing polymeric prodrug nanocarrier. Theranostics 10(6):2463–2478

Manjili HK, Malvandi H, Mousavi MS, Attari E, Danafar H (2018) In vitro and in vivo delivery of artemisinin loaded PCL–PEG–PCL micelles and its pharmacokinetic study. Artif Cells Nanomed Biotechnol 46(5):926–936

Yu F, He C, Waddad AY, Munyendo WLL, Lv H, Zhou J et al (2014) N-octyl-N-arginine-chitosan (OACS) micelles for gambogic acid oral delivery: preparation, characterization and its study on in situ intestinal perfusion. Drug Dev Ind Pharm 40(6):774–782

Liu Y, Fu S, Lin L, Cao Y, Xie X, Yu H et al (2017) Redox-sensitive Pluronic F127-tocopherol micelles: synthesis, characterization, and cytotoxicity evaluation. Int J Nanomedicine 12:2635–2644

Tran TQM, Hsieh MF, Chang KL, Pho QH, Nguyen VC, Cheng CY et al (2016) Bactericidal effect of lauric acid-loaded PCL-PEG-PCL nano-sized micelles on skin commensal propionibacterium acnes. Polymers (Basel) 8(9):321

Li W, Xue J, Xu H (2019) Combined administration of PTX and S-HM-3 in TPGS/solutol micelle system for oncotarget therapy. Int J Nanomedicine 14:1011–1026

Zhong P, Gu X, Cheng R, Deng C, Meng F, Zhong Z (2017) Avβ3 integrin-targeted micellar Mertansine prodrug effectively inhibits triple-negative breast cancer in vivo. Int J Nanomedicine 12:7913–7921

Li X, Zhang Z, Li J, Sun S, Weng Y, Chen H (2012) Diclofenac/biodegradable polymer micelles for ocular applications. Nanoscale 4(15):4667–4673

Gulfam M, Matini T, Monteiro PF, Riva R, Collins H, Spriggs K et al (2017) Bioreducible cross-linked core polymer micelles enhance: in vitro activity of methotrexate in breast cancer cells. Biomater Sci 5(3):532–550

Stimuli-Sensitive Polymeric Micelles for Biomedical Applications

7

Kaushal K. Jain, Akanksha Yogesh Kadam, Yashika Tomar, and Gautam Singhvi

Abstract

Stimuli-responsive polymeric micelles (PMs) have emerged not only as a promising technology to achieve on-demand, targeted, and regulated drug release for therapeutic applications but also for diagnostic purposes, emerging collectively as a theranostic toolkit. The stimuli can be exogenous (extrinsic) such as light, temperature, ultrasound, electric and magnetic field, or endogenous (intrinsic) such as pH, glucose, reactive oxygen species, redox, hypoxia, and enzymes. Stimuli-responsive PMs find extensive application in the field of cancer research. Tumor cells as well as tumor sites undergo extensive changes during their development and progression including the accumulation of acidic metabolites which reduces their pH in comparison to normal tissues, the elevation of several enzymes as well as proteins, increased reactive oxygen species, reduced oxygen levels, altered redox potential in certain tumor tissues, and increased temperature around certain tumor sites. These unique features encourage the development and fabrication of stimuli-responsive systems. In addition, these systems can also be engineered to encapsulate a wide variety of bioactive molecules and tuned in a way to overcome biological barriers by external stimuli to improve the efficiency of accumulation at the target site or within the tumor cells. In recent years stimuli-responsive systems are capturing the attention in research on the controlled delivery of drugs. In this review, an emphasis on the recent developments of the external, internal, dual and multi-stimuli-responsive polymeric micelles is made along with its future perspectives.

K. K. Jain · A. Y. Kadam · Y. Tomar · G. Singhvi (✉)
Industrial Research Laboratory, Department of Pharmacy, Birla Institute of Technology and Science (BITS)—Pilani, Pilani, Rajasthan, India
e-mail: gautam.singhvi@pilani.bits-pilani.ac.in

S. K. Singh et al. (eds.), *Polymeric Micelles: Principles, Perspectives and Practices*,
https://doi.org/10.1007/978-981-99-0361-0_7

Keywords

Polymeric micelles · Stimuli-responsive · Drug delivery · Cancer targeting

7.1 Introduction

Nanotechnology-based systems play an important role in the delivery of drugs by improving solubility, bioavailability, safety, efficacy, and stability as well as reducing dose and drug resistance along with various biomedical applications such as targeting and imaging. One of these systems has gained considerable interest among scientists over the past few decades which includes polymeric micelles (PMs). The name is so given because of the presence of amphiphilic polymeric materials in their design (Kaur et al. 2022; Qiu et al. 2021; Jain et al. 2019). Amphiphilic block copolymers (ABCs) can mimic the biological system by self-assembling within a specified solvent by the virtue of covalent bonding between the hydrophobic and hydrophilic segments encompassing them. In polymeric micelles, these segments organize themselves into a micellar structure possessing a supramolecular configuration and are referred to as the amphiphilic polymeric micelles (APMs) (Kaur et al. 2021; Kumar et al. 2020). Possessing a shell-core structure they can play an important role in overcoming the solubility issues of hydrophobic drugs by encapsulating them within the hydrophobic core. They also prevent the elimination from the reticuloendothelial system (RES) improving the circulation time. Targeting capability as well as bioavailability is improved due to leaky vasculature which enhances accumulation in solid tumors through enhanced permeation and retention effect (EPR) (Qiu et al. 2021; Kim et al. 2021). A distinctive advantage of polymeric micelles is their stability in extremely dilute conditions due to low values of critical micellar concentration (10^{-6}–10^{-7} M) (Majumder et al. 2020). However, premature release of drug in circulation and inadequate drug release from the micelles into the desired site is a major drawback. To overcome the excessive retention of drug in the core of micelles and confer spatiotemporal control which is an added drawback of the conventional system, exogenous and endogenous triggers can be used to form a stimuli-responsive system which plays a crucial role in modulating the delivery of the drug (Zhao et al. 2021; Rapoport 2007). These stimuli which are explored include various external factors such as magnetic and electric fields, ionizing radiation, light, and temperature as well as internal factors such as enzymes, pH, and redox potential as depicted in Fig. 7.1 (Zhao et al. 2021; Sana et al. 2022).

In comparison to normal tissues, tumor cells and tissues exhibit differences in both physiological as well as biochemical properties due to changes in cell growth and metabolism. These differences can be observed concerning oxidation, pH, and some enzymes. The key principle behind these smart responsive delivery systems involves compositional and/or structural changes on exposure to the specific stimuli. High efficiency has been observed in tumor treatment as well by alterations in the external environment such as irradiation of light, magnetic field application, and changes in the temperature conditions. Stimuli-responsive polymeric micelles are

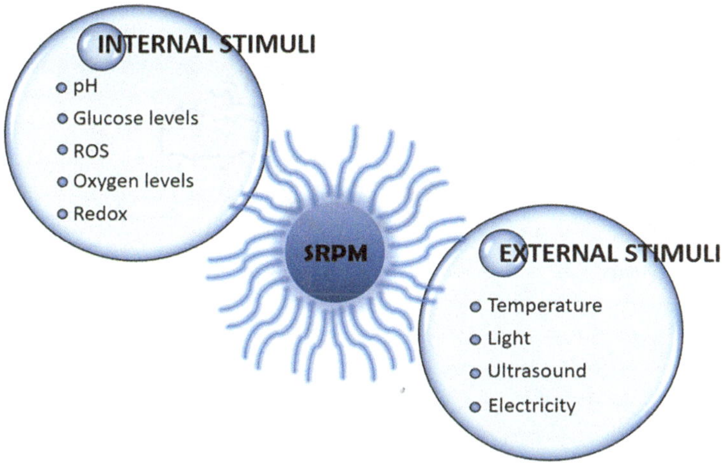

Fig. 7.1 Various stimuli on which the stimuli-responsive polymeric micelles (SRPM) are based

advantageous over conventional polymeric micelles in being highly specific in the identification of tumor tissues and subsequent response to various biological and physicochemical signals or exogenous conditions (Li et al. 2018; Hejmady et al. 2020). This specificity is responsible for producing maximal efficacy at target sites, preventing unnecessary exposure to healthy tissues and hence reducing toxic side effects (Nakayama et al. 2014; Nitheesh et al. 2021). However, stimuli-responsive nanocarriers are facing several challenges such as their possibility to produce toxicity and limitations in the absorption of drugs, among others. A need for further characterization of these nanocarriers and their clinical transition are also some added challenges associated with them (Chang et al. 2021).

7.2 pH-Responsive Polymeric Micelles for Drug Delivery and Theranostics

There exists a range of pH gradient throughout the human body, from acidic pH in the gastric region to alkaline in the intestinal region. Various tissues and regions in the body have this disparity in pH as well, such as lysosomal pH of 4–5, endosomal pH of 5–6, etc. In certain disease states pH varies too such as slightly acidic pH of tumor tissues (~6.5) and inflamed tissue being slightly less than physiological pH (7.4). This gradient could be utilized for targeted as well as specific delivery of therapeutic agents. pH-responsive polymeric micelles are one of many extensively studied drug delivery systems for targeted drug delivery (Makhlouf and Abu-Thabit 2018).

The pH-responsive polymeric micelles are nanocarriers made up of block copolymers consisting of ionizable pendant type groups linked to the hydrophobic polymer chain of the micelles. These pendant groups can be weak acids like

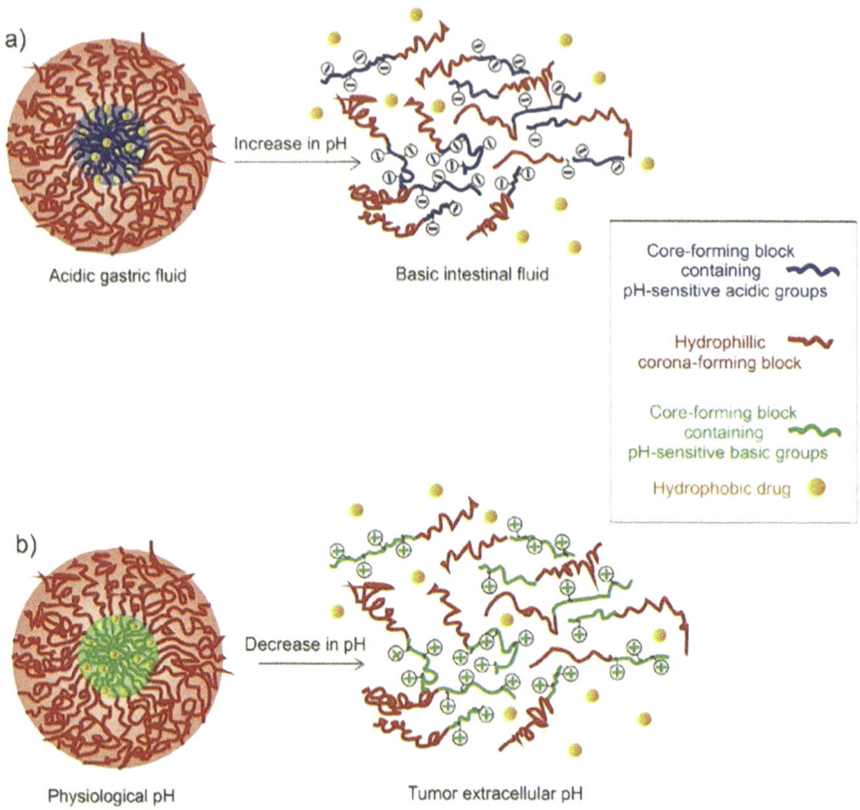

Fig. 7.2 Schematic representation of pH-sensitive mechanism of (**a**) anionic group containing polyacid and (**b**) cationic group containing polybasic. (Reprinted with permission from Bazban-Shotorbani et al. 2017 © 2017 Elsevier)

carboxylic or sulfonic acids or basic groups like amines or pyridines, etc. The ionizable pendant groups can undergo protonation or deprotonation in response to the surrounding pH changes. This leads to structural and conformational changes along with hydrodynamic volume changes in the polymer chains which helps to release the drug at a specified site (Makhlouf and Abu-Thabit 2018; Bazban-Shotorbani et al. 2017).

Some pH-sensitive polymers which are commonly used include poly[methacrylic acid], poly[vinyl pyridine], poly[vinyl imidazole], poly[L-histidine], poly[glutamic acid], poly[β-amino ester], chitosan, etc. Based on the pendant group present on the polymer chain the polymers can be of two types, namely anionic polymers (consisting of carboxyl or sulfonic group) and cationic polymers (consisting of amine group); The pH-sensitive mechanisms are depicted in Fig. 7.2. Anionic polymers are majorly used in pH-responsive delivery including polyacids such as poly[methacrylic acid], poly[acrylic acid], poly[glutamic acid], etc. At physiological pH of 7.4 the -COOH group ionizes (deprotonates) and releases its protons. This makes the -COOH group negatively charged COO^- and renders the backbone hydrophilic. But in acidic pH there is an abundance of protons in the surrounding medium and the COO^- group changes back to the COOH group by accepting (H^+) protons, thus rendering the backbone hydrophobic. This change in structure of the backbone results in destabilization (due to swelling/collapsing) or a solubility change causing polymer disassembly and releasing of the therapeutic moiety. Having a pKa of 4.28 the poly-acrylates are negatively charged at pH >4.28 and neutral below its pKa (pH <4.28). The poly[sulfonic acids] have sulfonate groups (SO_3H) with pKa around 2–3 and hence can be ionized over a varied range of pH conditions. Poly[sulfonic acids] are made up of monomers like 2-methoxy sulfonic acid, ethylene sulfonic acid, sulfoxy-ethyl methacrylates, styrene-sulfonic acid, among others (Makhlouf and Abu-Thabit 2018; Bazban-Shotorbani et al. 2017). V.P. Sant and co-workers designed self-assembling pH-sensitive polymeric micelles consisting of poly(ethylene glycol)-block poly(alkyl acrylate-co-methacrylic acid). This copolymer formed micelles at pH <4.7 and the micelles were disaggregated at pH >4.7. This system increased the oral bioavailability of fenofibrate, a poorly water-soluble drug. In acidic pH of the stomach, the micelles remained aggregated and prevented drug release. While in the intestinal alkaline pH the micelles dissociated releasing the molecularly dispersed form of entrapped drug, thus maximizing the oral bioavailability (Sant et al. 2005). The pH-sensitive cationic polymers owing to their modifiable positive charge show enhanced cellular uptake which can be utilized for cancer cell treatment. They include polyamines such as poly(N, N′-dimethyl aminoethyl methacrylate), poly(4- vinyl pyridine), poly(L-histidine), poly(β-amino ester), etc. Most polycations have amine groups (primary/secondary/tertiary) in their structure and are ionized at low values of pH. They undergo deprotonation at higher alkaline pH and protonation at lower acidic pH. This shift in charge destabilizes the structure leading to the release of the therapeutic moiety. Monomers particularly acrylamide, N, N′ dimethyl aminomethyl acrylamide, N, N′-diethyl aminoethyl methacrylate can be used for synthesizing pH-sensitive polyamines. The tertiary amine functional groups in poly(N, N′-dimethyl aminoethyl methacrylate) having pKa of 8 positively ionize at pH < pKa by accepting (H^+) protons from the surrounding medium and lose their positive charge at pH > pKa by releasing these (H^+) protons. Poly(N, N′-diethyl aminoethyl methacrylate) (pKa ~ 7.3) undergo deprotonation in alkaline pH. Due to reduction in several ionic amino functional groups and increased hydrophobic interactions the polymer abruptly precipitates over pH 7.4. Poly(2-aminoethyl methacrylate) having pKa ~7.6 remains stable in an acidic as well as neutral environment while in alkaline pH due to deprotonation of the primary amine group showed slow chemical degradation. Linear chain poly(ethyleneimine) with a pKa value of 4.5 related to the primary amine group and 6.7 related to the secondary amine group along with its proton buffering capacity renders this polymer suitable for gene delivery applications. Polypyridines like poly(2-vinyl pyridine) (pKa 5.9) and poly(4-vinyl pyridine) (pKa 5.39) have pyridine in their side chain available for protonation. Polyimidazoles like poly(N-vinyl imidazole) (pKa ~ 6) can become protonated under acidic conditions. Chitosan, a positively charged polysaccharide is composed of D-glucosamine and N-acetyl-D-glucosamine units. Its degree of deacetylation decides the number of amine groups in its structure which in turn affects its pH sensitivity. Poly-histidine is a polycationic peptide with a pKa value of 6 owing to the imidazole ring present in its side chain which exhibits pH sensitivity (Makhlouf and Abu-Thabit 2018). Lipeng Qui and colleagues developed a self-assembling pH-responsive micellar system with hyaluronic acid-g-poly(L-histidine) for the delivery of doxorubicin, an anticancer agent to MCF-7 breast cancer cells. The copolymer showed rapid disassembly at lysosomal pH due to poly(L-histidine) protonation and effective delivery of doxorubicin (Qiu et al. 2014).

Another mechanism for pH-responsive delivery includes pH-cleavable bonds. It consists of linkers which are acid labile such as ketal, acetal, oxime, orthoester, hydrazone, etc., which help release the drug in an acidic environment (Makhlouf and Abu-Thabit 2018). Rupei Tang and co-workers synthesized, developed, and characterized pH-sensitive self-assembling polymeric micellar formulation consisting of a copolymer of poly(ethylene glycol)-b-poly (2-ethoxytetrahydrofuran-2-yloxyethyl methacrylate) which contains acid labile orthoester side chain. Doxorubicin was successfully loaded at pH 7.4 in this system and released in biphasic manner at a slightly lower pH of 5. It demonstrated increased uptake by glioma cells and effective pH-responsive delivery to cancer cells (Tang et al. 2011).

In a study involving the utilization of pH-responsive polymeric micelles for drug delivery of a small molecule, short interfering RNA or siRNA molecules silencing specific genes and preventing protein expression by degrading the mRNAs which code for them were explored. This siRNA is up taken into the cell by endocytosis and ultimately after internalization by endosome it is degraded in lysosomes. To overcome its degradation pH-responsive polymeric micelles have been designed which transition from the ionized structure at normal physiological pH, i.e., ~7.4 to a conformation that disrupts membrane at endosomal pH of ~6.6. It transitions from hydrophilic structure to hydrophobic structure for facilitating the delivery of siRNAs. A. J. Convertine and co-workers used a di-block copolymer of poly(N, N-dimethyl aminoethyl methacrylate) (polyDMAEMA) block for siRNA binding purpose and pH-responsive block made up of DMAEMA (positively charged), propylacrylic acid (negatively charged), and butyl methacrylate (hydrophobic moiety) for membrane destabilizing purpose; 2 μg/mL as critical micelle concentration was used for the successful delivery of siRNA. This system showed 90% uptake in cells and as compared to standard lipid transfection agent 3 times increase in siRNA per cell (Convertine et al. 2010).

In a study for the diagnostic application of pH-responsive polymeric micelles, magnetite nanoparticles (Fe_3O_4) are commonly used as contrast agents in MRI scans. Incorporation of these magnetite nanoparticles in pH-responsive polymeric micelles could be utilized for the detection of acidic regions such as tumors. In a study by G.H.Gao and his colleagues a polymer system consisting of methoxy poly (ethylene glycol) [PEG]–hydrophilic portion and poly(b-amino ester) [PAE]–pH-sensitive portion was used in polymeric micelle synthesis. The PAE segment ionizes and solubilizes in acidic pH, thus releasing the magnetite nanoparticles in an acidic environment. This is further used for the detection of cancerous tissues in the body (Gao et al. 2010).

Another study for theranostic application was conducted by Kyung Hyun Min and colleagues who formulated and characterized methyl ether poly(ethylene glycol)-poly(β-amino ester) based pH-responsive polymeric micelles. This system was self-assembling and showed sharp micellization/demicellization at tumoral pH of 6.4. It was incorporated with fluorescence dye for cancer cell imaging, i.e., tetramethyl rhodamine isothiocyanate (TRITC) and camptothecin, an anticancer agent. The micellar system depicted substantially enhanced delivery of both the dye and drug to cancer cells with reduced side effects (Min et al. 2010).

7.3 Glucose-Responsive Polymeric Micelles for Drug Delivery

Glucose being an essential biomolecule is necessary as a ubiquitous fuel for our body and is overexpressed in case of diseases such as diabetes. In diabetics, the blood glucose level is high which can serve as an ideal stimulus for the delivery of antidiabetic agent insulin. Insulin delivery has been an extensively studied subject of research. Glucose-responsive polymeric micelles show the potential to be an ideal candidate for its delivery. These systems are composed of an insulin carrier and a glucose sensitive moiety. In order to regulate the rate of release of insulin many modifications like hydrophilicity, crosslinking, and pH are made on the carrier. The studies show three basic approaches for glucose-responsive delivery which include the utilization of enzymes such as glucose oxidase, utilizing proteins which bind carbohydrates like lectins as naturally occurring receptor mediated glucose detecting materials and use of polymers which show glucose sensitive properties (Das et al. 2020). Polymers depicting glucose-responsive properties have been extensively studied. The boronic acid-derived polymers have been an area focusing on development of glucose sensitive platforms for drug delivery. It has several advantages like high glucose sensitivity, storage for a longer time, and better stability. It is able to form reversible covalent complexes with hydroxyl groups containing molecules. Among the derivatives of boronic acid, phenylboronic acid (PBA) is widely researched. In an aqueous solution it exists as two forms in equilibrium which are a hydrophobic, uncharged (triangular) form and a hydrophilic, charged (tetrahedral) form (boronate). The charged form undergoes firm covalent bonding with molecules containing diol groups such as glucose and leading to phenyl boronic acid–glucose conjugate or complex. The apparent pKa of PBA is approximately 8.2–8.6 which is slightly higher than the physiological pH of 7.4. This indicates that at a slightly higher pH than the pKa of PBA (~9) or in presence of PBA-glucose complex it will be responsive to glucose. The incorporation of PBA into polymers will render the polymer glucose sensitivity (Das et al. 2020). In a study by Hao Yang and co-workers, complex form of glucose-responsive polymeric micelles was fabricated by complexing a PBA block copolymer PEG-b-P(Asp-co-AspPBA) and a glycopolymer P(Asp-co-AGA) which displayed notable glucose-responsiveness, good stability, biodegradability, and biocompatibility (Yang et al. 2013). Debashish

Roy and co-workers developed a boronic acid block copolymer, PDMA-b-PAEBB [poly(N, N-dimethylacrylamide-b-poly(N-(2-acryloylamino-ethyl)-4-borono-benzamide)-b-poly(N-(2-acryloylamino-ethyl)-4-(4,4,5,5-tetramethyl-[1,3,2]-dioxaborolan-2-yl)-benzamide)], which in response to changes in solution pH or glucose concentration can self-assemble and dissociate (Roy and Sumerlin 2012).

Lectins are defined as a group of protein molecules that naturally bind to carbohydrates like glucose, mannose, etc. Concanavalin A (Con A) is a lectin that binds specifically and reversibly to glucose molecules. Hence, it is widely studied for glucose-responsive drug delivery systems (Das et al. 2020). When in presence of divalent cations, four Con A molecules bind to one another and form a tetramer which in turn binds to four glucose molecules. Thus, it functions as a crosslinker of macromolecules with pendant saccharide molecules. When free glucose is present, the Con A moiety binds preferentially to free the glucose molecule, thus weakening the linkage between itself and saccharide moiety rendering the incorporated drug moiety free (Wang et al. 2020). A bio-affinitive, polymeric micelle with glucosamine in its structure was prepared by Yong-Chen Hu and co-workers. The self-assembling micelles were made of di-block copolymer, poly(p-nitro-phenyl acrylate)-block-polystyrene (PNPA-b-PSt) and were further crosslinked. Glucosamine interacts and binds with Con A. In presence of free glucose, Con A bonds with glucose and thus weakens the crosslinked structure of polymeric micelle and releases the incorporated drug (Hu and Pan 2005).

Glucose oxidase is an enzyme that oxidizes oxygen and glucose to hydrogen peroxide and gluconic acid, respectively. This property of the enzyme has been utilized for developing a variety of glucose-responsive delivery systems. As the by-product is gluconic acid there is expected to be a slight decrease in pH which can be further customized for pH-sensitive delivery. There are also swelling mechanisms established for glucose oxidase based delivery. These systems have been employed for the delivery of insulin (Das et al. 2020). Akifumi Kawamura and co-workers developed a comb-type polyelectrolyte, poly(ethylene glycol)-graft-poly(allyl amine) (PEG-g-PAA) for fabricating poly ion complex (PIC) micelle. Glucose oxidase was incorporated into the core of this PIC micelle. Once the micelles are in presence of glucose in the surrounding, the glucose can diffuse inside easily and can get oxidized to gluconic acid. This system was used for the detection of glucose through a reaction leading to a supramolecular architecture (Matsumura et al. 2008).

7.4 ROS: Responsive Polymeric Micelles for Theranostics

National Cancer Institute defines reactive oxygen species or (ROS) as "a type of unstable free radicals containing oxygen that easily react with other molecules in a cell." It includes molecules like (-OH) hydroxyl radical, (O^{-2}) superoxides, (-$ONOO^{-}$) peroxynitriles, (H_2O_2) hydrogen peroxides, and many more. As compared to pathological cells, normal cells produce low levels of ROS. Low ROS levels relate to normal functions of cells like cell growth modulation, migration, secretion, inflammation, programmed cell death (apoptosis), etc. Higher levels of ROS are

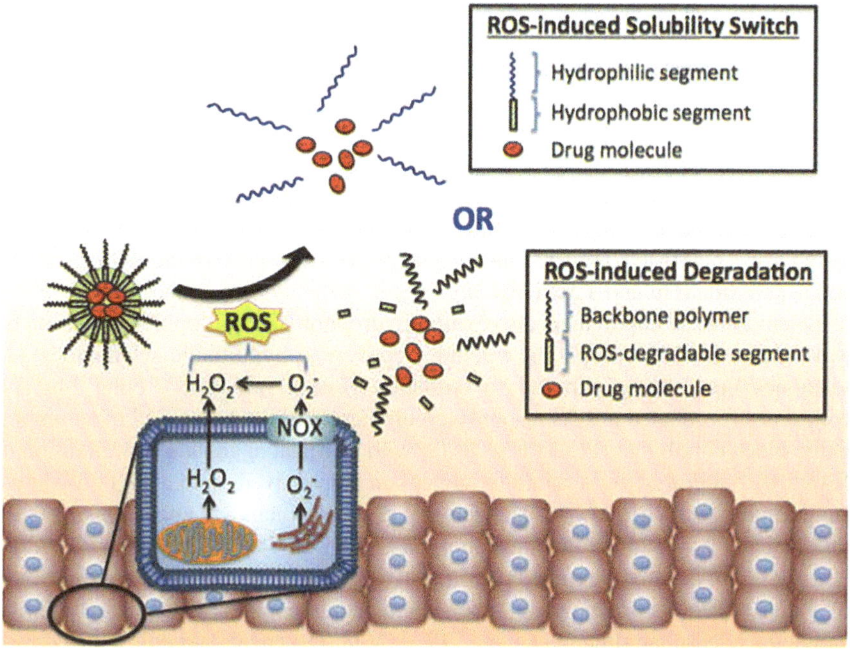

Fig 7.3 A generalized schematic diagram of reactive oxygen species (ROS)-responsive drug delivery and drug release via (*1*) solubility switch and (*2*) degradation mechanisms. (Reprinted with permission from (Lee et al. 2013) © 2013 John Wiley and Sons)

related to vascular pathogenesis such as hypertension, diabetes, cancer, atherosclerosis, aging, etc. The same reactive oxygen species which modulate normal cellular functions at lower concentrations; at a higher concentration causes oxidative stress and damages critical cellular components like DNA, proteins, lipids, etc. which may lead to dysfunctional proteins, mutated DNA, and other undesirable effects. Hence, ROS can be utilized as an internal stimulus or indicator that differentiates pathological and normal sites for enhanced and improved diagnosis and therapy. The development of ROS responsive polymeric micelles will enable site-specific targeting of therapeutic and diagnostic agents. The ROS responsive polymeric micelles release the incorporated agent by either of the two mechanisms, namely switching of solubility or degradation (Lee et al. 2013). Figure 7.3 schematically describes these mechanisms of releasing the payload from micelles. In the solubility switch mechanism, the polymer used in reacting with reactive oxygen species converts to hydrophilic form from hydrophobic form or vice versa and releases the incorporated agent. Some of the polymers which undergo solubility switch mechanism include polypropylene sulfide, selenium-based copolymers, polythioether ketal, etc.

Polypropylene sulfide undergoes a phase transition reaction under an oxidative environment and converts from hydrophobic sulfide to sulfoxide or sulfone which is hydrophilic. On being oxidized by the ROS it dissolves and releases the drug.

Mukesh K Gupta and his colleagues designed a polymeric micellar system consisting of a di-block copolymer of propylene sulfide (PS) and N, N-dimethylacrylamide (DMA) (poly(PS-b-DMA)) for inflammation site-specific release of hydrophobic drugs. The CMC of this di-block copolymer was estimated to be 0.09 mg/mL. These micelles were found to be sensitive to H_2O_2, peroxynitrile, and 3-morpholinosydnoimine (Gupta et al. 2012). Tirelli et al. synthesized and developed a ROS responsive polymeric micelles consisting of polysulfide/polyethylene glycol block copolymer which was conjugated to an enzyme superoxide dismutase. This system rendered the micelle responsive to two of the ROS: peroxide and superoxide (Hu and Tirelli 2012).

Selenide-based copolymers also exhibit hydrophobic to hydrophilic transition on exposure to ROS. Hydrophobic selenides convert to hydrophilic selenoxides and selenones upon being exposed to oxidation (Lee et al. 2013). Ning Ma and colleagues formulated a ROS responsive polymeric micelle composed of an amphiphilic block copolymer (PEG-PUSe-PEG) with a critical aggregate concentration of 1.4×10^{-4} mg/mL. It was found to have good responsiveness to a slight oxidative environment (0.1% H_2O_2 v/v). The delivery of doxorubicin was enabled using these micelles and it was found that the copolymer (PEG-PUSe-PEG) released 72% drug after 10 h of exposure to oxidation (Ma et al. 2010). The diselenide bond (Se-Se) in oxidative situations is cleaved to either selenic acid or selenol which is hydrophilic and helps release the drug (Lee et al. 2013). V.G. Deepagan and colleagues developed a polymeric micelle composed of a triblock copolymer of polyethylene glycol and polypeptides with diselenide crosslinking. Doxorubicin was encapsulated into its core. In an environment with higher levels of ROS the micelle delivered 3.73-fold more doxorubicin to tumor cells than free drug (Deepagan et al. 2016).

Polythioether ketal polymers have a similar mechanism as polypropylene sulfides, which involves conversion from hydrophobic to hydrophilic state. On exposure to ROS the ketal group hydrolyzes into biocompatible by-products like acetone and diols (Lee et al. 2013). Elena Gardey and colleagues developed thioether based micelles for inflammatory bowel disease (IBD) using the polymer pNAM-b-pNAT. pNAM is the water-soluble component poly(N-acryloylmorpholine), while the pNAT is the hydrophobic component poly(N-acrylthiomorpholine). These polymeric micelles were found to be sensitive to H_2O_2 and released the incorporated agent in the surroundings with higher levels of ROS in primary human monocytes (Gardey et al. 2022). In the degradation mechanism the involved polymers degrade or break down to by-products in presence of reactive oxygen species and the polymer structure breaks down releasing the incorporated agent in ROS rich environment. Some of the polymers following this degradation mechanism are explained consequently.

Boronic esters are one such class of polymers that undergo oxidation by addition of oxygen molecules as a linkage between themselves and the target moiety. On further hydrolysis the bond is cleaved. The boronic ester groups have been used as ROS degradable protecting groups (Lee et al. 2013). Yi Yuan and co-workers developed a ROS responsive polymeric micelle with amphiphilic poly(aspartic acid) and phenyl borate-serine [PASP-BSer]. The pendant group of phenyl borate-serine acted as ROS responsive component. This system was used for encapsulating the anticancer drug doxorubicin. It depicted H_2O_2 triggered the release and relatively higher uptake in A549 cancer cells than L929 normal cells (Yuan et al. 2020).

Polythioketal is another polymer which aids in the ROS responsiveness of micelles. The thioketal group degrades to acetone and thiols in superoxide rich surroundings (Lee et al. 2013). In a study conducted by Changzhen Sun and colleagues, a polymeric micellar system which was ROS responsive for the delivery of doxorubicin was designed. They utilized thioketal moiety as the ROS sensitive agent and the polymer used was methoxy poly (ethylene glycol)-thioketal-poly (ε-caprolactone) [mPEG-TK-PCL]. The micelle presented a high drug loading capacity of 12.8%, relatively less toxicity to normal cells and improved antitumor efficacy (Sun et al. 2017). Ruixin Li and colleagues developed a polymeric micelle for the efficient use of pesticides. They utilized the polymer methoxy poly(ethylene glycol)-thioketal-2-mercaptothiazoline [mPEG-TK-MTL] and loaded it with validamycin. These micelles were tested for antifungal activity against *Rhizoctonia solani.* During plant–pathogen interaction at an early stage, there is an outburst release of ROS which is used as a stimulus for the release of pesticides (validamycin). It was found that on the fifth day inhibition rate of validamycin loaded polymeric micelles at 1.56 mg/L was 50.37% (Li et al. 2020).

Another class of polymers that respond to ROS are aryl oxalate containing polymers. On reacting with H_2O_2 they form 1,2-dioxetanediones which are further converted rapidly to CO_2 and phenols. Thus, on being exposed to ROS rich environment polymeric micelles degrade to release the incorporated agent. These have high specificity to H_2O_2 and are therefore often used for their detection. The by-products produced by its degradation are high energy intermediates which on interaction with a fluorophore may lead to fluorescent emission (Gao and Xiong 2021; Song et al. 2014). Dongwon Lee and co-workers developed a polymeric micelle consisting of amphiphilic peroxalate based polynorbornene copolymers, fluorescent dye rubrene along with a pegylated corona for stealth effect. These polymeric micelles were able to detect as low as 50 nM concentration of H_2O_2 through chemiluminescence and can be utilized as imaging and contrasting agent for H_2O_2 detection (Lee et al. 2008).

Ferrocene is an organometallic compound with two cyclopentadienyl rings. It is capable of redox activity and gets oxidized in presence of ROS. The ferrocene group (hydrophobic) oxidizes to the ferricenium group (hydrophilic) triggering the ROS responsive mechanism (Nakayama et al. 2014). In a study by Feng Xu and co-workers a ferrocene based polymeric micelle composed of poly (N-acryloylmorpholine) (PACMO) and poly(2-acryloyloxyethyl ferrocene carboxylate) (PAEFC) [PACMO-b-PAEFC] was developed and used for delivery of paclitaxel, an anticancer agent. These micelles depicted 61.4% encapsulation efficiency of paclitaxel as well as oxidation-based release behavior (Xu et al. 2017).

7.5 Redox Responsive Polymeric Micelles for Drug Delivery and Theranostics

Redox is one of the many conditions that can be used as internal stimuli for targeted delivery, due to the varied redox states of the intracellular and extracellular environment in the body. Glutathione is a tripeptide (γ-L-glutamyl-L-cysteinyl-glycine) synthesized in cells. It has antioxidant activity and is involved in reduction and conjugation reactions providing means for removal of peroxides and xenobiotic compounds. As is established in the literature glutathione (GSH) levels are 100–1000 times higher in intracellular spaces (~2–10 mM) than in extracellular spaces (~2–10 μM) (Forman et al. 2009; Zhou et al. 2018). GSH is also found abundantly in tumorous cancer tissues which is quadruple times more than in normal tissues. This redox capacity difference is utilized for targeted delivery through redox responsiveness. For designing such systems, a redox responsive unit is incorporated into the amphiphilic polymer blocks which then forms a micellar structure on self-assembling in an aqueous medium. Such redox responsive units include disulfide bonds, diselenide bonds, ditellurium bonds, etc. The polymeric micelles thus formed remain stable in blood circulation but disaggregate in tumoral sites (Zhang et al. 2017; Yang et al. 2020).

The most common bond used in the redox responsive category is the disulfide bond. When the surrounding environment is reducing in nature and contains GSH, the disulfide bonds in the micellar delivery system undergo reduction to free thiol groups. Thus, the payload is released at the targeted site within minutes to hours. In a study performed by Sarvin Shirani and colleagues, a redox sensitive micelle consisting of abietic acid-cystamine-gellan gum (AB-ss-GG) linked by disulfide bond of cystamine group was synthesized for delivery of ribociclib, a CDK4 and CDK6 selective inhibitor to treat breast cancer cells. The estimated CMC was 40.15 mg/mL and demonstrated greater cytotoxicity (IC50 = 47.86 μmol/L) than blank micelles as well as a free drug (Shirani et al. 2022). In another study conducted by Jun Hu and his team, a redox sensitive polymeric micelle for a theranostic application consisting of amphiphilic copolymer TPE-conjugated poly(aspartic acid)-block-poly(2-methacryloyloxyethyl phosphorylcholine) with disulfide linkage (TPE-SS-PLAsp-b-PMPC) was designed. These micelles successfully incorporated the anticancer drug doxorubicin during self-assembling process and demonstrated quick disassembly in response to the greater concentration of GSH. Additionally, they showed aggregation induced emission (AIE) imaging which facilitated bioimaging of cancer cells. These micelles exhibited excellent antitumor efficacy, lesser side effects in comparison to free drug and active imaging (Hu et al. 2018). A team led by Huanli Sun developed a biodegradable micelle system consisting of disulfide linked poly(ethylene glycol)-b-poly(3-caprolactone) (PEG-SS-PCL) with the ability to shed PEG shells. These micelles had a CMC of ~6.6 mg/L and were utilized for delivery of doxorubicin while demonstrating 60% loading efficiency (Sun et al. 2009).

Diselenide bond is another bond used in the redox responsive category. As compared to the disulfide bonds, selenium has a larger radius and weaker electronegativity resulting in lower bond energy (172 kJ/mol) than S-S (240 kJ/mol). Thus, diselenide bonds are more sensitive to being cleaved under redox conditions (Shi et al. 2020). These diselenide bonds are found to be responsive to oxidation as well as reduction forming selenic acid and selenol, respectively (Zhang et al. 2017). Liangjie Shi and colleagues developed a redox responsive polymeric micelle consisting of diselenide consisting non-ionic gemini polymeric micelles of poly (ethylene glycol) (PEG-G12), where 12 indicates the alkyl chain length containing carbon atoms. They exhibited greater drug loading efficiency (77%) and low CMC. Indomethacin was loaded in these micelles and it showed redox responsive release with 27% release in 0.02 mM 1,4-dithiothreitol and 82% release in 10 mM 1,4-dithiothreitol (DTT) (Shi et al. 2020).

Similar to selenium, tellurium is another element that has been explored for redox responsiveness as it shows glutathione peroxidase mimicking activity. In a study performed by Wei Cao and co-workers, a polymeric micelle composed of triblock copolymer of poly(ethylene glycol)–poly(urethane)tellurium–poly(ethylene glycol) (PEG−PUTe−PEG) was synthesized. These were used for the delivery of drugs containing platinum like oxaliplatin and cisplatin. It was observed that redox responsive sustained release of platinum was achieved for a duration of more than a month (Cao et al. 2014).

Ferrocene as explained in Sect. 4.3 is an organometallic compound exhibiting redox activity. On being exposed to oxidizing environment ferrocene (Fc-uncharged) converts to ferrocenium (Fc^+-charged) (Fabbrizzi 2020). In an experiment conducted by Anchao Feng and co-workers, they were able to develop a brush like new type of supramolecular copolymer which is capable of self-assembling into polymeric micelles exhibiting redox responsive behavior. The micelle was made of two kinds of polymers: poly(ethylene glycol)-block-poly (glycidyl methacrylate) decorated with β-CD pendants (GA-CD) and poly (caprolactone) containing end-capping ferrocene (Fc) moieties (L-Fc). In normal conditions, the ferrocene is uncharged and can form non-covalent bonds with the cyclodextrin cavity resulting in an inclusion complex. On being exposed to a redox environment, the ferrocene converts to ferrocenium and gets charged, thus cleaving the non-covalent bond leading to reversible disassembling of the structure and release of the payload (Feng et al. 2014).

Another example of the redox responsive unit is trimethyl-locked quinone propionic acid (QPA) groups. It gets converted to lactone on being reduced and is sensitive to dithionite diaphorase an enzyme highly expressed in many cancer tissues. Jungeun Bae and co-workers developed a redox responsive amphiphilic polymer made of poly(QPA)-methoxy poly(ethylene glycol) (PolyQPA-mPEG) with CMC of 0.039%w/v. These were used for delivery of paclitaxel and were found to release 65.73% of paclitaxel at 36 h. These micelles depicted IC50 values of 217.6 μM as compared to 323.3 μM of the free drug for MDA-MB-231 cell lines (Bae et al. 2015).

7.6 Hypoxia Responsive Polymeric Micelles for Drug Delivery and Theranostics

The environment of tumor tissue tends to have low oxygen levels due to rapidly proliferating cells along with poor blood supply and thus is hypoxic ($<1.4\%$ of oxygen). There is a significant difference in oxygen levels in normal tissue and tumor tissue varying from ~5 mmHg in hypoxic tumor tissues to ~70 mmHg in normoxic tissues. This creates scope for developing drug delivery systems that are responsive to hypoxia. The controlled release obtained from these delivery systems is due to bio-reductive units which induce hydrophobic–hydrophilic transitions of the polymers and subsequently lead to disassembly of the delivery system, in this case polymeric micelles. Due to the blood supply being curbed in the innermost regions of tumor tissue, the delivery of traditional contrast agents and drugs becomes difficult; interference being accumulated in such hypoxic regions. Hypoxia responsive polymeric micelles provide a solution for overcoming such theranostic obstacles. There are many hypoxia sensitive bio-reductive molecules which are used for imaging and delivery of therapeutic moieties such as nitroaromatic compounds, derivatives of azobenzenes, N-oxide compounds, quinone compounds, etc. (Sun et al. 2018; Zhang et al. 2021).

The nitroaromatic compounds and their derivatives exhibit hypoxia sensitivity due to single-electron reduction to amines by nitro reductases and bio-reducing agents present in tissues. This bio-reduction step is oxygen dependent and reversible in nature and therefore under normal oxygen levels it is inhibited. But under hypoxic oxygen levels the nitro group can undergo reduction to a free radical anion and alter the behavior of the molecule from hydrophobic to hydrophilic or cleavage (Sun et al. 2018; Zhou et al. 2021). N-oxide compounds and quinone compounds have hypoxia responsive properties. These undergo one or two electron bio-reductions in hypoxic conditions and thus produce radical species (Zhou et al. 2021). In a study performed by Xin Shun Sun and colleagues, a polymeric micelle for the delivery of cytochrome c was designed. Cytochrome c is a metalloprotein whose release from mitochondria is one of the initial steps for the initiation of cell death pathways. Hence, a polymeric micelle consisting of methoxy PEG-block-poly(diethylene-triamine-graft-4-nitrobenzyl chloroformate)-L-glutamate [mPEG-b-P(DETA-NBCF)-LG] was synthesized to incorporate cytochrome c, with the nitro-benzyl group acting as the hypoxia sensitive unit. At a concentration of 100 μg/mL, these micelles exhibited 70% mortality of HepG2 cancer cells (Sun et al. 2020). Yinlu Deng and his team developed a polymeric micelle consisting of poly(ethylene glycol)-block-poly (methacrylic acid-co-2-nitroimidazole methacrylate) [PEG-b-P(MAA-co-NIMA)]. The 2-nitroimidazole group displayed the hypoxia responsive property. These micelles were loaded with anticancer agent doxorubicin and the release rate was checked in normoxic and hypoxic tissues. The micelles exhibited a 35% release of doxorubicin over 16 h in normoxic conditions while about 79% in hypoxic conditions (Deng et al. 2019).

An azo group containing compounds such as azobenzene is another category of hypoxia sensitive unit. It is incorporated as a bio-reductive linker in the polymer chains. Azo groups under hypoxic conditions are reduced to amine derivatives and thus lead to change in polymer structure and release of the payload (Zhou et al. 2021). In a study conducted by Mengmeng Long and co-workers, for the treatment of bone metastatic prostate cancer, a polymeric micelle with azobenzene as hypoxia sensitive linker and alendronate as a bone-targeting moiety was designed. The polymeric micelle consisted of alendronate-poly(ethylene glycol)-azobenzene-poly (L-lactic acid) [ALN-PEG-AZO-PLL] and was loaded with doxorubicin. The micelles were able to suppress the tumor growth and also preserve bone structure by promoting osteoblastic activity and suppressing osteoclastic activity and hence achieving an enhanced therapeutic outcome (Long et al. 2021). In another study conducted by Li Li and his colleagues, polymeric micelle consisting of Fucoidan (hydrophilic segment), azobenzene (linker), and linolenic acid (hydrophobic segment) [FAL-DOX] were synthesized and doxorubicin was incorporated in it. The micelles exhibited improved antitumor performance of doxorubicin with a (FAL-DOX) IC_{50} of 1.685 μg/mL as compared to IC_{50} of 7.847 μg/mL of doxorubicin and demonstrated approximately a 4.6-fold enhanced effect on hypoxic tumor cells (Li et al. 2022a). Shiya Chen and co-workers conducted a study in which they synthesized amphiphilic polymeric micelles consisting of polymer,1,2-distearoyl-sn-glycero-3-phosphoethanolamine-azo-N-[maleimide (polyethylene glycol-2000)] (Mal-PEG2000-Azo-DSPE) with an azo bond as hypoxia sensitive linker. These micelles were loaded with Mito rHP; a fluorescent moiety targeting mitochondria. This fluorescent micellar probe designed was able to image hypoxia induced mitophagy (Chen et al. 2020). F. Perche and colleagues developed a polymeric micelle for siRNA delivery to hypoxic tumor tissue for gene silencing. The micelle consisted of poly(ethylene glycol)-azobenzene-poly(ethyleneimine)-1,2-dioleyl-sn-glycero-3-phosphoethanolamine [PEG-Azo-PEI-DOPE]. The micelles achieved moderate in vitro and in vivo silencing activity (Perche et al. 2014).

7.7 Enzyme-Responsive Polymeric Micelles for Drug Delivery and Theranostics

Enzymes perform a decisive function biologically due to their specific and superior catalytic nature. In many diseases, there is an abnormal expression of enzymes and hence they can be manipulated endogenously to design a stimuli-responsive system for the delivery of drugs. Enzymes possess several significant advantages as a stimuli-responsive system for the disassembly and degradation of polymeric micelles, such as their natural presence in all living organisms, exceptional specificity and efficiency, peptides substrates offer good biodegradability and biocompatibility as well. Some promising biological triggers include proteases, peptidases, phospholipases, and oxidoreductases (Yang et al. 2020; Dai et al. 2019). A lysosomal cysteine protease called Cathepsin B and a large family of endopeptidases called matrix metalloproteins (MMPs) are expressed in large amounts in tumorous cells and at the sites of tumor, respectively. Thus, they possess the potential to be designed as enzyme-responsive polymeric micelles (Dai et al. 2019). Yan et al.

designed polymeric micelles encapsulating doxorubicin (DOX) emitting blue fluorescence in UV light. The functional hydrophobic chain consisted of a tetraphenylethene (TPE) segment and was responsible for the self-assembly of the block copolymer, inducing luminescence and offering favorable stability to this polymer-based nanostructure. The polymeric micelles were disassembled in the ubiquity of esterase, consequently quenching its fluorescence by the release of the drug (Yan et al. 2020).

Cabazitaxel an up-coming second-generation-taxane treats castration-resistant prostate cancer which is metastatic in nature. Barve et al. worked on a targeted, enzyme-responsive and biodegradable polymeric micelle to overcome the issues of poor solubility and limited targeting ability of cabazitaxel. Two amphiphilic block copolymers made up the micelle. One of the block copolymers was made of cholesterol and PEG (peptide sensitive to enzymes and cleavable by MMP-2). The other block copolymer included PEG, cholesterol, and DUPA (targeting ligand). The micelle combined with DUPA showed greater fluorescence intensity, i.e., a twofold in comparison to micelle which was not modified and threefold in comparison to free dye. In the evaluation of DUPA as a prostate-specific membrane antigen (PSMA) targeting ligand, free cabazitaxel and polymeric micelles were incubated with PSMA-positive prostate cancer C4–2 cells for a duration of 0.5 h and 1 h. In comparison to free cabazitaxel a three and fivefold uptake and in comparison, to unmodified micelle a 1.5- and 2.1-fold higher uptake were observed at 0.5 h and 1 h, respectively, when the micelle was coupled with DUPA. The micelle coupled with DUPA also demonstrated the highest antitumorigenic activity in xenograft tumors of mice (Barve et al. 2020).

Nitro-reductase has been overexpressed in tumor cells and is crucial in the reduction of nitro compounds. Harnoy and his co-workers designed nitro-reductase (NTR) sensitive micelles based on a polymer TNP, amphiphilic in nature consisting of the 4-nitrobenzyl group, hydrophilic polyethylene (PEG) moieties and hydrophobic AIE tetraphenyl-ethylene (TPE) encapsulating the hydrophobic drug doxorubicin (DOX). The simple mechanism involves the reduction of the 4-nitrobenzyl group followed by decomposition of the micellar structure and finally the drug discharge as observed in Fig. 7.4. The fluorescence of the TPE derivatives aids in tracking the disassembly of the micelle. CLSM and FCM help in visualizing drug release. An obvious toxicity against cancer cells was shown by TNP@DOX micelles in MTT assays (Sun et al. 2021). Wei and his co-workers researched amphiphilic-linear dendritic block copolymers (LDBCs) that self-aggregated into a micellar structure in an aqueous environment, encapsulating Nile red which is a hydrophobic dye or even doxorubicin. The block copolymers are made up of hydrophobic dendritic phenylalanyl-lysine (Phe-Lys) dipeptides and hydrophilic linear poly (N-vinylpyrrolidone) (PNVP). The disassembly of the micelle is triggered by an enzyme serine protease trypsin. The stability of the micelles can be enhanced by expanding the generation of the dendrons, i.e., by growing the branch units of a dipeptide, (Phe-Lys). The rate of release of the dye or the drug can also be modified depending on the generation of the dendrons. MTS assays were performed which

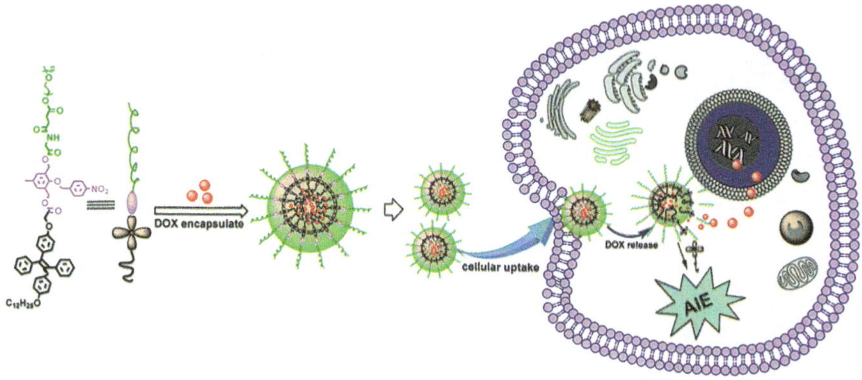

Fig. 7.4 Schematic diagram of DOX transport and cargo release of TNP@DOX micelles. (Reprinted with permission from Sun et al. 2021 © 2010 Royal Society of Chemistry)

confirmed cytocompatibility with the cell lines of human liver cancer cells (SMMC-7721) and lung epithelial cells (BEAS-2B) (Wei et al. 2019).

NAD(P)H: quinone oxidoreductase-1 (NQO-1) expression is almost 50 times greater in lungs and liver cancer cells when compared with the normal cells. In the research conducted by Park and his colleagues amphiphilic block copolymer, QPA-P was synthesized containing hydrophilic polyethylene glycol (PEG) and hydrophobic enzyme triggered QPA (trimethyl-locked quinone propionic acid)-locked polycaprolactone (PCL). A two-step cyclization process is responsible for the depolymerization of QPA-locked PCL by NQO1, followed by disassembly of the micelles, releasing the encapsulated drug doxycycline (DOX). In the tumor mouse model, greater deposition was seen at the target site by QPA-PM-DOX when correlated with free DOX. Tumor growth inhibition was also high for QPA-PM-DOX than benzene-PM-DOX (Park et al. 2021).

7.8 Dual-Responsive Polymeric Micelles for Drug Delivery and Theranostics

Researchers have used a mixture or combination of different sensitive polymers (e.g. pH and temperature) to impart dual-responsiveness. One of the trends in dual-responsive polymeric micelles includes the creation of new polymers by using different existing monomers. Dual-responsive systems are used not only for controlling but also enhancing the release of the drug at the targeting site (Kalhapure and Renukuntla 2018). Wang along with his colleagues studied a ROS (reactive oxygen species) and esterase dual-stimuli responsive system for the delivery of drug moiety. In designing this system, the combined effect of N-isopropylacrylamide (NIPAM) and boronic esters was exploited via the RAFT (reversible addition-fragmentation chain transfer) polymerization technique. In water solution the

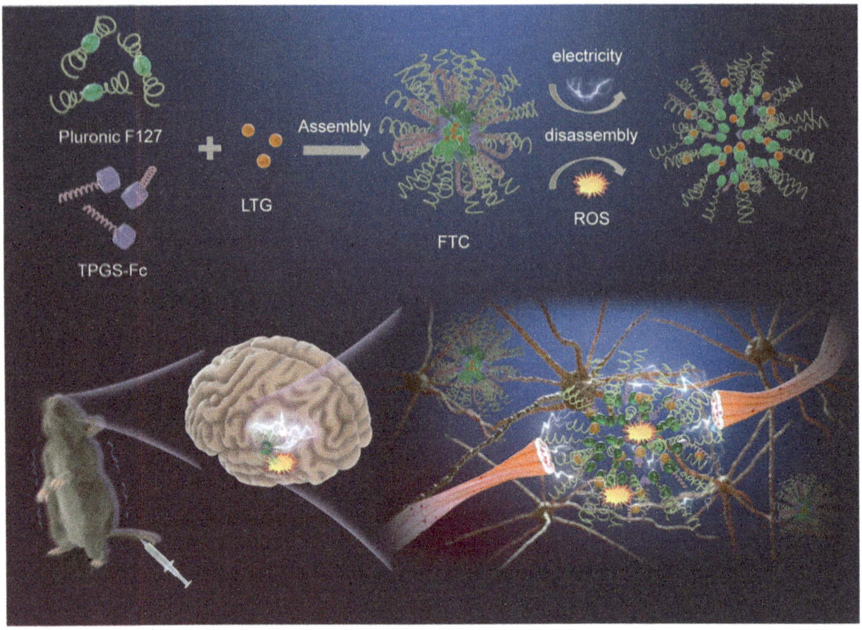

Fig. 7.5 Schematic of an electric and reactive oxygen species dual-responsive nanocarrier in the treatment of epilepsy. (Reprinted with permission from Li et al. 2022b © 2022 Elsevier)

copolymer self-assembled leading to the formation of polymeric micelles. In vitro release of doxycycline (DOX) from mPEG$_{2000}$-PNIPAM-co-PBABE (P3) polymeric micelles under physiological conditions showed that at blank conditions, the cumulative drug release was only 30% for 54 h, however, in 10 mM H_2O_2 (ROS) the drug release increased by 30% compared to blank conditions. The release of the drug was expedited in the presence of both H_2O_2 and esterase, with 90% release observed within 54 h. In vitro cytotoxicity testing was carried out using the MTS assay method in both HL-7702 and HeLa cells. Results showed that the copolymer is less toxic since the viability of cells was high even at greater concentrations of polymer (200 mg/L) and is highly biocompatible (Wang et al. 2019).

Li and his colleagues developed an electroresponsive micelle, targeting the brain for the delivery of lamotrigine (LTG) an anti-seizure drug, the delivery of which can be controlled by an H_2O_2 (ROS) as seen in Fig. 7.5. These mixed micelles contain pluronic F127/ferrocene (Fc) modified D-α-tocopherol polyethylene glycol succinate referred to as the FTC micelles. In vitro release study of the drug found that the cumulative release of drug from the micelles at 1 mM and 10 mM, 1 h after coincubation was found to be 1.5 times and 1.9 times greater in comparison to the control group which contained drug-loaded FTC micelles without H_2O_2. The findings also suggested that as the concentration of ROS increased, the rate and

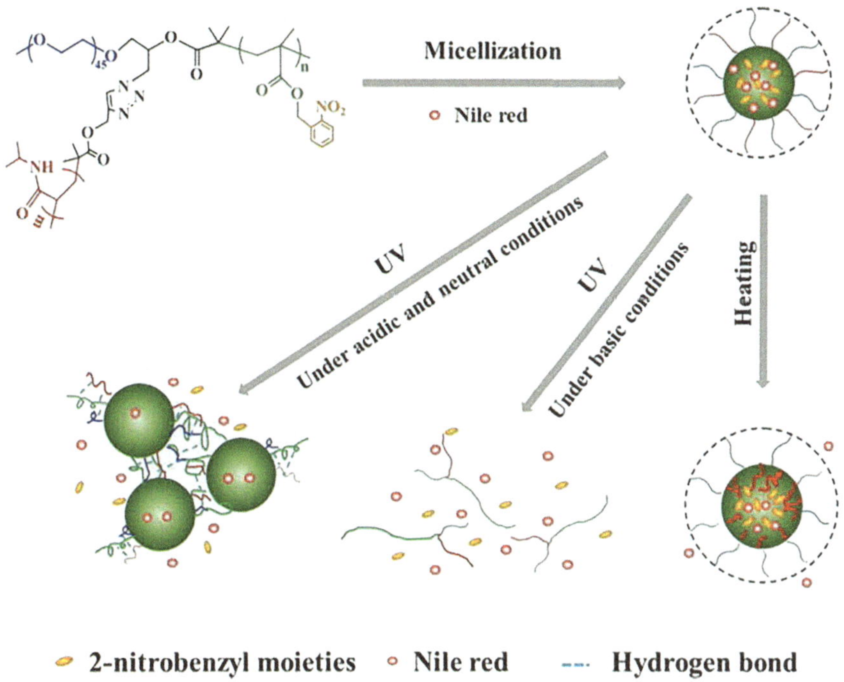

Fig. 7.6 Schematic illustration of the micellization of the miktoarm star terpolymer and possible morphology changes under the stimuli of temperature and the combined stimulations of pH and UV light. (Reprinted with permission from Huo et al. 2017 © 2017 Elsevier)

the amount of release of the drug also accelerated. The anti-seizure effect was investigated on PTZ rats for preventing epilepsy and suppressing seizures. This in vivo evaluation between the free LTG and non-responsive FT-LTG showed greater efficacy against seizures in the latter. Moreover, findings suggested that the potential side effects due to LTG are decreased and the drug dose required to attain the anti-epileptic effect is also reduced (Li et al. 2022b).

Huo and his colleagues designed a dual-responsive system consisting of amphiphilic ABC miktoarm star terpolymer sensitive toward light and temperature. In an aqueous solution, the terpolymer self-assembles into poly(2-nitrobenzyl methacrylate) (PNBM)-core micelles with hydrophilic polyethylene glycol (PEG) sensitive to light and poly(N-isopropylacrylamide) (PNIPAM) coronas sensitive toward temperature as shown in Fig. 7.6. Nile red was encapsulated within the micelles and fluorescence measurements revealed its release, when exposed to change in temperature as well as UV light. An interesting outcome was that the triggering by UV light weakened the release of encapsulated Nile Red in acidic conditions; however, the release was promoted in basic conditions (Huo et al. 2017). Zhuang and his co-workers designed intelligent polymeric micelles with the theranostic application for the treatment and diagnosis of cancer. These polymeric micelles were dual-responsive synthesized using the mPEG-P (TPE-co-AEMA) copolymer consisting of tetraphenylethene (TPE) which is an AIE-based fluorogen and 2-azepane ethyl methacrylate (PAEMA), a pH-sensitive polymer encapsulating the drug Doxorubicin (DOX) within them. Additionally, the release of the drug was increased marginally when an excess of glutathione (GSH) is present.

An in vitro study was performed on 4T1 and HeLa cells. The outcome showed that DOX-loaded mPEG-P (TPE-co-AEMA) micelles had greater antitumor efficacy and less toxicity as compared to free DOX. Blue fluorescence produced by TPE and red fluorescence produced by DOX are observable in the cytoplasm when co-cultured with micelles loaded with DOX for 1 h. The uptake of these micellar structures by 4T1-cells can be observed by an overlap of blue and red fluorescence, which becomes stronger with time indicating more uptake (Zhuang et al. 2018).

Liu 2020 and his team of researchers designed a glucose and H_2O_2 dual-responsive micellar system for insulin delivery in the treatment of diabetes. The polymeric micelles consisted of a hydrophilic section made of PEG and a hydrophobic segment made of poly(amino phenylboronic ester) (PAPBE) sensitive to H_2O_2 and glucose. Sensitivity toward glucose was responsible for the release of insulin which can be further facilitated by co-encapsulation of glucose oxidase (GO_x). GO_x produces H_2O_2 as the by-product in the process of formation of gluconic acid from glucose which in turn acts as a stimulus for the release of insulin. Streptozotocin (STZ) was used to induce diabetes in mice to carry out the in vivo studies. The hypoglycemic effect observed in insulin/GO_x is superior in contrast to free insulin or micelles containing only insulin (Liu et al. 2020).

7.9 Multi-Responsive Polymeric Micelles for Drug Delivery and Theranostics

Targeting the release of a drug by a carrier at a specific site of disease by incorporating systems responding to multiple stimuli is becoming a trend in recent years. Multi-responsive systems like these are helpful in avoiding side effects and achieving a greater control over the release of drugs (Ruiz et al. 2022). Though these systems are rarely reported, they present greater possibilities of achieving more functionalities considering various parameters. Dong and his colleagues developed multiple stimuli-responsive polymeric micelles using a pyrene-functionalized poly (dimethylaminoethyl methacrylate) polymer responsive to pH, light, and temperature. The drug encapsulated for the controlled release study was Nile Red. On exposure to UV rays the micellar structure undergoes photolysis leading to the formation of a unimer. At pH 3, the core is protonated, hence the micelle swells or undergoes dissociation. At pH 10, the shells collapse because the PDMAEMA segments undergo deprotonation, followed by shrinkage of the micelle and finally aggregating to form a complex micelle. At high temperature conditions, the shrinkage of the micelle can be observed, when the solution containing micelles was heated from 20° to 60°. Though controlled release can be obtained by this triple-responsive system, complete release of Nile Red is demonstrated by UV light and low pH conditions (Dong et al. 2013).

Dong et.al 2018 synthesized quadruple responsive copolymers forming micelles in an aqueous solution with a disulfide bridge linking the hydrophobic cores made of [poly (dimethylaminoethyl methacrylate)-copoly(2-nitrobenzyl methacrylate) (PDMAEMA-co-PNBM) and the hydrophilic coronas, i.e., PEG. The controlled release profile was evaluated using the hydrophobic molecule Nile Red via fluorescence intensity of the system. Results showed that the micelles were responsive to various stimuli such as the presence of UV rays, DTT (dithiothreitol), temperature as well as pH. Stimulus dependent behavior can be observed in these micelles such as a change in the pH leading to alteration in the release pattern of the encapsulated Nile Red and temperature changes affect the size of the micelles. Such responses are crucial and help in the construction of complex logic gates and dealing with multi-tasks (Dong et al. 2018).

Xu and his colleagues designed novel amphiphilic copolymers based on azobenzene via addition-fragmentation chain transfer (RAFT) polymerization method alongside multi-stimuli-responsiveness. An azo group with dual-responsiveness toward light and reduction, super hydrophobic polyhedral oligomeric silsesquioxane (POSS) moiety and tertiary amine group with the pH-responsive property was integrated into the polymer chain to generate a system with multiple responsive property. Reduction in the hydrophobic azo pendant leads to the formation of corresponding hydrophilic-substituted anilines which consequently is responsible for the disassembly of the micellar structure. Variations in the micelle size can be seen due to cis-trans isomerization of the azo groups when exposed to UV rays. Tertiary amine groups are responsible for imparting pH-sensitive property to the polymeric micelles. This multiple stimuli-responsive property of P(POSSMA-co-AZOMA-co-DMAEMA) polymer can be utilized in controlling the release of the encapsulated drug moiety for a targeted therapy (Xu et al. 2018).

Gurgain 2012 designed polymeric micellar system responsive to quadruple stimuli by designing poly(N-isopropylacrylamide-b-sodium 2-(acrylamido)-2-methylpropane sulfonate) tagging it with a dimer of spiropyran at the poly (N-isopropylacrylamide) end (SP2-b-$NIPAM_{154}$-b-$AMPS_{148}$). It can be seen that the block copolymer so designed was responsive to multiple stimuli which included pH, metal ion, temperature, and light. $AMPS_{148}$ showed stimuli-responsive behavior toward metal ions and $NIPAM_{154}$ block copolymer toward temperature. Light, pH, and thermal stimuli sensitivity are shown by the block polymer due to a spiropyran moiety (SP2) photochromic in nature. Both size and color changes can be observed by irradiation of visible light or UV light (due to the responsiveness of the SP block) or by changing the pH of the solution. With an increase in temperature significant rise in the size of the polymeric micelles can be observed due to the $NIPAM_{158}$ block. The micellar system also undergoes a change in conformation from extended to shrunken on the addition of Fe^{3+} ion, due to the properties of $AMPS_{148}$ block (Guragain et al. 2012).

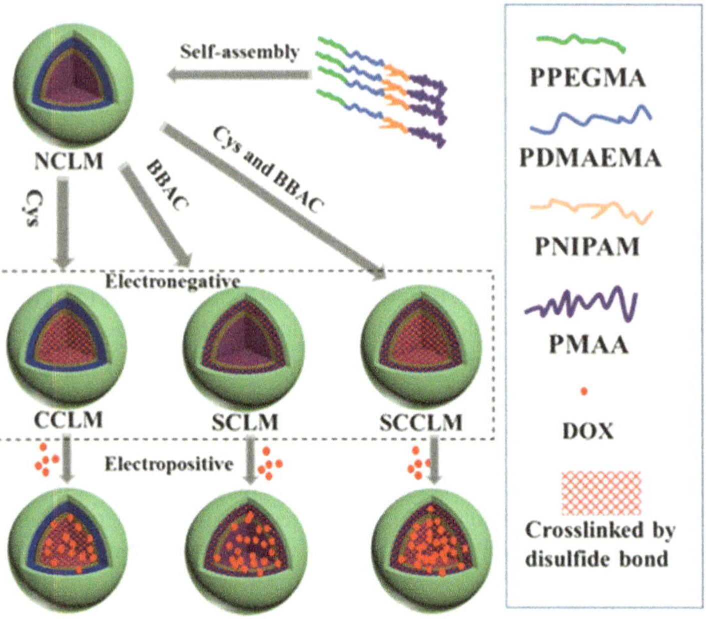

Fig. 7.7 Illustration of the self-assembly, crosslinking process, and preparation of three types of DOX-loaded crosslinked micelles by electrostatic interaction. (Reprinted with permission from Zhang et al. 2018 © 2018 John Wiley and Sons)

Zhang along with his team developed a triple stimuli-responsive system with sensitivity toward pH, temperature, and reduction using a block copolymer poly (polyethylene glycol methacrylate)–poly[2-(dimethylamino) ethyl methacrylate]–poly(N-isopropylacrylamide)–poly(methylacrylic acid) (PPEGMA-PDMAEMA-PNIPAM-PMAA). This tetra block copolymer in acidic solution self-assembles into noncross-linked micelles (NCLM). These micelles were crosslinked via carbodiimide chemistry reaction or quaternization reaction forming core-crosslinked micelles (CCLM), shell-crosslinked micelles (SCLM), and shell-core dilayer-crosslinked micelles (SCCLM) as shown in Fig. 7.7.

These crosslinked micelles have the advantage of greater stability below CMC (critical micelle concentration), presence of high salt concentration, and shearing forces. The moiety enclosed within these crosslinked micelles is doxorubicin (DOX). Studies on the release pattern revealed that release of DOX was highest when micelles involved the trigger of all stimuli at pH 5, 37 °C, and 10 mM DTT (Dithiothreitol). In vitro cytotoxicity assays were conducted by MTT assay for free DOX, three variations of DOX-loaded crosslinked micelles, and four types of blank micelles against HepG2 cells. The inhibitory effect of free DOX on the HepG2 cells was higher compared to micelles crosslinked and DOX-loaded. The DOX-loaded micelles also provided excellent antitumor action in the MTT assay. Hence, it was concluded considering various factors that SCCLM was the best suited nanocarrier for the delivery of DOX (Zhang et al. 2018).

7.10 Extrinsic Stimuli-Responsive Polymeric Micelles for Drug Delivery and Theranostics

One of the classifications of biological stimuli for the release of payloads from the nanocarriers is physical stimulus with external application which includes heat, mechanical pressure, light, and electric or magnetic field strengths (Fleige et al. 2012). External stimuli-responsive nanocarriers offer the benefits of temporal, spatial as well as dose control in the release of the drug considering a remote apparatus that can be controlled by a switch ("On" and "Off") at will. However, with the stimuli-responsive systems, depending on the stimulus used, an exact spot for setting off the release of the drug and the response rate thereof can cause variations in the degree of drug release and therapeutic activity (Cheng et al. 2013).

Cheng et al. designed light-sensitive polymeric micelles (LRPMs) by combining poly(ethylene glycol) with hydrophilic properties and hydrophobic poly (caprolactone) forming the block copolymer and maleimide-anthracene linkers acting as the light-sensitive molecular groups. Doxycycline (DOX) was encapsulated within the 3PEG-PCL micelles followed by irradiation with ultraviolet light at 254 nm for about 10 s which was responsible for the complete release of the drug into the intracellular environment, i.e., the cell nucleus, exerting strong cytotoxic effects on the tumor cells affecting their growth as well as proliferation. The in vitro cytotoxic effects of the developed photosensitive micelle were carried on SAS cells. The comparison between doxycycline loaded 3PEG-PCL micelles and the free DOX involved conduction of SRB (Sulforhod-amine-B) assay. The IC50 values of the polymeric micelles were greater compared to the free DOX which indicated similar results to that of MTT assay, i.e., dose dependent cell toxicity. Further, staining of cells was carried out using DAPI (4,6-diamidino-2-phenylindole) and observation was carried out using CSLM microscopy. DAPI-stained cell nuclei were indicated by blue areas and red areas represent DOX fluorescence. The pink fluorescent color represents the internalization of DOX by cells. When the cells were irradiated with UV light for a duration of about 10 s, gradual transfer of the pink fluorescence from the membrane of the cell to its nucleus is observed. Hence, it can be said that these micelles possess significant potential for treating superficial basal-cell carcinoma of the skin as well as the oral cavity (Cheng et al. 2019).

Liu and his colleagues developed a photodynamic theranostic micelle encapsulating indocyanine green (ICG) with self-quenchable properties along with folic acid behaving as the targeting moiety (FA-ICG-micelles). Folic acid plays a crucial role in improving the effectiveness of photodynamic therapy (PDT) as well as the intracellular uptake of the micelles. ICG was incorporated within the star-shaped poly(lactide-co-glycolide) (PLGA) and polyethylene glycol (PEG) (4s-PLGA-PEG) block copolymers which acts as the hydrophobic core of micellar structure. The cytotoxic effects of FA-ICG micelles were studied by CCK assay, culturing MDA-MB-231 human breast adenocarcinoma cells. Fluorescence analysis showed that average intensity was 2.8-fold greater in FA-ICG-micelles in tumors compared to free ICG indicating tumor specific accumulation. In vivo antitumor studies were carried out on nude mice with tumor-bearing MDA-MB-231 cells to study the PDT effect of the photosensitive micelles. Compared to day 0, a 16 and 12-fold increase in tumor volume was observed on day 14 in saline and free ICG group, respectively. However, significant inhibition of tumor was observed in FA-ICG-micelles comparatively. This can be due to sufficient accumulation of ICG in the tumor site responsible for production of singlet oxygen (Liu et al. 2016).

Liang and his colleagues conducted research on the breakdown of the micellar structure by high intensity focused ultrasound (HIFU). The micelles were synthesized by a supramolecular di-block copolymer with the presence of metal–ligand bond at the intersection of both the blocks. The di-block consisted of poly (ethylene glycol) (PEG) and poly(propylene glycol) (PPG) along with HIFU responsive, mechano-labile bis (terpyridine)-Cu (II) bond at the juncture producing a (PPG-[Cu]-PEG) system. The micelles formed using PPG-[Cu]-PEG released the encapsulated payload within several minutes as compared to PEG-[Ru]-PPG (stronger Ru(II)−Tpy bond) and PEO−PPO−PEO [poly(ethylene glycol)-block-poly (propylene glycol)-bloc– poly(ethylene glycol)] (covalent bond) mainly due to disruption of the weak Cu (II)−Tpy bonds in the PEG-[Cu]-PPG chain when exposed to HIFU. This opened up opportunities for the further designing of delivery systems mediated by ultrasound.

Kim et al. designed a polymeric micelle conjugated with magnetic nanoparticle (MNP-PMs) to produce a combined effect of hyperthermia and chemotherapy in treating cancer. Doxorubicin (DOX) was incorporated within these micellar structures made of PEG-PLA (poly(ethylene glycol)-poly(lactide)) and nanoparticles of iron oxide. Alternating magnetic field (AMF) is used for inducing hyperthermia (41–46 °C) with an intensity of about 1.5 kA/m and a 200 kHz frequency, which improves the release of the drug in comparison to that in the absence of AMF. Human lung adenocarcinoma cells A549 cells are used for carrying out hyperthermia and in vitro cytotoxicity experiments. It has been seen that DOX-MNP-PMs are responsible for causing a significant reduction in the proliferation of cells from 0.52- to 0.22-fold in comparison to free DOX via hyperthermia induced by magnetism consequently enhancing the apoptosis of cancerous cells (Kim et al. 2015).

Hassanzadeh and his colleagues designed thermo-responsive nano-micelles using a succinylated form of surface-modified poly(ethylene-co-vinyl alcohol) with pegylation by methoxy polyethylene glycol succinate (5000 Da) along with retinoic acid (RA) to get a targeted conjugate to deliver epirubicin for treating carcinoma of hepatic cells. Cell viability study was performed using HepG2 cells by MTT assay. At a higher temperature (45 °C) significant growth inhibition was observed indicating a thermal responsive behavior and the optimum formulation released 40% of the drug within 400 h which signifies a controlled release behavior (Hassanzadeh et al. 2017).

7.11 Conclusion

Stimuli-responsive polymeric micelles have shown several advantages and overcome many challenges in the area involving delivery of drugs as well as theranostic application. Due to their stimuli-responsiveness, they can be delivered to the targeted site enhancing the effectiveness and specificity of the therapy. Various internal and external stimuli cater control the release profile of the targeted moiety. The polymeric micelles responsive to internal stimuli like pH, ROS, oxygen levels, glucose levels, presence of enzymes, etc. take advantage of the physiological phenomena occurring in vivo, while, the ones responsive to external stimuli employ some external source like temperature, light, ultrasound, electricity, etc., for biomedical purposes. These have proven to be useful in cases of various diseases and disorders like cancer and diabetes to name a few. By altering the polymeric chain by a stimuli-responsive molecule, the mechanism of release is made highly specific which in turn improves the targeting potential, reduces the wastage of drug as well as side effects. Majorly research is being carried out on the utilization of stimuli-responsive polymeric micelles for cancerous tumors. Since the tumor microenvironment has significant differences from the normal cell environment, this property is exploited by the SRPMs for delivery as well as diagnostic purposes. They have proven to be more efficacious than other conventional delivery designs by requiring lower doses, having a longer period of release, etc., but still need work in areas of clinical transitions and limiting toxicities. Stimuli-responsive polymeric micelles have opened new gateways for drug delivery and diagnosis which were previously inaccessible and made treatment better.

7.12 Future Prospects

Polymeric micelles became a boon for compounds experiencing problems due to reduced solubility and also improved the possibility of conquering multi-drug resistance. Stimuli-responsive polymeric particles are becoming a trend in the delivery of gene/drug considering a range of internal and external stimuli with greater emphasis on cancer research because of conclusive results in therapeutic responses, diagnosis monitoring, and reduced side effects in comparison to traditional approaches. A step ahead in this technology can be an improvement in the biodegradable or biocompatible stimuli-responsive copolymers that encapsulate a diverse range of therapeutic moieties with a stable micellar system. Though there is an edge over the current stimuli-responsive systems not many have been explored in vivo and hence in vivo stability as well as drug release studies need extensive research. It is also important to address the degradation kinetics of polymers as well as aspects related to their toxicity. The complexity of multiple stimuli systems is a significant drawback in its translation from the experimental phase to clinical applications. Hence, most useful systems can aim at a simple polymer or hybrid nanocarrier system which can be synthesized easily and be cost effective as well, to overcome the challenges during scale-up. The future trend can also focus on combination therapy wherein immune or gene therapy can be combined with nanomedicine to provide a synergistic or additive effect.

References

Bae J, Maurya A, Shariat-Madar Z, Murthy SN, Jo S (2015) Novel redox-responsive amphiphilic copolymer micelles for drug delivery: synthesis and characterization. AAPS J 17(6):1357–1368

Barve A, Jain A, Liu H, Zhao Z, Cheng K (2020) Enzyme-responsive polymeric micelles of cabazitaxel for prostate cancer targeted therapy. Acta Biomater 113:501–511. https://doi.org/10.1016/j.actbio.2020.06.019

Bazban-Shotorbani S, Hasani-Sadrabadi MM, Karkhaneh A, Serpooshan V, Jacob KI, Moshaverinia A et al (2017) Revisiting structure-property relationship of pH-responsive polymers for drug delivery applications. J Control Release 253:46–63

Cao W, Gu Y, Meineck M, Li T, Xu H (2014) Tellurium-containing polymer micelles: competitive-ligand-regulated coordination responsive systems. J Am Chem Soc 136(13):5132–5137

Chang D, Ma Y, Xu X, Xie J, Ju S (2021) Stimuli-responsive polymeric nanoplatforms for cancer therapy. Front Bioeng Biotechnol 9:1–26

Chen S, Liu J, Li Y, Wu X, Yuan Q, Yang R et al (2020) Hypoxia-responsive fluorescent nanoprobe for imaging and cancer therapy. Trends Anal Chem 131:116010

Cheng R, Meng F, Deng C, Klok HA, Zhong Z (2013) Dual and multi-stimuli responsive polymeric nanoparticles for programmed site-specific drug delivery. Biomaterials 34(14):3647–3657. https://doi.org/10.1016/j.biomaterials.2013.01.084

Cheng CC, Huang JJ, Lee AW, Huang SY, Huang CY, Lai JY (2019) Highly effective photocontrollable drug delivery systems based on ultrasensitive light-responsive self-assembled polymeric micelles: an in vitro therapeutic evaluation. ACS Appl Bio Mater 2(5):2162–2170

Convertine AJ, Diab C, Prieve M, Paschal A, Hoffman AS, Johnson PH et al (2010) pH-responsive polymeric micelle carriers for siRNA drugs. Biomacromolecules 11(11):2904–2911

Dai Y, Chen X, Zhang X (2019) Recent advances in stimuli-responsive polymeric micelles: via click chemistry. Polym Chem 10(1):34–44

Das SS, Bharadwaj P, Bilal M, Barani M, Rahdar A, Taboada P et al (2020) Stimuli-responsive polymeric nanocarriers for drug. Polymers (Basel) 12:1–45

Deepagan VG, Kwon S, You DG, Nguyen VQ, Um W, Ko H et al (2016) In situ diselenide-crosslinked polymeric micelles for ROS-mediated anticancer drug delivery. Biomaterials 103:56–66

Deng Y, Yuan H, Yuan W (2019) Hypoxia-responsive micelles self-assembled from amphiphilic block copolymers for the controlled release of anticancer drugs. J Mater Chem B 7(2):286–295

Dong J, Wang Y, Zhang J, Zhan X, Zhu S, Yang H et al (2013) Multiple stimuli-responsive polymeric micelles for controlled release. Soft Matter 9(2):370–373

Dong Y, Ma X, Huo H, Zhang Q, Qu F, Chen F (2018) Preparation of quadruple responsive polymeric micelles combining temperature-, pH-, redox-, and UV-responsive behaviors and its application in controlled release system. J Appl Polym Sci 135(35):1–11

Fabbrizzi L (2020) The ferrocenium/ferrocene couple: a versatile redox switch. ChemTexts 6(4):1–20

Feng A, Yan Q, Zhang H, Peng L, Yuan J (2014) Electrochemical redox responsive polymeric micelles formed from amphiphilic supramolecular brushes. Chem Commun 50(36):4740–4742

Fleige E, Quadir MA, Haag R (2012) Stimuli-responsive polymeric nanocarriers for the controlled transport of active compounds: concepts and applications. Adv Drug Deliv Rev 64(9):866–884. https://doi.org/10.1016/j.addr.2012.01.020

Forman HJ, Zhang H, Rinna A (2009) Glutathione: overview of its protective roles, measurement, and biosynthesis. Mol Aspects Med 30(1–2):1–12

Gao F, Xiong Z (2021) Reactive oxygen species responsive polymers for drug delivery systems. Front Chem 9:1–17

Gao GH, Im GH, Kim MS, Lee JW, Yang J, Jeon H et al (2010) Magnetite-nanoparticle-encapsulated pH-responsive polymeric micelle as an MRI probe for detecting acidic pathologic areas. Small 6(11):1201–1204

Gardey E, Sobotta FH, Quickert S, Bruns T, Brendel JC, Stallmach A (2022) ROS-sensitive polymer micelles for selective degradation in primary human monocytes from patients with active IBD. Macromol Biosci 2100482:1–6

Gupta MK, Meyer TA, Nelson CE, Duvall CL (2012) Poly(PS-b-DMA) micelles for reactive oxygen species triggered drug release. J Control Release 162(3):591–598

Guragain S, Bastakoti BP, Ito M, Yusa SI, Nakashima K (2012) Aqueous polymeric micelles of poly[N-isopropylacrylamide-b-sodium 2-(acrylamido)-2-methylpropanesulfonate] with a spiropyran dimer pendant: quadruple stimuli-responsiveness. Soft Matter 8(37):9628–9634

Hassanzadeh F, Farzan M, Varshosaz J, Khodarahmi GA, Maaleki S, Rostami M (2017) Poly (ethylene-co-vinyl alcohol)-based polymeric thermo-responsive nanocarriers for controlled delivery of epirubicin to hepatocellular carcinoma. Res Pharm Sci 12(2):107–118

Hejmady S, Pradhan R, Alexander A, Agrawal M, Singhvi G, Gorain B et al (2020) Recent advances in targeted nanomedicine as promising antitumor therapeutics. Drug Discov Today 25(12):2227–2244. https://doi.org/10.1016/j.drudis.2020.09.031

Hu YC, Pan CY (2005) Bioaffinitive and nanosized polymeric micelles based on a reactive block copolymer. Macromol Rapid Commun 26(12):968–972

Hu P, Tirelli N (2012) Scavenging ROS: superoxide dismutase/catalase mimetics by the use of an oxidation-sensitive nanocarrier/enzyme conjugate. Bioconjug Chem 23(3):438–449

Hu J, Zhuang W, Ma B, Su X, Yu T, Li G et al (2018) Redox-responsive biomimetic polymeric micelle for simultaneous anticancer drug delivery and aggregation-induced emission active imaging. Bioconjug Chem 29:1897–1910

Huo H, Ma X, Dong Y, Qu F (2017) Light/temperature dual-responsive ABC miktoarm star terpolymer micelles for controlled release. Eur Polym J 87:331–343. https://doi.org/10.1016/j.eurpolymj.2016.12.038

Jain S, Cherukupalli SK, Mahmood A, Gorantla S, Rapalli VK, Dubey SK et al (2019) Emerging nanoparticulate systems: preparation techniques and stimuli responsive release characteristics. J Appl Pharm Sci 9(8):130–143

Kalhapure RS, Renukuntla J (2018) Thermo- and pH dual responsive polymeric micelles and nanoparticles. Chem Biol Interact 295:20–37. https://doi.org/10.1016/j.cbi.2018.07.016

Kaur J, Mishra V, Kumar S, Gulati M, Kapoor B (2021) Harnessing amphiphilic polymeric micelles for diagnostic and therapeutic applications: breakthroughs and bottlenecks. J Control Release 334:64–95. https://doi.org/10.1016/j.jconrel.2021.04.014

Kaur J, Gulati M, Kapoor B, Kumar N, Prasher P, Ansari S et al (2022) Advances in designing of polymeric micelles for biomedical application in brain related diseases. Chem Biol Interact 361: 109960. https://doi.org/10.1016/j.cbi.2022.109960

Kim HC, Kim E, Jeong SW, Ha TL, Park SI, Lee SG et al (2015) Magnetic nanoparticle-conjugated polymeric micelles for combined hyperthermia and chemotherapy. Nanoscale 7(39): 16470–16480

Kim KN, Oh KS, Shim J, Schlaepfer IR, Karam SD, Lee J (2021) Light-responsive polymeric micellar nanoparticles with enhanced formulation stability. Polymers 13:1–11

Kumar H, Navya D, Shivendu PN, Nandita R, Eric D (2020) Nanoscience in medicine. Vol. 1. Anticancer Res 40:51–88

Lee D, Erigala VR, Dasari M, Yu J, Dickson RM, Murthy N (2008) Detection of hydrogen peroxide with chemiluminescent micelles. Int J Nanomedicine 3(4):471–476

Lee SH, Gupta MK, Bang JB, Bae H, Sung HJ (2013) Current progress in reactive oxygen species (ROS)-responsive materials for biomedical applications. Adv Healthc Mater 2(6):908–915

Li Y, Yu A, Li L, Zhai G (2018) The development of stimuli-responsive polymeric micelles for effective delivery of chemotherapeutic agents. J Drug Target 26(9):753–765. https://doi.org/10.1080/1061186X.2017.1419477

Li R, Xie H, Zhang C, Sun Y, Yin H (2020) ROS-responsive polymeric micelle for improving pesticides efficiency and intelligent release. J Agric Food Chem 68(34):9052–9060

Li L, Wang B, Zhang Q, Song P, Jiang T, Zhao X (2022a) Hypoxia responsive fucoidan-based micelles for oxidative stress-augmented chemotherapy. Eur Polym J 174:111340

Li J, Meng X, Men Z, Chen X, Shen T (2022b) Electric and reactive oxygen species dual-responsive polymeric micelles improve the therapeutic efficacy of lamotrigine in pentylenetetrazole kindling rats. Colloids Surf A Physicochem Eng Asp 642:128628. https://doi.org/10.1016/j.colsurfa.2022.128628

Liu L, Ma G, Zhang C, Wang H, Sun H, Wang C et al (2016) An activatable theranostic nanomedicine platform based on self-quenchable indocyanine green-encapsulated polymeric micelles. J Biomed Nanotechnol 12(6):1223–1233

Liu X, Li C, Lv J, Huang F, An Y, Shi L et al (2020) Glucose and H_2 O_2 dual-responsive polymeric micelles for the self- regulated release of insulin. ACS Appl Bio Mater 3:1598

Long M, Liu X, Huang X, Lu M, Wu X, Weng L et al (2021) Alendronate-functionalized hypoxia-responsive polymeric micelles for targeted therapy of bone metastatic prostate cancer. J Control Release 334:303–317

Ma N, Li Y, Ren H, Xu H, Li Z, Zhang X (2010) Selenium-containing block copolymers and their oxidation-responsive aggregates. Polym Chem 1(10):1609–1614

Majumder N, Das NG, Das SK (2020) Polymeric micelles for anticancer drug delivery. Ther Deliv 11(10):613–635

Makhlouf ASH, Abu-Thabit NY (2018) Stimuli responsive polymeric nanocarriers for drug delivery applications, volume 1: types and triggers, vol 59. Woodhead Publishing, Sawston, pp 1–971

Matsumura S, Hlil AR, Lepiller C, Gaudet J, Guay D, Shi Z et al (2008) Stability and utility of Pyridyl disulfide functionality in RAFT and conventional radical polymerizations. J Polym Sci Part A Polym Chem 46:7207–7224

Min KH, Kim JH, Bae SM, Shin H, Kim MS, Park S et al (2010) Tumoral acidic pH-responsive MPEG-poly(β-amino ester) polymeric micelles for cancer targeting therapy. J Control Release 144(2):259–266

Nakayama M, Akimoto J, Okano T (2014) Polymeric micelles with stimuli-triggering systems for advanced cancer drug targeting. J Drug Target 22(7):584–599

Nitheesh Y, Pradhan R, Hejmady S, Taliyan R, Singhvi G, Alexander A et al (2021) Surface engineered nanocarriers for the management of breast cancer. Mater Sci Eng C 130:112441. https://doi.org/10.1016/j.msec.2021.112441

Park J, Jo S, Lee YM, Saravanakumar G, Lee J, Park D et al (2021) Enzyme-triggered disassembly of polymeric micelles by controlled depolymerization via cascade cyclization for anticancer drug delivery. ACS Appl Mater Interfaces 13(7):8060–8070

Perche F, Biswas S, Wang T, Zhu L, Torchilin VP (2014) Hypoxia-targeted siRNA delivery. Angew Chem Int Ed Engl 126(13):3430–3434

Qiu L, Li Z, Qiao M, Long M, Wang M, Zhang X et al (2014) Self-assembled pH-responsive hyaluronic acid-g-poly(l-histidine) copolymer micelles for targeted intracellular delivery of doxorubicin. Acta Biomater 10(5):2024–2035

Qiu N, Du X, Ji J, Zhai G (2021) A review of stimuli-responsive polymeric micelles for tumor-targeted delivery of curcumin. Drug Dev Ind Pharm 0(0):1–18. https://doi.org/10.1080/03639045.2021.1934869

Rapoport N (2007) Physical stimuli-responsive polymeric micelles for anti-cancer drug delivery. Prog Polym Sci 32(8–9):962–990

Roy D, Sumerlin BS (2012) Glucose-sensitivity of boronic acid block copolymers at physiological pH. ACS Macro Lett 1(5):529–532

Ruiz AL, Ramirez A, McEnnis K (2022) Single and multiple stimuli-responsive polymer particles for controlled drug delivery. Pharmaceutics 14(2):421

Sana B, Finne-wistrand A, Pappalardo D (2022) Recent development in near infrared light-responsive polymeric materials for smart drug-delivery systems. Mater Today Chem 25: 100963. https://doi.org/10.1016/j.mtchem.2022.100963

Sant VP, Smith D, Leroux JC (2005) Enhancement of oral bioavailability of poorly water-soluble drugs by poly(ethylene glycol)-block-poly(alkyl acrylate-co-methacrylic acid) self-assemblies. J Control Release 104(2):289–300

Shi L, Jin Y, Du W, Lai S, Shen Y, Zhou R (2020) Diselenide-containing nonionic gemini polymeric micelles as a smart redox-responsive carrier for potential programmable drug release. Polymer 198:122551

Shirani S, Varshosaz J, Rostami M, Mirian M (2022) Redox responsive polymeric micelles of gellan gum/abietic acid for targeted delivery of ribociclib. Int J Biol Macromol 215 (June):334–345

Song CC, Du FS, Li ZC (2014) Oxidation-responsive polymers for biomedical applications. J Mater Chem B 2(22):3413–3426

Sun H, Guo B, Cheng R, Meng F, Liu H, Zhong Z (2009) Biodegradable micelles with sheddable poly(ethylene glycol) shells for triggered intracellular release of doxorubicin. Biomaterials 30(31):6358–6366

Sun C, Liang Y, Hao N, Xu L, Cheng F, Su T et al (2017) A ROS-responsive polymeric micelle with a π-conjugated thioketal moiety for enhanced drug loading and efficient drug delivery. Org Biomol Chem 15(43):9176–9185

Sun M, Wang K, Oupický D (2018) Advances in stimulus-responsive polymeric materials for systemic delivery of nucleic acids. Adv Healthc Mater 7(4):1–17

Sun XS, Jang MS, Fu Y, Lee JH, Lee DS, Li Y et al (2020) Intracellular delivery of cytochrome C using hypoxia-responsive polypeptide micelles for efficient cancer therapy. Mater Sci Eng C 114:114

Sun XY, Liang YX, Wu CY, Tang Q, Liu R, Lu ZL et al (2021) Nitroreductase-responsive polymeric micelles based on 4-nitrobenzyl and AIE moieties for intracellular doxorubicin release. Polym Chem 12(17):2618–2626

Tang R, Ji W, Panus D, Palumbo RN, Wang C (2011) Block copolymer micelles with acid-labile ortho ester side-chains: synthesis, characterization, and enhanced drug delivery to human glioma cells. J Control Release 151(1):18–27

Wang N, Chen XC, Ding RL, Yang XL, Li J, Yu XQ et al (2019) Synthesis of high drug loading, reactive oxygen species and esterase dual-responsive polymeric micelles for drug delivery. RSC Adv 9(5):2371–2378

Wang J, Wang Z, Yu J, Kahkoska AR, Buse JB, Gu Z (2020) Glucose-responsive insulin and delivery systems: innovation and translation. Adv Mater 32(13):1–19

Wei J, Lin F, You D, Qian Y, Wang Y, Bi Y (2019) Self-assembly and enzyme responsiveness of amphiphilic linear-dendritic block copolymers based on poly(N-vinylpyrrolidone) and dendritic phenylalanyl-lysine dipeptides. Polymers (Basel) 11(10):1625

Xu F, Li H, Luo YL, Tang W (2017) Redox-responsive self-assembly micelles from poly (N-acryloylmorpholine-block-2-acryloyloxyethyl ferrocenecarboxylate) amphiphilic block copolymers as drug release carriers. ACS Appl Mater Interfaces 9:5181–5192

Xu Y, Cao J, Li Q, Li J, He K, Shen T et al (2018) Novel azobenzene-based amphiphilic copolymers: synthesis, self-assembly behavior and multiple-stimuli-responsive properties. RSC Adv 8(29):16103–16113

Yan K, Zhang S, Zhang K, Miao Y, Qiu Y, Zhang P et al (2020) Enzyme-responsive polymeric micelles with fluorescence fabricated through aggregation-induced copolymer self-assembly for anticancer drug delivery. Polym Chem 11(48):7704–7713

Yang H, Sun X, Liu G, Ma R, Li Z, An Y et al (2013) Glucose-responsive complex micelles for self-regulated release of insulin under physiological conditions. Soft Matter 9(35):8589–8599

Yang H, Khan AR, Liu M, Fu M, Ji J, Chi L et al (2020) Stimuli-responsive polymeric micelles for the delivery of paclitaxel. J Drug Deliv Sci Technol 56:101523. https://doi.org/10.1016/j.jddst.2020.101523

Yuan Y, Zhao L, Shen C, He Y, Yang F, Zhang G et al (2020) Reactive oxygen species-responsive amino acid-based polymeric nanovehicles for tumor-selective anticancer drug delivery. Mater Sci Eng C 106:1101159

Zhang X, Han L, Liu M, Wang K, Tao L, Wan Q et al (2017) Recent progress and advances in redox-responsive polymers as controlled delivery nanoplatforms. Mater Chem Front 1(5):807–822

Zhang K, Liu J, Ma X, Lei L, Li Y, Yang H et al (2018) Temperature, pH, and reduction triple-stimuli-responsive inner-layer crosslinked micelles as nanocarriers for controlled release. J Appl Polym Sci 135(40):1–10

Zhang P, Li M, Xiao C, Chen X (2021) Stimuli-responsive polypeptides for controlled drug delivery. Chem Commun 57(75):9489–9503

Zhao X, Bai J, Yang W (2021) Stimuli-responsive nanocarriers for therapeutic applications in cancer. Cancer Biol Med 18:319

Zhou Q, Zhang L, Yang TH, Wu H (2018) Stimuli-responsive polymeric micelles for drug delivery and cancer therapy. Int J Nanomedicine 13:2921–2942

Zhou H, Qin F, Chen C (2021) Designing hypoxia-responsive nanotheranostic agents for tumor imaging and therapy. Adv Healthc Mater 10(5):2001277

Zhuang W, Xu Y, Li G, Hu J, Ma B, Yu T et al (2018) Biological and medical applications of materials and interfaces redox and pH dual-responsive polymeric micelle with aggregation-induced emission feature for cellular imaging and chemotherapy redox and pH dual-responsive polymeric micelle with aggregation. ACS Appl Mater Interfaces 10(22):18489–18498

Nucleic Acid-Based Micellar Therapy for the Treatment of Different Diseases

8

Subhasri Bogadi, Divya Pamu, Lavanya Mude,
Madhukiran Parvathaneni, Pavan Kumar Chintamaneni,
and Veera Venkata Satyanarayana Reddy Karri

8.1 Background

The theory of genetics is that branch of biology that deals with the study of genetic variation, genes, and heredity, and Gregor's medals discovered it worked in 1940. The main role of genetics is deoxyribonucleic acid (DNA). James Watson and Francis crick discovered the structure of DNA in 1953. In 1967, the enzyme DNA ligase was isolated; this enzyme joined 2 strands of DNA together to constrict the recombinant molecule. Some enzymes, referred to as restriction enzymes, are primarily responsible for DNA sequences and making fragments. Recombinant deoxyribonucleic acid (r DNA), a construct used to treat numerous diseases, is created by joining pieces of DNA sequences. Recombinant DNA, developed in 1972, aids in combining double-standard DNA molecules.

A cell is the fundamental building block of an organism, and it is formed of a cell membrane, which serves as the major barrier between the inside of the cell and the outside world. There are numerous cell organelles in a cell that are useful for the storage of genes. The genome is nothing more than a collection of genes. Cells that

S. Bogadi · D. Pamu · L. Mude · V. V. S. R. Karri (✉)
Department of Pharmaceutics, JSS College of Pharmacy, JSS Academy of Higher Education and Research, Ooty, Tamil Nadu, India

M. Parvathaneni
Department of Biotechnology, Harrisburg University of Science and Technology, Harrisburg, PA, USA

Arni Medica, South Plainfield, NJ, USA

CRC Pharma LLC, Parsippany, NJ, USA

P. K. Chintamaneni
Department of Pharmaceutics, GITAM School of Pharmacy, GITAM Deemed to be University (HYD-Campus), Medak, Telangana, India

S. K. Singh et al. (eds.), *Polymeric Micelles: Principles, Perspectives and Practices*,
https://doi.org/10.1007/978-981-99-0361-0_8

carry genetic information are given this information to function correctly in the body. These data are transmitted by the nitrogen bases found in DNA and RNA, such as adenine, guanine, cytosine, and thymine (or) uracil. Unlike genetic information found in RNA, which has a single standard structure, genetic information is found in DNA as a double standard structure. Usually, RNA comes in three different forms: ribosomal RNA, transfer RNA, and messenger RNA. Ribosomal RNA, which makes up 85% of all cellular material and is primarily responsible for the collection of genes, is present in every cell. Methods for isolating DNA or RNA include gene modification experiments for the nucleic acid source in DNA or RNA. During cloning research, nucleic acid was employed in modest amounts (Gutiérrez-Castrellón et al. 2007). Nucleic acid is a naturally occurring compound comprised of sugars, phosphoric acids, purines, and pyramids. Nucleic acids represent a special and mostly untapped target for medical therapy.

DNA and RNA are contained in the monomers known as nucleotides. When monomers are joined to create a polynucleotide, that is what the term means. These polynucleotides comprise a nitrogen base, a sugar molecule with five carbons, and a phosphate group. Micelles are amphiphilic molecules with easier functionalization and can encapsulate both hydrophilic and hydrophobic substances. It is a nanocarrier for transporting drugs and genes that responds to stimuli. Immunomicelles, a different targeting method, exhibit excellent binding specificity and target ability. So in this book chapter, we highlight the benefits of nucleic acid-based micellar therapy for treating numerous illnesses like cancer, diabetes mellitus (DM), acquired immunodeficiency syndrome (AIDS), and other genetic abnormalities.

8.2 Nucleic Acid Based Micellar Therapy

Nucleic acid is an essential component of all living cells. It is produced in three forms: phosphoric acid, sugars, and organic bases such as purines and pyramids. The primary role of nucleic acid is to transform genetic information into the cell (Patra et al. 2018). It is the makeup of some polymers, which are acidic monomeric subunits known as nucleotides, and control the production of proteins. Nucleic acids are of 2 types, one is DNA (deoxy ribonucleic acid), and another one is RNA (riboxy nucleic acid). For medicinal treatments, nucleic acids constitute a unique and primarily untapped target. The monomers known as nucleotides make up DNA and RNA; these are combined to form the three components: nitrogen base, which contains the nitrogen ring, five-carbon sugar molecules, and one phosphate group. The nucleotides lose two phosphate groups as it expands the chains of DNA (or) RNA.

An ATP phosphate group creates deoxyribonucleoside triphosphate, which is the first step in the formation of DNA (De Vries et al. 2013). The nucleotides contain four nitrogen bases, such as adenine, guanine, cytosine, and thymine, that form the polymer known as DNA, whereas RNA contains uracil instead of thymine. These nitrogen-containing bases are complementary to each other. Adenine is always complementary to thiamine, whereas guanine is always complementary to Uracil.

While complementary for this nitrogen basis of adenine and thymine using two hydrogen bonds, guanine and cytosine use the three hydrogen bases. This structure and clinical stability of molecules create genetic material.

Additionally, complementary bonding nucleotides offer a mechanism for DNA replication and genetic information transmission. The tiny circular DNA molecules known as plasmids, found in some bacteria and archaea and typically only include a few genes, are another molecule (Barnaby et al. 2014; Yin et al. 2014; Wu et al. 2014; Chinen et al. 2015, 2016, 2017). A large number of plasmids are easily transferred between cells. In all cells, DNA is found as a protein-coated structure termed chromatin rather than free in solution. DNA is a double helix in structure, but when a segment is present between 140 and 200 base pairs, it shows a spherical structure around eight positively charged proteins called histones. As more histones are added, consecutive pieces of DNA are wrapped around them, forming nucleosomes that resemble beads strung together (Xu et al. 2012a; Grigoryev and Woodcock 2012; Bamshad et al. 2011; Rivera and Ren 2013; Rakyan et al. 2011; Deng et al. 2014). Genetic material can change due to chemically altered DNA (Deng et al. 2014). Inducing COT transitions, anions like bisulfite can deaminate C to produce U. Adjacent pyrimidines can dimerize when exposed to UV radiation. F. Sanger and W. Gilbert invented sequence determination techniques in the 1970s; for their work, they were awarded the Nobel Prize in 1980. Sanger's approach depends on the enzymatic production of DNA, whereas Gilbert-M depends on the chemical activities of nucleotide bases. These two techniques are used to calculate the distance between each instance of specific base pairs, A, C, G, or thiamine (T), and a fixed place on the DNA.

Unlike DNA, RNA is typically single-stranded. A, U, G, or C is one of the four nitrogenous bases that make up a nucleotide in an RNA chain. In this chain, an enzyme known as RNA polymerase is responsible for this process, which is known as transcription (Liu et al. 2005). RNA is a volatile nucleic acid that performs several roles in cells, including serving as a template for protein production and giving structure to higher-order complexes. RNA nucleic acids are of three forms; they are messenger RNA (mRNA), ribosomal RNA (rRNA), and transfer RNA (tRNA). About 80% of cellular RNA comprises rRNA, with tRNA and m RNA making up the remaining 3–5%. However, R. Robert and Philip Sharp discovered in the 1970s that the coding sequences in eukaryotic genes are frequently "split" in the genome. They won the 1993 Nobel Prize. Before mRNAs can function, these coding sequences must be brought together. Exons are the regions of DNA or RNA that code for proteins, whereas introns are the non-coding regions found between exons. RNA splicing is the procedure that involves removing the introns and then re-joining the exons. Each intricate piece of enzyme-powered equipment known as a spliceosome eliminates each intron through a unique series of processes.

A messenger mRNA delivers the message between a protein and its particular gene. Consider a scenario where a cell must produce a specific protein (Xiong et al. 2012). The protein-coding gene will then go "on, "which causes an RNA-polymerizing enzyme to deliver the gene's sequence.

An mRNA will join a ribosome once created. Proteins are put together by this molecular assembly line using amino acids. A group of mRNA are called codons, and each codon produces proteins through appropriate amino acids. Prokaryotic mRNA typically degrades quickly, whereas eukaryotic mRNA's (Li et al. 2015). rRNA binds both mRNA and tRNA to facilitate translating the mRNA's codon sequence into amino acids. They aid mRNA in binding to the appropriate location. Ribozymes are RNA molecules that function as enzymes. The prokaryote E. coli produces approximately 15,000 ribosomes per cell using these ribosomal genes. Individual amino acids are transported by transferring tRNAs and bringing amino acids to the ribosome. The transfer RNA molecule contains long-chain nucleotides; it usually requires 20 amino acids for protein synthesis. Another type of RNA is regulatory RNA, roughly 22 nucleotides long and used as small regulatory RNA molecules to diminish the stability or interfere with the translation of specific mRNA molecules (with wholly or partially complementary sequences). Not all cellular catalysis is performed only by proteins (Abdelhamid et al. 2014). Artificial miRNAs, also known as miRNA mimics, shRNA-miRs, or pri-miRNA-like shRNAs, have the most complex structures and undergo two steps of processing in cells to generate mature siRNAs, RNAi effectors.

A specialized component called a ribosome is used in translation to decode the information contained in RNA molecules. The discovery of the genetic code in the 1960s was one of the pinnacles of molecular biology. All 64 possible triplet codons were examined by Khorana and Nirenberg using artificial templates and in vitro protein synthesis systems. 61 codons encode 20 amino acids in the genetic code, so many codons frequently indicate one amino acid. This is the main characteristic of the genetic code. The genetic code's structure, encoded by the first two bases of each codon, strongly suggests that it evolved from a more rudimentary code that used 16 dinucleotides. An assembly of monomer surfactant molecules scattered in a liquid colloid is known as a micelle. Micelles are amphiphilic molecules with easier functionalization and can encapsulate both hydrophilic and hydrophobic substances. The hydrophobic "tail" sections are trapped in the micelle center by the hydrophilic "head" portions, which are in contact with the surrounding solvent. In inverse micelles, the head groups extend the tails from the center. Micellization is the process of creating a micelle (Chen et al. 2013). An average micelle has a spherical shape and 50–100 monomers. The aggregation number is the number of monomers required to form a micelle. Amphiphilic block/graft copolymers are hydrophobic and hydrophilic to the exterior self-aggregate to form polymeric micelles (shell).

Polymeric micelle (PM) belongs to the group of "nanocarriers." It has two portions hydrophobic and hydrophilic portions. Amphiphilic block or graft copolymers act similarly to typical amphiphiles and produce polymer micelles in an aqueous solution above the critical micelle concentration. The solubility of the original medicine can be improved by PM, which also enhances biocompatibility. The hydrophilic shell and the nanoscale size prevent mechanical clearance. Its inner core has a significant potential for drug loading. The receptor-mediated drug delivery mechanism can take advantage of it. It is a nanocarrier for transporting drugs and genes that responds to stimuli. Immunomicelles, a different targeting method, exhibit excellent binding specificity and target ability (Xu et al. 2012b).

8.3 Treatment Strategies for Nucleic-Acid-Based Micellar Therapy

Many treatment strategies of nucleic acid-based micellar therapy examine the actions of genes. This can change or transform cellular physiology. Gene therapy's effectiveness comes from its ability to access molecular pathways previously inaccessible by conventional pharmacological methods by altering cell physiology at the genetic and epigenetic levels. This makes it possible to target specific pathways and variables with an unmatched level of precision, greatly enhancing the efficacy of monotherapy and drastically lowering the adverse effects frequently connected to wide-spectrum pharmacological agents. Delivery is the main obstacle for gene treatments, as it is for all drug development processes. This main problem was initially solved by creating disarmed retroviruses. The creation of synthetic non-viral gene delivery systems has encountered several biological and technical difficulties (Xiong et al. 2012). Various cationic polymers and lipids with different transfection efficiencies have been investigated for this. Whenever the nucleic acid is delivered as a prodrug, it is expressed as a transgenic. It would be necessary to retool current techniques to ascertain the relationship between delivery effectiveness and therapeutic efficacy. Gene replacement therapy identifies a defective gene and applies a correct piece of DNA to it via a viral vector (also known as a carrier molecule), thereby replacing the identified defective gene with the correct copy. Gene suppression, gene editing, and gene repair are used to grow an understanding of genes. The biological result is highly dependent on identifying the targeted gene, even though the modes of genetic changes differ in their impact on the gene of interest. These alterations are not mutually exclusive and can be used singly or in combination to obtain the same therapeutic result (Chinen et al. 2016). There are four target treatment strategies present: (1) Gene replacement, (2) Gene addition, (3) Gene expression, (4) Gene editing.

In the replacement treatment strategy, first, identify the particular disease and replace it similarly to the original gene. But this therapy takes a long process to identify the specific gene. In the case of gene addition, they were adding a particular protein instead of a gene which goes wrong for making the affected. Whereas gene expression is by altering the DNA/RNA sequences using interfaces. Mammalian gene expression is a plasmid DNA usually obtained from bacteria. These plasmid DNA molecules contain coding sequences, promoters, enhancers, and polyadenylation sites (Chinen et al. 2015). These DNA molecules are essential for the production and mRNA sequences. These sequences are located beyond these mammalian elements. Because healthy, well-established procedures make it simple to insert. The pDNA is used in gene replacement therapy; it delivers the mRNA Sequences, ios, and is an alternative strategy for avoiding the need for nuclear entrance (Abdelhamid et al. 2014). The fact that RNA molecules are highly vulnerable to being broken down by extracellular and intracellular nucleases is a fundamental disadvantage of this method. However, recent chemical modification and in vitro transcription developments have significantly increased. Like pDNA, an artificial chromosome is also used in gene replacement therapy because this chromosome has the capacity for replication and expression maintenance over multiple generations. In addition, AC has a far greater capacity for the transgene size than any currently existing vector system. The gene expression is "knocked down," which downregulates the DNA and RNA sequences. Before any mechanism of action, such as gene expression knockdown, mRNA splicing change, transcriptional and epigenetic regulation, or genome editing, a specific intracellular localization (e.g., cytosol or nucleus) is frequently required.

RNA interfaces (double standard RNA (ds RNA), and siRNA are used in gene expression strategies because they connect to RNA-induced silencing complexes while in the interfacing mechanism. The ds RNA is present in the cytosol, which helps to RNA interface and has been driven to transcription by pDNA harboring either the RNA POI Il or POI Ill promoter for greater long-lasting silencing efficacy (Xiong et al. 2012). Small hairpin RNA (sn RNA) acts as a mediator between the RNA and si RNA; it may reduce the effectiveness of its silencing. siRNA molecules have been subjected to various chemical alterations to enhance their efficacy, potency, serum stability, specificity, and transport, control their immunogenicity, and lessen the off-target effect. Antisense oligonucleotides (AON) can be used in an RNA-independent way to shut down genes (ODN). Single-stranded DNA molecules with an 18–21 length, known as As-ODN, share complementary sequences with their target gene (Bamshad et al. 2011). Hybridizing the As-ODN to target regions in several ways decreases gene expression. These oligonucleotides, like siRNA, are chemically altered to improve their effectiveness and stability and protect them from nuclease attacks. The binding affinity, target selectivity, and nuclease resistance of the current generation of chemical moieties and stereoisomers are considerably improved over those of the previous generation. The ideal remedial strategy for gene repair is gene replacement; risks associated with ectopic integration frequently outweigh the advantages; restoring wild-type activities in dominantly harmful mutations using gene repair is quite exciting, removing miscoded regions, a modification carried in gene sequences.

In genome editing (also called gene editing) is a group of technologies that give scientists the ability to change an organism's DNA. These technologies allow genetic material to be added, removed, or altered at particular locations in the genome. Several approaches to genome editing have been developed. Cells use RNA editing, a naturally occurring mechanism, for posttranscriptional processing in the case of RNA sequence editing using the nucleic acid base pairs. Generally, the four nucleic acid base pairs are present: adenine, guanine, cytosine, and thymine. In these base pairs, adenine is always complementary to thymine, and guanine is always complementary to cytosine in the case of DNA. In contrast, RNA, instead of thymine uracil, is complementary to adenine. But here, especially in the gene editing technique, changes in the complementary base pairs are adenine binds with the guanine, and cytosine binds with thymine in the case of DNA.

In contrast, RNA and cytosine bind with the uracil. Another complementary base pair is adenosine, and cytosine deaminase rings acting on the RNA sequence. These two deaminases are viral vectors, viral RNA, non-coding miRNA, and pre-mRNA. As a result, it has the potential to treat a variety of disorders. Gene correction, however, can be permanently fixed through genome-level gene editing. However, gene correction can be permanently fixed through genome-level gene editing. As a result, it can potentially treat a wide range of disorders. It can be done with various DNA molecules, including short DNA fragments, RNA/DNA hybrid oligonucleotides, and triplex-forming oligonucleotides. In the case of DNA, plasmids are mainly used for editing the gene sequence in the treatment of nucleic acid micellar therapy. The plasmid DNA molecules act as prodrugs at the molecular level since they use the cell's DNA transcription and translation machinery to biosynthesize the therapeutic protein after internalization. The plasmid molecules must enter the cytoplasm and then reach the nucleus to carry out the P DNA's mode of action.

The effectiveness of gene expression is ultimately governed by nuclear access or lack thereof (Bamshad et al. 2011). It can also be used as a DNA vaccine for gene editing techniques and curing diseases. Short single-stranded DNA segments known as oligonucleotides can specifically block the production of a single protein when internalized by tissue applications; oligonucleotides engage and prevent translation or processing. Oligonucleotides act as antigens in the triplex form of DNA standards used for therapeutic illness. Human clinical studies for cancer are currently being conducted with AON, such as M G98 and ISIS 5132. Double-stranded nucleic acid fragments, known as DNA aptamers, mainly focus on the molecular actions of proteins and have the potential to inhibit the HIV-1 integrase enzyme's ability to synthesize harmful proteins. DNA enzymes are ribozyme analogues with superior biological stability. The DNA motifs confer the physical strength, which takes the place of the RNA backbone chemistry (Deng et al. 2014). Single-stranded nucleic acid segments known as RNA aptamers can directly interact with proteins; they do so by complementing the shapes of their targets.

Additionally, they have high affinities comparable to monoclonal antibodies. RNA aptamers have shown potential in inhibiting HIV-I transcriptase's ability to biosynthesize harmful proteins. Further, RNA aptamers have preferentially bind and render inactive growth factors. Eighty per cent of individuals in a clinical trial who received eye injections of anti-VEGF aptamers had no adverse side effects while maintaining or improving their vision. The RNA decoys aim to offer an alternative, competitive binding site for proteins that function as mRNA stabilizers or translational activators. Decoys obstruct translation or cause mRNA instability and eventual destruction (Surana et al. 2015). As decoys for transactivating proteins stop attaching to the viral Genome's corresponding cis-acting elements. Antisense medications are brief segments of the deoxyribonucleotide sequence. They can also process protein synthesis. These medications start to operate before a disease-causing protein is ever produced (Hossen et al. 2019). They can theoretically be used to treat various illnesses whose fundamental pathophysiology. An antisense oligonucleotide have enzyme moiety which is mainly responsible for RNA sequence and forms the triplex. RNA through RNA interference, siRNAs can be employed to suppress the expression of disease-causing genes. It stopped complementary to these brief dsRNA segments, which have 2123 nucleotides.

8.4 Diseases Associated with the Genetic Mutation

Only three nucleic acid-based treatments received global approval in early 2007; these therapies primarily revolve around gene therapy and antisense technology. Only three treatments based on nucleic acids were granted approval on a global scale at the beginning of 2007. These treatments primarily focus on gene therapy and antisense technology and include gene therapy, aptamer, antisense-based products, and transplantation of organs or tissues. Replacing a damaged gene with a new, healthy gene is the fundamental tenet of gene therapy. In vitro gene therapy is a feasible strategy that can be used to treat an inborn metabolic defect. This therapy removes target cells and incubates them with a nucleic acid-containing vector. Human cells are returned to the body once the vector has delivered the nucleic acid. The vector must be administered intravenously in vivo (Nasongkla et al. 2006). Only the target cells were meant to be recognized and bound by the vector. The nucleic acid is supplied to these cells. In the case of in situ delivery, the vector is directly injected next to the body's target cells. These treatments mainly address illnesses, e.g., cancer, DM, AIDS, and hereditary diseases (Fig. 8.1).

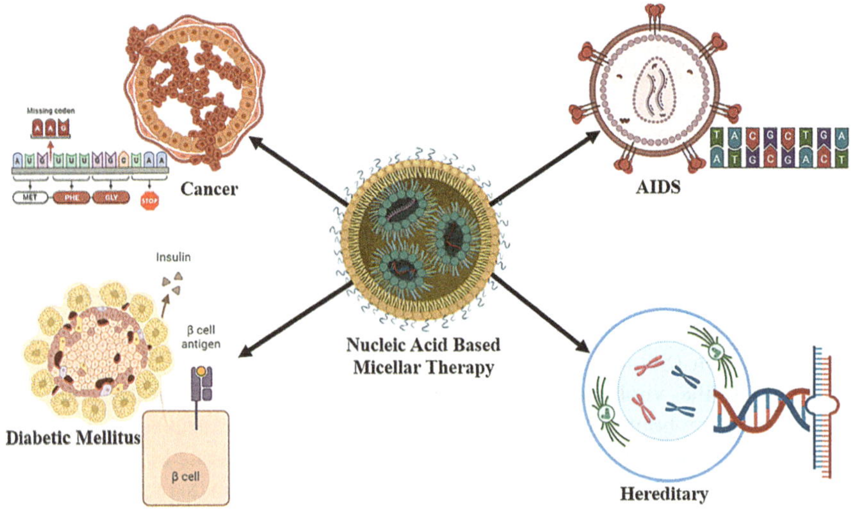

Fig. 8.1 Nucleic acid-based micellar therapy for the treatment of different diseases

8.4.1 Cancer

A malignant neoplasm, also known as cancer, is a general term for several different disorders. It is a body-wide abnormal proliferation of dead cells. Because the old cell does not perish, it develops (Gao et al. 2005). Instead, they develop atypically. The collection of recently created dead cells is referred to as a tumor. Gene therapy, a method for repairing the faulty genes responsible for cancer formation, can be used to treat it. It uses cancer as a vector to spread the gene. Virus vectors are the most prevalent type of vector utilized in gene therapy. Cancer gene therapy has two goals: help hasten recent improvements in cancer survival rates using proven cancer therapies. The second goal is maximum efficacy with minimal harm. When administering drug-nucleic acid combinations, many factors must be considered. Normal cells and tumor cells are both components of gene therapy. Mutant tumor suppressor genes can be replaced by tumor cells. Cell surface receptors that are overexpressed are downregulated, or their kinase activity is inhibited. It can suppress tumor angiogenesis, restore the cell's ability to experience senescence, and prevent abnormal intracellular signaling pathways (Jeong et al. 2005). The use of chemotherapy in a more dense-intense manner without sensitive, normal cells is possible in the case of normal cells. It is the effector mechanism of gene therapy to correct the molecular defects that result in malignant transformation and suicide genes.

Transgenes can be injected directly into the tumor to reach it in situ. They remove cells from the patient carrying the tumor, insert the transgene extracorporeally into the cells, and reintroduce the genetically altered cells into the host. Ex vivo adaptive transfer aids in the treatment of immunotherapy and hematological malignancies (peripheral bloodstream cells) (Lee et al. 2012). However, reliable tumors—an alternative method of delivery for superficial lesions when the malignancy has spread—do not fall under this category. The intra-tumoral injection of naked DNA can be done under direct observation while the endoscope views the tumor. It adds no value, particularly when the current local treatments, such as surgery or radiotherapy, may reach cure rates. The optimum distribution method for gene therapy is typically intravenous and sporadically into arteries, even though it is essential for treating advanced diseases with many metastases.

Nevertheless, systemic gene therapy has certain drawbacks, including extravasation and interaction with extracellular matrix (ECM), circulating antibodies and immune effector cells, non-specific adsorption in the liver, and circulating antibodies and immune effector cells. The trans-gene inside autologous cells, which have the innate propensity to home to sites of tumor growth, can be used to circumvent these issues. The cells are mesenchymal, endothelial, macrophage, and lymphocytes. The trans-gene is then dispersed throughout the tumor once released from a cellular carrier using antigen recognition and pharmacological stimulation.

Vectors are mainly responsible for producing genes, which are viral-based and non-viral-based types. Viral-based vectors are RNA and DNA. In RNA, viral-based vectors contain the three crucial genes: gap, pol, and env, DNA-based viral vectors contain adeno-associated viruses (AAV), Herpes viruses, and sends viruses. At the same time, non-viral vectors are naked DNA shown in Table 8.1.

Table 8.1 Different types of vectors and their properties

<table>
<tr><th>Vectors</th><th colspan="4">Properties</th><th>Advantages</th><th>Disadvantages</th></tr>
<tr><td rowspan="5">Adenoviruses</td><td colspan="4">• Relatively large
• Double-stranded DNA
• Large genome and complex</td><td rowspan="5">• Gene transfer to non-dividing cells is possible, it is simple to reproduce, and gene expression is high</td><td rowspan="5">• High immunogenic
• The duration of a transferred gene can vary
• Highly selective for cells</td></tr>
<tr><td>Generation</td><td>Genes</td><td>Modification</td><td>Implications</td></tr>
<tr><td>First</td><td>E1</td><td>Viral replication and activation of other ad transcriptional units prevented</td><td>Transgenes can replace E1 regions of viral vectors</td></tr>
<tr><td>Second</td><td>E2 or E4</td><td>Viral replication is prevented</td><td>Toxicity of E2 gene expressed protein prevented
E4- viral replication and late protein synthesis are prevented</td></tr>
<tr><td>Third</td><td>–</td><td>Development of high-capacity helper-dependent vector</td><td>Activity regulation possible</td></tr>
<tr><td>Adeno-associated viruses (AAV)</td><td colspan="4">• Nano pathogenic viruses
• Contain in single standard DNA genome of 4.7 kb
• It requires co-infection with helper viruses for efficient, site-specific integration
• It binds to human chromosome number 19, a highly desirable feature of a gene therapy vector</td><td>• An adenoviral vector may accommodate small genes, which facilitates long-term gene expression</td><td></td></tr>
<tr><td>Herpes virus</td><td colspan="4">• It has the following features:
• A neurotropic vector
• Gene delivery to the PNS and CNS
• Latent infection in non-dividing neurons
• 150 kb genome</td><td>• Large size
• Provide long-term CNS gene
• High titer</td><td>• Difficult to generate
• It is under development
• Currently,</td></tr>
</table>

			available vectors show transient expression • Minimal transduction effectiveness
Sindbis virus	• It belongs to the alphavirus family • It affects cells that are not dividing		
Non-viral vectors Naked DNA	• It is the simplest method for injecting naked DNA into humans • Relatively low expression	• The therapeutic genes' failure to integrate into the host chromosome • It is a simple method of non-viral transfection	
Lipoplexes	• Lipids may cover P-DNA in a structured manner, such as a compound of DNA and a liposome known as a lipoplex • Lipids come in three types: Anionic, cationic, and neutral		
Polyplexes	• Polyplexes are complexes of DNA and polymers • It comprises cationic polymers whose synthesis is controlled by ionic interactions • These are quickly taken out of circulation		

Table 8.2 Prominent strategies of gene therapy for cancer

Genes	Strategies
Gene cytokines	Pro-inflammatory genes, such as IL-12 and IFN-g, are introduced; local information is induced, and tumor suppression occurs Cell-mediated immunity destruction
Genes for suicide	Phosphorylation of ganciclovir by HSV-tk The "bystander effect"
Cancer genes	P53 gene introduction, which is typically mutated in most cancers – Anti-oncogenes
HSC protection	– MDR-1 gene introduction in HSC: Improves HSC tolerance to chemotherapy; permits more aggressive treatment

8.4.1.1 Characteristics Strategies for Gene Therapy for Cancer

- Enhancing the tumor's immunogenicity.
- Inhibiting the expression of oncogenes.
- Introducing a wild-type tumor suppressor gene.
- Protecting stem cells from chemotherapy's toxic side effects.
- Obstructing the tumors' ability to avoid being destroyed by the immune system.
- And killing tumor cells by introducing toxin genes under the control of a tumor-specific gene.

It mainly uses three strategies: RNA-directed plans, gene replacement, and suicide gene therapy (Table 8.2). Recessive suppressor genes are lost in both copies in many human cancers due to gene replacement techniques. Therefore, it introduces a plan for eliminating the cancer cell phenotype (Cheng et al. 2012). The genome guardian P53 is the most alluring candidate for gene substitution. Gene therapy can encode proteins that have direct cellular cytotoxicity, such as the A-subunit of Diphtheria toxins, in the case of suicide. However, the suicide gene product is typically not harmful by itself. The original prodrug activating enzyme is utilized in bacterial CD cytosine. Accelerating the triphosphorylation processes activates the prodrug ganciclovir (Koganti et al. 2013). Triphosphate is a DNA replication chain terminator. While endogenous kinases carry out the di- and tri-phosphorylation, the HSV-tk assists in the first phosphorylation. In RNA-directed targeting, these techniques focus on the mRNA rather than the DNA encoding a protein or the protein itself. Consider epidermal growth factor receptor (EGFR) mRNA targeting in malignancies with elevated EGFR expression.

There are 3 types

1. Technology that blocks sense.
2. Ribozymes.
3. RNA interference.

Single standard ODNs complementary to suppress the production using antisense technology. Either antisense ss-ODN and the targeted mRNA couple up in a specific base-pairing sequence that prevents translation, or ss ODN forms a duplex detected by the RNase H, causing mRNA cellular enzyme cleavage. In the case of ribozymes, these natural molecular derivatives can be found in various organisms. ODNs also bind to the target mRNA. AIDS and healthy volunteers are still research subjects, not cancer patients.

SiRNA, complementary to mRNA targets, is a 21–23 nucleotide base ssRNA effector used in RNA interference or gene silencing. The production of siRNA is carried out by the RISC intracellular multimolecular complex (RNA-induced silencing complex). Slamon et al. (2001) found that humanized monoclonal antibodies improved survival in breast and colon cancer subgroups. In addition, gastrointestinal stromal tumors and chronic myelogenous leukemia can be treated with receptor tyrosine kinase inhibitors. In the phase I clinical trial, Arie Belldegrun et al. (2001) used gene-based immunotherapy on 24 patients with locally advanced prostate cancer (CaP). Transrectal ultrasound delivered a functional DNA-lipid complex encoding the interleukin 2 (IL-2) gene into the hypoechogenic tumor lesion. After treatment with high-dose IL2, in vitro data showed a decrease in plasma serum antigen (PSA) mRNA expression in the LNCaP cell line (derived from a metastatic lymph node lesion of human prostate cancer), a normally high PSA producer. Thus, these differential mechanisms may explain the inconsistent mRNA-protein disparity. Finally, the author concluded that IL-2 gene therapy is safe and well tolerated by patients based on the findings.

2013 Jing Li: Nucleic acid therapies that have recently shown promise in cancer treatment include siRNA, AON, shRNA, and pDNA. This approach is effective for treating monogenic illnesses. Due to the complicated signaling pathways in cancer cells and the various compensatory mechanisms involved, effective treatment strategies are according to the current cancer therapy regimens. Suggested that nanocarriers that can transport drug-nucleic acid mixes will lead to the development of better-controlled, safer, and more effective cancer treatments.

Cancer cells are directly killed by chemotherapy medications (apoptosis). Assume that any apoptotic network faults will significantly reduce the effectiveness of these networks. P53, retinoblastoma protein, and MDA-7 are a few tumor suppressor genes. Kiwada et al. have described the xenograft model for treating human colon cancer (2011). This method used Bcl-2 siRNA produced in PEGylated lipoplexes and the chemotherapy drug 5-fluorouracil (5-FU). The combination greatly enhanced the capacity to inhibit cell proliferation and supported tumor suppression compared to either monotherapy (Li et al. 2012). Chu et al. used cisplatin and surviving siRNA in combination therapy to treat ovarian cancer (2012). When delivered using polyethyleneimine, the siRNA/drug combination efficiently caused apoptosis and suppressed tumor formation in vitro and in vivo (PEI).

8.4.2 DM

DM is a metabolic disorder of 2 types. One is insulin-dependent (type 1), and the other is insulin-resistant (type 2), which means insufficient insulin and more complications than type 1. According to the World Health Organization (WHO), diabetes affects 537 million people globally (Tressler et al. 2003). It is estimated that 643 million people will be impacted in 2030, rising to 783 million in 2045. India is slowly progressing to the top of the world with the most significant volume of diabetic subjects and is anticipated to be the "diabetes capital of the world." India will be the capital of people with diabetes. By 2025.69 million people in India were affected due to diabetes in 2019. It is estimated that over 101 million people will be involved in 2030 and 134 million in 2045.

Genes involved in DM are defective lipoprotein lipase (LpL) genes that may risk coronary artery disease and type 2 diabetes in people with it (Taylor 2012). Calpain-10, also known as CAPN10, is the protease from the calpain family that was found to be the first candidate susceptibility gene for type 2 diabetes mellitus (T2DM). Defective genes that regulate a molecule called peroxisome proliferator-activated receptor (PPAR) gamma may contribute to type 2 diabetes and high blood pressure in some patients. A defective gene that affects the beta3-adrenergic receptor has been detected, found in visceral fat cells (those occurring around the abdominal region). The result is a slowdown in metabolism and an increase in obesity (Samuel and Shulman 2012). A theory is that some cases of type 2 diabetes and obesity are derived from everyday genetic actions that were once important for survival—the existence of a so-called thrifty gene, which regulates hormonal fluctuations to accommodate seasonal changes. In specific nomadic populations, hormones are released during seasons when food supplied has traditionally been low, which results in resistance to insulin and efficient fat storage (Hirabara et al. 2012). The process is reversed in seasons when food is readily available. High-carbohydrate and fatty foods are functional all year long; the gene no longer serves a practical function and is now harmful because fat, initially stored for famine situations, is not used up (Table 8.3).

Increasing hyperglycemia is the aim of nucleic acid-based therapies, particularly siRNA-based medicines that concentrate on potential novel targets for effective and focused Type 2 DM (Tang et al. 2012). The antisense sequence of messenger RNA forms the small duplexes which aid in treating type II diabetes. This technique was discovered by Small fire and Mello, and they received the Nobel prize for this research. Developing RNAi-based therapeutics to treat diseases like DM and its effects is a top priority for many businesses (Gaggini et al. 2013). To cure obesity and T2DM, Cyt Rx Corp. and the University of Massachusetts Medical School started developing therapeutic compounds based on RNAi technology; human T2DM patients' skeletal muscles showed improved insulin signaling and glucose metabolism. Using RNA interface technology is a new technique for treating Type 2 DM; it controls the metabolism, insulin activities.

8.4.3 AIDS

AIDS is one of the immune-related diseases that is brought on by the immunodeficiency virus. (HIV). HIV weakens the immune system, making it challenging to fight viruses. It is easily contagious by contaminated vaginal secretions, semen, or

Table 8.3 Genes susceptible to type 1 and type 2 diabetes

Genes	Gene susceptible
Type 1	
PPAR	This gene mainly decreases insulin sensitivity and increases T2DM risk
ABCC8	It is an ATP binding cassette, a subfamily belonging to C, eight-member. This gene encodes the high-affinity sulfonylurea receptor
Gene treatment assistance with islet transplantation	These genes serve as carriers for transgenes that regulate the immune system. These genes exhibit pre- and post-transplantation states when given as an infusion. These genes, when administered in vivo, produce trophic factors for islets
Type 2	
HLA-DQB1	This gene belongs to the class II beta, which consists of alpha and beta, present on the membrane. Both chains play a crucial role in immune systems. Especially since these class II molecules represent the antigen-presenting cells. Particularly one phenotype, DQB1C 0201/*0302, which shows high toxicity in type 1 diabetes
INS	It is an insulin gene which contains the 1430 base pairs, which helps with the translation of proinsulin
CTLA4	Cytotoxic T-lymphocyte antigen-4 gene (CTLA-4) encoded on chromosome 2q3CTLA-4 produces negatively charged ions for inhibiting the T-cells
Possible intervention tactics	Insulin-sensitizing genes serve as vectors for the transduction and translation of genes and are primarily in charge of preventing insulin resistance
Islets of allograft/xenograft survival-promoting genes Gene anti-apoptotic	Primarily guard against cytokine-induced cell death Bcl-2, heme oxygenase-1, hsp70, delta-dominant protein kinase C, and manganese superoxide dismutase are a few examples
Genes that regulate immunity	For instance, indoleamine 2,3 dioxygenase
Cytokines	For example, the IL-1 receptor under clinical experimentation mainly focuses on obesity-related complications and satiety centers. Beta cells and a beta cell growth factor engineered into a vector could partially cure the disease

blood. Nowadays, drugs are only used to treat illness rather than to stave it off. Use nucleic acid base techniques employing two types of ssRNA standards, reverse transcriptase and integrase; integrase aids in the insertion of the human immunodeficiency virus (HIV) DNA into host DNA and infects macrophages and helper T cells (Van Herk et al. 2016). Making immune cells resistant to HIV through gene therapy (the AIDS virus) is possible. Enhancing the body's immunological response to these substances may be utilized to assist patients in destroying HIV and HIV-infected cells. The strategies for gene therapy for AIDS are shown in Table 8.4.

It aids in treating cancer, infectious disease, inherited diseases, and medical ailments. AIDS is treated using a variety of methods. Oncogene expression, cytokine overexpression, and angiotensinogen overproduction are the causes of illness.

Table 8.4 Gene therapy strategies for AIDS

Intracellular immunization	To make the engineered cells resistant to viral infection, genetic engineering produces an antibody intracellularly
Ribosomes	These RNA molecules have an enzymatic activity that cleans the RNA and an antisense sequence for specific recognition
Transdominant mutant	These proteins include changed amino acids, which enable the mutant protein to interfere with the actions of the protein in its wild-type form
Trojan horse	The HIV packaging signal and an anti-HIV nucleic acid are present in this anti-HIV RNA

Intracellular immunization techniques involve introducing genes into pathogen-susceptible cells, such as injecting gag-encoding genes into viral-sensitive cells. It is capable of preventing viral replication in its mutant form. A gene that codes for antibody fragments that can attach to the HIV envelope proteins is transferred to susceptible cells to disrupt the formation of the virus. The soluble CD4 antigen, which resembles the HIV cell surface receptor, can also be secreted by recombinant cells. The delivery of virions—DNA vectors containing the gene encoding for a surface antigen protein from the target pathogen—would bind with this soluble viral receptor. Any cell in the body may be the target. The antigenic protein that results is exported by the target cell. An immunological response can be triggered by transitory gene expression alone. Clinical studies for gene-based vaccinations against AIDS, hepatitis B, and malaria have begun (Kokil et al. 2015). When the line is known to cause disease, a nucleotide sequence in antisense technology attaches to mRNA or DNA. As a result, genes "switch off." Antisense nucleotides are brief and single-stranded nucleic acid sequences with a particular sequence of nucleotides, which are utilized in this technology. Nucleotides can be acquired from natural sources or chemically produced, whereas "oligo" denotes "few."

Short DNA or RNA molecules known as oligonucleotides are often created in the lab using solid-phase chemical synthesis. These oligomers have various uses in genetic research, testing, and forensics. The binding of oligonucleotides stops the transcription and translation processes to DNA or RNA (Taylor 2012). AON is found in the appropriate mRNA sequences. These oligonucleotides are administered intravenously, subcutaneously, and intradermally; charged oligos accomplish this by receptor-mediated endocytosis, while uncharged oligos accomplish this through passive diffusion (Nifontova et al. 2022). Most oligos are delivered through polymeric carriers that are mediated by liposomes. These oligos appear to be finally digested by nucleases, specifically 3-exonucleases, inside the cells. The urine pathway is then mainly used to excrete breakdown metabolic products.

8.4.4 Hereditary Diseases

A hereditary disease is a sickness present from birth and brought on by one or more defects in the genome (congenital). Most generic illnesses have a rating and affect one in a few thousand or millions. Gene therapies for hereditary diseases are shown in Table 8.5.

Table 8.5 Gene therapy for hereditary disease

Common hereditary diseases	Single gene inheritance	Multifractional inheritance	Chromosome abnormalities	Mitochondrial inheritance
	Hemochromatosis, Huntington's disease, Sickle cell anemia, Marfan syndrome, and cystic fibrosis	High blood pressure and heart disease Alzheimer's condition Arthritis diabetes Obesity and cancer	Klinefelter syndrome and the Turner syndrome	Leber's hereditary optic atrophy, mitochondrial encephalomyopathy, lactic acidosis, and stroke-like episodes are all examples of diseases that affect the eyes (MELAS)
Severe combined immune deficiency (SCID)	Children who were born lacking immune systems were impacted. In this case, an adenosine deaminase (ADA) gene was introduced into the bone marrow of particular children in labs, followed by transplantation of the same children			
Chronic granulomatous disease (CGD)	People with this disease cannot frequently combat bacterial and fungal infections. The ADA gene, which is primarily in charge of developing the immune system as well as the ability to fight against all infections, particularly those brought on by bacteria and fungi, is administered to treat this condition			
Hemophilia	The leading cause of this condition is the incorrect formation of blood clots. When administered into the liver of a specific patient, the therapeutic gene indicates the formation of blood clots			
Other genetic disorders	The various genetic disorders will soon be treated in numerous clinical trials			

8.5 Conclusion & Future Aspects

Cancer, diabetes mellitus, AIDS, and other inherited disorders may be treated with the help of nucleic acid base micellar treatments. They use various techniques, such as gene substitution, gene editing, modification, etc. By utilizing these many methods, it might be able to treat a specific illness, particularly for those that have no known treatments. Due to the need for some safety and effectiveness for particular targets, all strategies in this treatment may not yield correct outcomes. When a novel gene is administered to the body, it may behave not as an antibody but as an antigen. It must go through numerous clinical trials to succeed and become commercially available.

References

Abdelhamid D, Arslan H, Zhang Y, Uhrich KE (2014) Role of branching of a hydrophilic domain on physicochemical properties of amphiphilic macromolecules. Polym Chem 5(4):1457–1462

Bamshad MJ, Ng SB, Bigham AW, Tabor HK, Emond MJ, Nickerson DA, Shendure J (2011) Exome sequencing as a tool for Mendelian disease gene discovery. Nat Rev Genet 12(11): 745–755

Barnaby SN, Lee A, Mirkin CA (2014) Probing the inherent stability of siRNA immobilized on nanoparticle constructs. Proc Natl Acad Sci 111(27):9739–9744

Belldegrun A, Tso C-L, Zisman A, Naitoh J, Said J, Pantuck AJ, Hinkel A, deKernion J, Figlin R (2001) Interleukin 2 gene therapy for prostate cancer: phase I clinical trial and basic biology. Hum Gene Ther 12(8):883–892. https://doi.org/10.1089/104303401750195854

Blanco E, Kessinger CW, Sumer BD, Gao J (2009) Multifunctional micellar nanomedicine for cancer therapy. Exp Biol Med 234(2):123–131

Chen L, Sha X, Jiang X, Chen Y, Ren Q, Fang X (2013) Pluronic P105/F127 mixed micelles for the delivery of docetaxel against Taxol-resistant non-small cell lung cancer: optimization and in vitro, in vivo evaluation. Int J Nanomedicine 8:73

Cheng D, Cao N, Chen J, Yu X, Shuai X (2012) Multifunctional nanocarrier mediated co-delivery of doxorubicin and siRNA for synergistic enhancement of glioma apoptosis in rat. Biomaterials 33(4):1170–1179

Chinen AB, Guan CM, Mirkin CA (2015) Spherical nucleic acid nanoparticle conjugates enhance G-Quadruplex formation and increase serum protein interactions. Angew Chem 127(2): 537–541

Chinen AB, Ferrer JR, Merkel TJ, Mirkin CA (2016) Relationships between poly (ethylene glycol) modifications on RNA–spherical nucleic acid conjugates and cellular uptake and circulation time. Bioconjug Chem 27(11):2715–2721

Chinen AB, Guan CM, Ko CH, Mirkin CA (2017) The impact of protein corona formation on the macrophage cellular uptake and biodistribution of spherical nucleic acids. Small 13(16): 1603847

De Vries JW, Zhang F, Herrmann A (2013) Drug delivery systems based on nucleic acid nanostructures. J Control Release 172(2):467–483

Deng X, Berletch JB, Nguyen DK, Disteche CM (2014) X chromosome regulation: diverse patterns in development, tissues, and disease. Nat Rev Genet 15(6):367–378

Gaggini M, Morelli M, Buzzigoli E, DeFronzo RA, Bugianesi E, Gastaldelli A (2013) Non-alcoholic fatty liver disease (NAFLD) and its connection with insulin resistance, dyslipidemia, atherosclerosis, and coronary heart disease. Nutrients 5(5):1544–1560

Gao ZG, Fain HD, Rapoport N (2005) Controlled and targeted tumor chemotherapy by micellar-encapsulated drug and ultrasound. J Control Release 102(1):203–222

Grigoryev SA, Woodcock CL (2012) Chromatin organization—the 30 nm fiber. Exp Cell Res 318(12):1448–1455

Gutiérrez-Castrellón P, Mora-Magana I, Díaz-García L, Jiménez-Gutiérrez C, Ramirez-Mayans J, Solomon-Santibánez GA (2007) Immune response to nucleotide-supplemented infant formulae: systematic review and meta-analysis. Br J Nutr 98(S1):S64–S67

Hirabara SM, Gorjao R, Vinolo MA, Rodrigues AC, Nachbar RT, Curi R (2012) Molecular targets related to inflammation and insulin resistance and potential interventions. J Biomed Biotechnol 2012:1

Hossen S, Hossain MK, Basher MK, Mia MN, Rahman MT, Uddin MJ (2019) Smart nanocarrier-based drug delivery systems for cancer therapy and toxicity studies: a review. J Adv Res 15:1–8

Jeong YI, Seo SJ, Park IK, Lee HC, Kang IC, Akaike T, Cho CS (2005) Cellular recognition of paclitaxel-loaded polymeric nanoparticles composed of poly (γ-benzyl L-glutamate) and poly (ethylene glycol) diblock copolymer endcapped with galactose moiety. Int J Pharm 296(1–2): 151–161

Koganti S, Jagani HV, Palanimuthu VR, Mathew JA, Rao MC, Rao JV (2013) In vitro and in vivo evaluation of the efficacy of nanoformulation of siRNA as an adjuvant to improve the anticancer potential of cisplatin. Exp Mol Pathol 94(1):137–147

Kokil GR, Veedu RN, Ramm GA, Prins JB, Parekh HS (2015) Type 2 diabetes mellitus: limitations of conventional therapies and intervention with nucleic acid-based therapeutics. Chem Rev 115(11):4719–4743

Lee MJ, Albert SY, Gardino AK, Heijink AM, Sorger PK, MacBeath G, Yaffe MB (2012) Sequential application of anticancer drugs enhances cell death by rewiring apoptotic signaling networks. Cell 149(4):780–794

Li JM, Wang YY, Zhao MX, Tan CP, Li YQ, Le XY, Ji LN, Mao ZW (2012) Multifunctional QD-based co-delivery of siRNA and doxorubicin to HeLa cells for reversal of multidrug resistance and real-time tracking. Biomaterials 33(9):2780–2790

Li J, Wang X, Zhang T, Wang C, Huang Z, Luo X, Deng Y (2015) A review on phospholipids and their main applications in drug delivery systems. Asian J Pharm Sci 10(2):81–98

Liu SQ, Tong YW, Yang YY (2005) Thermally sensitive micelles are self-assembled from poly (N-isopropyl acrylamide-co-N, N-dimethyl acrylamide)-b-poly (D, L-lactide-co-glycolide) for controlled delivery of paclitaxel. Mol Biosyst 1(2):158–165

Nasongkla N, Bey E, Ren J, Ai H, Khemtong C, Guthi JS, Chin SF, Sherry AD, Boothman DA, Gao J (2006) Multifunctional polymeric micelles as cancer-targeted, MRI-ultrasensitive drug delivery systems. Nano Lett 6(11):2427–2430

Nifontova G, Tsoi T, Karaulov A, Nabiev I, Sukhanova A (2022) Structure-function relationships in polymeric multilayer capsules designed for cancer drug delivery. Biomater Sci 10(18): 5092–5115

Patra JK, Das G, Fraceto LF, Campos EV, Rodriguez-Torres MD, Acosta-Torres LS, Diaz-Torres LA, Grillo R, Swamy MK, Sharma S, Habtemariam S (2018) Nano-based drug delivery systems: recent developments and prospects. J Nanobiotechnol 16(1):1–33

Rakyan VK, Down TA, Balding DJ, Beck S (2011) Epigenome-wide association studies for common human diseases. Nat Rev Genet 12(8):529–541

Rivera CM, Ren B (2013) Mapping human epigenomes. Cell 155(1):39–55

Samuel VT, Shulman GI (2012) Mechanisms for insulin resistance: common threads and missing links. Cell 148(5):852–871

Slamon DJ, Leyland-Jones B, Shak S, Fuchs H, Paton V, Bajamonde A, Fleming T, Eiermann W, Wolter J, Pegram M, Baselga J, Norton L (2001) Use of chemotherapy plus a monoclonal antibody against HER2 for metastatic breast cancer that overexpresses HER2. N Engl J Med 344 (11):783–792. https://doi.org/10.1056/NEJM200103153441101

Surana S, Shenoy AR, Krishnan Y (2015) Designing DNA nanodevices for compatibility with the immune system of higher organisms. Nat Nanotechnol 10(9):741–747

Tang WH, Martin KA, Hwa J (2012) Aldose reductase, oxidative stress, and diabetic mellitus. Front Pharmacol 3:87

Taylor R (2012) Insulin resistance and type 2 diabetes. Diabetes 61(4):778–779
Tressler RL, Ramstack MB, White NR, Molitor BE, Chen NR, Alarcon P, Masor ML (2003) Determination of total potentially available nucleosides in human milk from Asian women. Nutrition 19(1):16–20
Van Herk A, Forcada J, Pastorin G (eds) (2016) Controlled release systems: advances in nanobottles and active nanoparticles. CRC, Baco Raton
Wu XA, Choi CH, Zhang C, Hao L, Mirkin CA (2014) Intracellular fate of spherical nucleic acid nanoparticle conjugates. J Am Chem Soc 136(21):7726–7733
Xiong XB, Binkhathlan Z, Molavi O, Lavasanifar A (2012) Amphiphilic block copolymers: preparation and application in nano drug and gene delivery. Acta Biomater 8(6):2017–2033
Xu X, Costa A, Burgess DJ (2012a) Protein encapsulation in unilamellar liposomes: high encapsulation efficiency and a novel technique to assess lipid-protein interaction. Pharm Res 29(7): 1919–1931
Xu Q, Liu Y, Su S, Li W, Chen C, Wu Y (2012b) Anti-tumor activity of paclitaxel through dual-targeting carrier of cyclic RGD and transferrin conjugated hyperbranched copolymer nanoparticles. Biomaterials 33(5):1627–1639
Yin H, Kanasty RL, Eltoukhy AA, Vegas AJ, Dorkin JR, Anderson DG (2014) Non-viral vectors for gene-based therapy. Nat Rev Genet 15(8):541–555

Polymeric Micelles in the Delivery of Therapeutic Phytoconstituents

9

Jayesh S. Unde and Rahul Shukla

Abstract

Since ancient times, medicinal plants and their phytochemical constituents have been utilized to cure a number of diseases. Currently, various herbal formulations are available on market but the problem is many of the phytoconstituents of herbal product are hampered by their less bioavailability, the reason behind that is poor aqueous solubility, limited absorption, and hygroscopicity leads to degradation of the phytoconstituents. These problems can be overcome by combining the phytoconstituent with nanocarriers such as liposomes, polymeric micelles, dendrimers, organic nanoparticles, nanoemulsion, emulsomes, phytosomes, and nanocrystals. These nanocarrier systems help to increase the pharmacokinetic and bioavailability of phytoconstituent.

Polymeric micelles are nanosized core-shell shape, self-assembly of amphiphilic macromolecules with lipophilic core and hydrophilic shell containing polymer including block and graft copolymer for favorable drug delivery. Polymeric micelles are between 10 and 100 nm in size They have unique characteristics such as increased bioavailability, biocompatibility, enhanced solubility of water-insoluble drug and enhance its absorption by minimizing the degradation rate. Polymeric micelles act as a targeted drug delivery because they target certain parts of body, for example, encapsulation of curcumin by polymeric micelles for the treatment of cancer.

Traditional micelles have hydrophilic head and hydrophobic tail, they do not contain polymers. In polymeric micelles different polymers are used for the preparation. The stearic acid and chitosan are grafted with copolymers like

J. S. Unde · R. Shukla (✉)
Department of Pharmaceutics, National Institute of Pharmaceutical Education and Research (NIPER)—Raebareli, Lucknow, India
e-mail: rahul.shukla@niperraebareli.edu.in

S. K. Singh et al. (eds.), *Polymeric Micelles: Principles, Perspectives and Practices*,
https://doi.org/10.1007/978-981-99-0361-0_9

diblock (poly(ethylene glycol), polystyrene) and triblock (poly(ethylene oxide)) copolymers respctively. These polymers protect the drug from degradation and provide in vivo stability, enhanced efficiency and biocompatibility.

Polymeric micelles are more valuable than liposomes comparing their size, stability, and accumulation at target site. Polymeric micelles has highest percentage of entrapment efficiency and loading capacity. Polymeric micelles deliver various poorly soluble phytoconstituents like artemisinin and curcumin (anticancer), Piper methysticum G. Forst. (Kava-kava) for local anesthetic activity, berberine for anticancer and antibacterial activity.

Keywords

Polymeric micelles · Therapeutic phytoconstituents · Polymers · Solubilization · Targeted delivery

9.1 Introduction

Herbal medicine is used since ancient times for the treatment of various diseases. Plant-derived phytoconstituents show multiple pharmacological actions and chances of fewer side effects. Developed countries increase the demand for various natural phytoconstituent due to their safety and efficacy. The extract and isolated phytoconstituent play a key role in therapeutic action including anticancer, antibacterial, antidiabetic, antifungal, and antimalarial treatment (Karar and Kuhnert 2017). In the nineteenth century the importance of herbal medicine decreased due to newer synthetic drugs and their quick action helps to get immediate relief. The derived phytoconstituent with synthetic drug shows synergistic action with a decrease in toxicity and resistance (Pal and Shukla 2003) that required carrier formulation for targeted delivery with minimal interaction between therapeutic active ingredients. Phytoconstituent require a significant delivery system to deliver drug in sustain release manner with less adverse effects. The nanocarriers enhance patient compliance and reduce toxicity by increasing therapeutic value. Some phytoconstituents have poor water solubility that leads to decreased bioavailability, therapeutic potential, and stability. To overcome that problem, the development of nanocarriers like nanocapsules, nanospheres liposomes, phytosomes, nanoemulsion, polymeric micelles, colloidal nanogels, solid lipid nanoparticles (SLN) help to enhance solubility, pharmacological effect, and stability. Thus, nanocarriers support the development of novel carriers to overcome problem-related delivery of phytoconstituent (Handa et al. 2022).

In nanocarriers, polymeric micelles are mostly used to improve the water solubility of phytoconstituent with weak solubility by using various copolymers. Mostly amphiphilic block copolymers contain both hydrophilic and hydrophobic block that help to achieve a small size (less than 100 nm). Various parameters affect the size of polymeric micelles including length and molecular weight of copolymers, the ratio of drug and copolymer, surface charge, temperature, and synthesis approach. The small size of polymeric micelles easily crosses the systemic barrier without clogging blood vessels. Polymeric micelles can enhance blood circulation time, increase permeability and absorption, targeting capability with high drug accumulation It supports the delivery of different monoclonal antibodies, proteins, as well as peptides and prevents them from enzymatic breakdown or the outside environment. Polymeric micelles are an effective drug delivery strategy for improving stability and helping to establish a good pharmacokinetic profile of phytoconstituent (Kaur et al. 2021a).

9.2 Micelles

Micelles are colloidal systems formed by the self-assembly of amphiphilic nanocarriers containing hydrophilic heads and hydrophobic tails in aqueous solvents. The hydrophilic shell dissolves in the aqueous phase which helps to deliver a drug that is poorly soluble in water present inside the hydrophobic core. The size of micelles is less than 50 nm (Aguilar 2013). Majorly non-ionic type of surfactant is used for delivery of drug through oral or parental route with less adverse effects than ionic surfactant. Non-ionic surfactant produces clear as well as stable micellar solution and has better targeting ability. The aggregates of surfactant rise different shapes of micelles and they are classified as spherical shape micelles, rod-shaped micelles, hexagonal phases, and lamellar phases (refer Fig. 9.1). Only the lamellar phase has the property to form reverse and normal orientation. The surfactant concentration changes the shape of micelles. In oil-in-water, they show the normal lamellar micelles, while the water-in-oil phase shows reverse lamellar micelles. Solvents like hexane, cyclohexane, and benzene show reverse orientation of micelles formation. The dilution of surfactant solution may alter the shape of micelles like spherical micelles change into lamellar phase, these create dose dumping related problems in the body. In polymeric micelles, the polymer creates specific cross-linking that helps to stabilize the micelles. They form polymeric aggregates with micelles using a polymerization reaction. Reverse micelles show aggregation in non-polar media. These phenomena are used for the reparation of therapeutic aerosols and metered dose inhalers (MDI) (Lawrence 1994).

Most of the internal body preparation contains a non-ionic surfactant that polyoxymethylated non-ionic surfactant has greater importance in ophthalmic preparation. J. Jiao describes the role of non-ionic surfactants in the delivery of water-insoluble compounds through the micellization process. The water-insoluble drug is partitioned into the micelles forming surfactant molecules. The micelle's core polarity plays important role in the shape as well as the size of a micelle but they do not alter the surfactant arrangement. Generally, the core polarity is inversely proportional to micelle solubilization. Polyethoxylated non-ionic surfactants like Cremophor EL has smaller micelle core polarity of about 1.05 and a bigger one is Triton X-100 that has 1.40 (Jiao 2008).

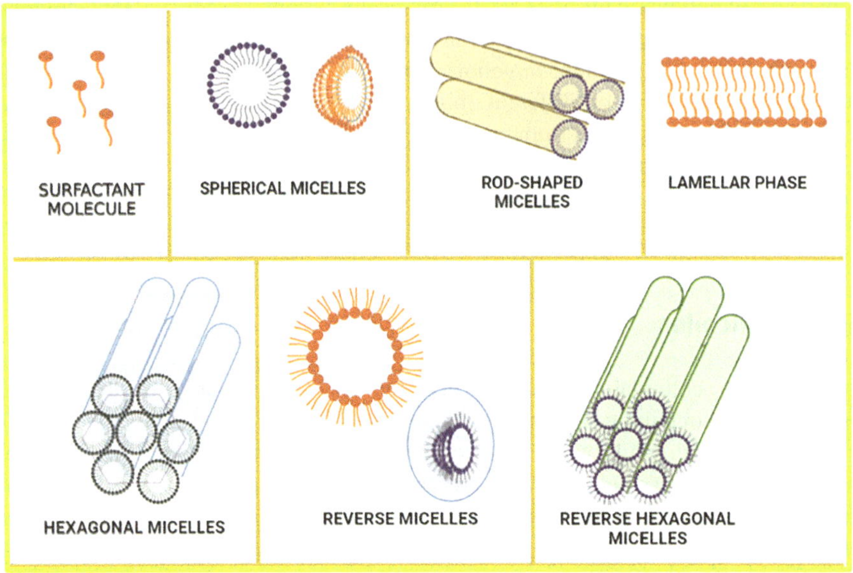

Fig. 9.1 The different shapes of micelles formed above the critical micellar concentration (CMC)

9.3 Polymeric Micelles

Polymeric micelles are a novel approach in nano delivery and contain core shells that have self-association formed by amphiphilic block copolymer inside water and the capacity to retain hydrophobic phytoconstituent within the core of micelles. In 1984, Bader et al. proposed different polymeric systems for the transport of a biologically active substance. They mentioned hydrophilic or hydrophobic cores are formed by micellar structure (Bader et al. 1984). Polymeric micelles have a size of about 10–100 nm. Depending on copolymer molecular weight, the chain length of the hydrophilic and hydrophobic copolymer affects the size. Polymeric micelles have a powerful tool for the delivery of less water-soluble drugs and also help in solubilization improvement. Other properties shown by polymeric micelles are attaining sustained release profile, protect the drugs from the activity of various enzymes, and help to accumulate the drug to the targeted location. Additional qualities include high biocompatibility, low toxicity, micellar association, core-shell, and high stability. Due to poor solubility problems, various phytoconstituent did not achieve the marketing potential and show toxicity problems by drug or excipient or both in the formulation. S. Croy et al. estimated that in 1999 total market sale of poorly water-soluble drugs was about $37 billion, even the formulation-related problems affect the market potential (Aliabadi and Lavasanifar 2006). There are various techniques to enhance the solubility of poorly or less water-soluble drugs using crystal modification, salt formation, and cocrystal formation are used, but using altering the pH technique decrease the chances of forming stable moiety. The counterions present in salt form also affect the dissolution rate of very poor water-soluble drugs. Reduction in particle size is another approach to enhance solubility. Also micronization of active ingredient done through using instruments like a jet mill or ball mill, nanocrystal formation by wet milling or high-pressure homogenization these methods mostly used. The major drawback of is affect the stability of heat liable phytoconstituent and API (Kawabata et al. 2011). Chemical entities obtained from a natural source or semisynthetic derivative show toxic action above a certain limit. Certain vehicles and cosolvents are used for the solubilization of water-insoluble drugs. In unionized drugs, cosolvents like ethanol, polyethylene glycol, propylene glycol, and dimethylacetamide show a toxicity profile by the increasing amount in the formulation. H. Gelderblom et al. studied the drawbacks as well as the advantage of Cremophor EL as a vehicle. Cremophor EL improves the solubility of water-insoluble as well as very less water-soluble drugs including various anticancer agents. Cremophor EL is a heterogeneous non-ionic surfactant that acts as a formulation carrier and solubility enhancer. An anticancer drug like the Paclitaxel derivative of Taxol has hardly water soluble and has a solubility of less than 0.03 mg/mL. It is slightly soluble in octanol, butanol, and propylene glycol and freely soluble in methanol, acetone, ethanol, chloroform, ether, and Cremophor EL. United States Pharmacopeia (USP) choose the 1:1 ratio mixture of Cremophor EL and dehydrated ethanol used to dissolve 30 mg paclitaxel in a 5 mL mixture. The research study found that Cremophor EL with paclitaxel shows severe anaphylactoid hypersensitivity reactions. Other symptoms of Cremophor EL like red rashes on body, skin flushing, dyspnoea, hypotension, and chest pain. One major side effect of Cremophor EL is neurotoxicity, peripheral neuropathy with axonal degeneration and demyelination. Therefore, a novel version of paclitaxel that is free of Cremophor EL is currently being developed. In that Cremophor EL is replaced by cosolvents like ethanol, Tween 80 and polymeric micelles help to increase bioavailability and decrease side effects (Gelderblom et al. 2001).

9.4 Polymers Used in the Preparation of Polymeric Micelles

Different polymers are mainly used in the preparation of polymeric micelles. The graft copolymers like chitosan on vinyl monomers and stearic acid, di-block copolymers like polyethylene glycol (PEG), and polystyrene and triblock polymer including polyethylene oxide and polypropylene oxide (refer Fig. 9.2). These polymers protect the drug from degradation and provide in vivo stability, enhanced efficiency, and biocompatibility. The block copolymer role in polymeric micelles is to enhance the solubility of hydrophobic molecules in a hydrophobic core stabilized by a hydrophilic aura. Block copolymers also play an important role in targeted delivery in both active and passive ways. They help to enhance circulation time and decrease the lysosomal activity in the presence of a hydrophilic shell. Polymeric micelle's nano range size makes it easier for them to pass through numerous barriers. Block copolymers are classified according to intermolecular interaction in the core

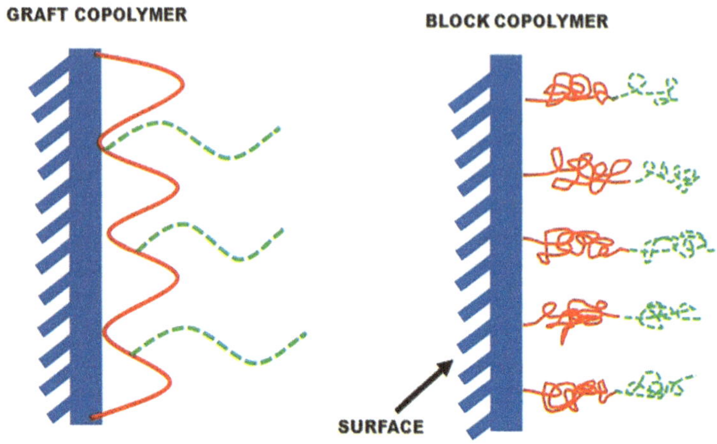

Fig. 9.2 Diagrammatic representation of graft and block copolymers attached on a surface

segment. They are classified as polyion complex micelles, amphiphilic micelles, and micelles stemming from metal complexation. In that polyion complex micelles formed by electrostatic interaction, while amphiphilic micelles resulting from hydrophobic interactions (Nishiyama et al. 2001). In block copolymer the shapes of polymeric micelles are spherical, and the chain length in the core region increases showing the direct effect on the shape of polymeric micelles. As the long core polymeric micelles become rod and lamellae shape. In polyion complex micelles are formed by segregation and neutralization of oppositely charged ions. The property of electrostatic interaction is due to the interaction between polycation and polyanion. Di-block copolymers mostly contain PEG's primary role in the hydrophilic segment. The polymerization reaction starts by using initiators like alpha-methoxy, hydroxyl group, and omega amino group. Here ethylene oxide is the functional initiator to achieve anionic polymerization and methoxy group as growing block polymer at the end terminal. If two or more than two di-block copolymers are coupled to produce multiblock copolymers. The coupling reaction produces a side product that should be avoided. The multifunctional core formed by coupling helps to initiate polymerization in the shell. This mechanism applied for the synthesis of poloxamers is a triblock copolymer that incudes polyethylene oxide (PEO), polypropylene oxide (PPO), and polyethylene oxide (PEO) (Gaucher et al. 2005; Singh et al. 2022).

Graft copolymers are large-sized macromolecules in which one or more blocks are attached to the main chain. Mostly monomolecular micelles formed from graft polymer contain multiple branches of macromolecules with covalent linkage. The bonds formed by side connecting the block to the main chain behave as a side chain showing constitutional or configurational features. The structure of graft polymer is comb-shaped. The preparation of graft copolymer is simple than block copolymer. The graft polymers prepared by long chain of one monomer is attached to the main chain backbone polymer through the process of polymerisation. The large-sized monomers of macromolecules best way to synthesize the graft copolymer. The graft polymerization helps to modify the physicochemical properties of cellulose. In cellulose, the surface creates certain grafts of synthetic polymers. It helps to impart specific properties on cellulose surfaces without altering intrinsic properties. The various grafting approaches are used for polymerization of cellulose, in that "grafting from" approach is mostly used. In the "grafting from" technique, the development of the polymer chain takes place at the cellulose's beginning point. The advantage of this method is the high graft density of polymer observed on cellulose derivative (Roy et al. 2009) (refer Table 9.1).

9.5 Mechanism of Micelle Formation

Two forces play important role in micelle formation: one is the attractive force used for the association of molecules and the other is repulsive for controlling the growth of micelles. Mostly amphiphilic copolymers play a selective role in hydrophilic as well as hydrophobic copolymers. In polymeric micelles, the formation process is the same as for micelles formed by surfactants. The micellization process starts with a low concentration of polymer, they form a single chain of the polymer. The critical micelle concentration (CMC) is concentration in which below that polymer exist at single molecule and above CMC they self assemble into micelles. The micelles are formed in a way that hydrophobic regions avoid the aqueous phase and tail-like structures are shown by hydrophilic regions in which the polymer is diluted (refer Fig. 9.3). Loose aggregates in micelles are formed due to the increased size of micelles at high concentrations (Jones and Leroux 1999). In comparison to phospholipids and surfactants, amphiphilic block copolymers often display greater stability and durability. It encapsulates the drug and enhances its stability and bioavailability. Critical micelle temperature (CMT) of the copolymer also archives micellar formation same way as CMC. The detailed mechanism of polymeric micelles from the growth of micelles to the encapsulation of drug or phytoconstituent by copolymer is recently unknown (Li et al. 2019).

9.6 Phytoconstituents

Ayurveda is one of the ancient and traditional medicine systems that originated over thousands of years. Ayurveda comes from a combination of two Sanskrit words *Ayur* means life and *Veda* means science. It describes the science of life. Ayurveda is also known as the "Mother of All Healing." It describes knowledge about herbal medicine and its use for curing various diseases. Ayurveda listed more than 700 plants containing various phytoconstituents used for treating various diseases. At present time advanced techniques are available that help to separate phytochemicals from different herbs. A very small amount of potent active phytochemicals are obtained from the bulky raw materials of herbals (Parasuraman et al. 2014).

The phytoconstituents are non-nutritive and contain chemical compounds obtained in the plant for their protective action. These are secondary metabolites

Table 9.1 Type and structures of amphiphilic copolymer

Types of copolymers	Characteristics	Structure	Example	Ref
Graft copolymers	High molecular weight, high degree of branching, complex structure, both 2D (planar) and 3D (cylindrical or spherical) surface plane, flexible, with tight binding, stable	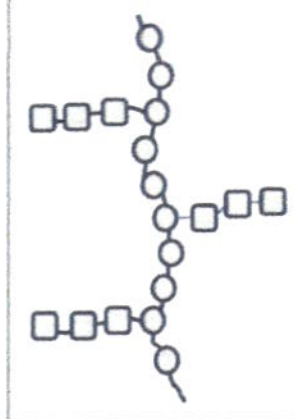	Polyethylene-g-poly(n-butyl acrylate)	Biswal and Singh (2004) and Feng et al. (2011)
			Poly(2-hydroxyethyl acrylate)-*g*-poly (ethylene oxide)	
			Carboxymethyl cellulose-g-polyacrylamide	
Block copolymer	Liner arrangement of block, mostly spherical or star shape block, covalent linkage, tensile strength and elastic properties. In- vitro stabilization	Di-block (AAABBB) 	Poly(p-dioxanone)-b-poly (epsiloncaprolactone)	Gaucher et al. (2005), Tamboli et al. (2013) and Hasegawa et al. (1996)
		Triblock (AABBCC) 	PCL-b-poly (2,4-dinitrophenylthioethyl ethylene phosphate)-b-PEG	
		Tetra-block (AABBCCAA) 	Poly(ethylene glycol)-b-poly (L-histidine)-b-poly(L-lactic acid)-b-poly(ethylene glycol)	
		Penta-block (ABCBA) 	(PAE-PCL-PEG-PCL-PAE)	
		Star-block (four arm) 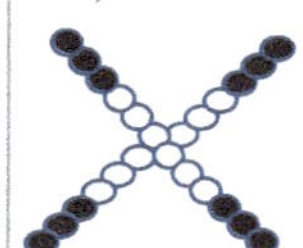	Poly(ethylene oxide)-block-poly (ε-caprolactone) [PCL-b-PEO]	

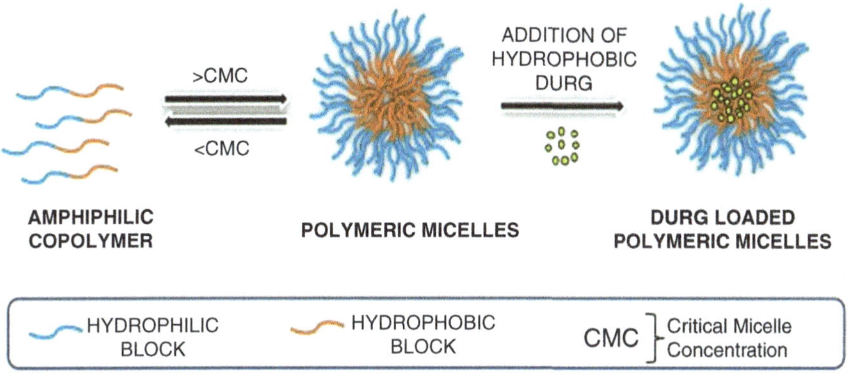

Fig. 9.3 Schematic representation shows formation of drug loaded polymeric micelles

of plants with active therapeutic properties. Phytochemistry directs the qualitative as well as quantitative activities of phytoconstituents and also checks the biological activities. The active drug constituent in plants plays important role in pharmacological activity. Inert nondrug constituent is preferred as an excipient such as bulking agent (Alamgir 2018). The phytoconstituents contain phenolics, alkaloids, cryogenics, saponins, glycosides, terpenes, tannins, anthraquinones, steroids, and various essential oils. In plants, they play a role in protection against various infection, pathogens, and microbes These phytoconstituent has potent inhibitor activity against enzyme that leads to desired pharmacological actions. The collecting of phytoconstituent from the herbal plant in high percentage yield requires special collection skills. Ayurveda describes the specific period from a selection of land to collection time, they affect yield percentage and potency of phytoconstituent. Chemo-profiling is an advanced way for standardization of several groups and classes in phytoconstituent obtained and fingerprinting used for evaluation purposes. Additionally, a variety of chromatographic methods, including gas chromatography (GC), high-performance liquid chromatography (HPLC), thin layer chromatography (TLC), and high-performance thin layer chromatography (HPTLC), are used to separate phytoconstituent from herbal for both lab and industrial scales (Akinyemi et al. 2018).

Different phytoconstituents from herbal plants are employed in nutraceuticals and the treatment of many ailments. Akinyemi et al. mentioned the global market for herbal medicine is about US $80 billion annually and grows daily. In some West African countries like Nigeria, Ghana, Mali, and Zambia about 60% of children show primary symptoms of malaria like fever treated at home using herbal medicines. WHO has declared that 11% of synthetic drugs obtained from plant origin mostly for treating diseases like cancer, bacterial, and viral about 60% of drugs in a clinical trial are from natural origin. A daily increment is observed in the usage of life-saving medications made from herbal plants and their phytoconstituents (Akinyemi et al. 2018).

9.7 Role of Nanocarriers for Delivery of Phytoconstituent

Herbal phytoconstituents are a safer alternative for achieving therapeutic delivery systems. The modification in plant-derived medicines is required to enhance stability and solubility. These approaches give rise to nanocarriers for the delivery of various phytoconstituents and behave as novel drug delivery systems (NDDS). The major limitation of phytoformulation is shown toxicity due to the low therapeutic index of phytoconstituents determined by performing toxicity analysis. Different nanocarriers help to enhance the therapeutic index and decrease cytotoxicity. Other factors as entrapment efficacy and drug loading are considered important parameters in nanodelivery systems. Therapeutic phytoconstituents are delivered via a variety of nanocarriers including liposomes, polymeric micelles, dendrimers, phytosomes, nanocrystals, dendrimers, nanoemulsion, cerasomes, and nanocrystals. Various nanoparticles like silica-based nanoparticles, solid lipid nanoparticles, and magnetic nanoparticles have an adjuvant role in the delivery of phytoconstituent (Ng et al. 2020).

V. D. Leo et al. prepared liposomes containing curcumin for targeted drug delivery to the colon. In that liposomes are coated with Eudragit S100 and act as a pH-responsive polymer and a characterization study is performed. Curcumin is a polyphenol derivative naturally obtained from roots and rhizomes of *Curcuma longa* and belongs to family Zingiberaceae. For centuries, there are potential advantages of curcumin like antibacterial, antiviral, antioxidant, anticancer, and anti-inflammatory. Curcumin has low solubility, poor bioavailability, high metabolism, and clearance as a phytoconstituent. These limitations are overcome by the preparation of curcumin liposomes. The small unilamellar vesicles of curcumin (size up to 100 nm) were prepared by solvent-free micelle-to-vesicle transition method (MVT) and then coated with Eudragit S100 polymer. The small unilamellar vesicles of curcumin (sized up to 100 nm) are prepared by the solvent-free micelle-to-vesicle transition method (MVT) and then coated with Eudragit S100 polymer. These liposomes help to enhance release rate, high encapsulation efficacy, better stability, and enhanced bioavailability (De Leo et al. 2018). Y.-M. Tsai et al. describe the kinetic study of curcumin and its ability to cross BBB using Curcumin-loaded PLGA nanoparticles for cancer treatment. The nanoparticles are prepared by emulsification solvent evaporation technique and further characterized by performing pharmacokinetic studies on different organs containing the brain, spleen, lung, liver, and kidney performed. A significant change in parameters half-life, average residence time, and area under the concentration curve (AUC) in PLGA nanoparticles of curcumin is higher than normal curcumin. This nanoformulation increases the accumulation of drugs at the target organ (Tsai et al. 2011).

Phytosomes are nanodelivery systems made up of phyto-phospholipid complex structurally similar to liposomes. The lipid complex contains a phospholipid head group and two long fatty chains work to encapsulate the polar part of the chain from the lipophilic surface. Phyto-phospholipid complex structurally works to enhance permeability and bioavailability. Most of the constituents derived from plants are polyphenols and biologically active with a high affinity toward the water. But some of them have poor permeability and even do not cross biological membranes like hesperidin as same in poorly water-soluble phytoconstituent like curcumin and rutin. The phytosomes most promising approach is to enhance the solubility of lipophilic polyphenol phytoconstituent and enhance the permeability of hydrophilic ones. Instead of that phytosomes protect the active constituent from degradation and other chemical factors like oxidation, hydrolysis, and photolysis (Lu et al. 2019).

Volatile oils are the secondary metabolite of plants derived from various oil glands and secretory cells. They have a regulatory role and complex mechanism to treat disease. About 60 families contain volatile oils including Zingiberaceae, Alliaceae, Umbelliferae, Lauraceae, Dipterocarpaceae, Compositae, etc. (Zhang et al. 2021). R.K. Harwansh et al. reported the formulation for delivery of active phytoconstituent in the form of nanoemulsion. Nanoemulsion droplets range in size from 20 to 200 nm. It may vary depending on the method of preparation and homogenization method used. The smaller is the size of globules in emulsion, leads to enhance surface area and it help to improve the solubility profile of drug. To create a clear and transparent formulation, a precise ratio of oil, water, and a combination of surfactant and co-surfactant is needed (Harwansh et al. 2019). The preparation of Quercetin containing nanoemulsion-based gel and its characterization is reported by J.P. Gokhale et al. for the management of rheumatoid arthritis (RA) using quercetin as a disease-modifying anti-rheumatic drug (DMARDs). Quercetin shows anti-inflammatory activity by inhabiting nitric oxide, IL-6. With high plasma protein binding, low permeability, and solubility their availability is low in the body. For enhancement in permeability for topical application prefer to choose a suitable excipient and optimized batch. Various types of oils are selected to examine the solubility of quercetin and it is found that the solubility of quercetin in arachis oil is 11.14 mg/ml, while in oleic acid is found to be 9.10 mg/ml. Both Tween 20 (surfactant) and PEG-400 (co-surfactant) added concentration were kept at 6% concentration. The accurate ratio of quercetin, oil, surfactant and co-surfactant is mixed in screw capped bottle and further vortex at optimum speed to form S_{mix}. The formed S_{mix} is added dropwise in magnetically stirred aqueous phase to obtain clear and transparent nanoemulsion (Gokhale et al. 2019).

9.8 Application of Polymeric Micelles for Delivery of Phytoconstituent

Polymeric micelles promise strategy for the delivery of various phytoconstituent with a problem related to stability and solubility. The block copolymer and graft copolymers play a key role in the stability aspect of polymeric micelles.

9.8.1 Anticancer Agent

Cancer is a leading disease that causes abnormal growth of organs or tissue. The major root cause of cancer is metastasizing, the spreading of cancer from one organ to another. According to World Health Organization (WHO) survey in 2018 the second leading cause of death is cancer. Approximately 9.6 million deaths occur annually and the death rate is about 1:6. There are various types of cancer like lung cancer, skin cancer, breast cancer, colorectal cancer, and blood cancer especially leukemia and non-Hodgkin lymphoma. In women mostly breast cancer, cervical and thyroid cancer are commonly occurring cancer (WHO n.d.). The multiple mutation in gene that contribute to cancer cancer development, such as germline mutation in women leads breast cancer. Various herbal plants preparation and their phytoconstituents are used for the treatment of anticancer agents (refer Table 9.2).

In Ayurveda, Vedic literature of India, it is mentioned that plant-derived products are nontoxic to human health and produces less tolerance. Different phytoconstituent have different mechanisms to treat cancer some of them kill rapidly dividing cells, decrease oxidative stress, stop gene mutation as well as the alteration in the gene, inhibit cell proliferation, and induce apoptosis. Polyphenols like resveratrol and gallocatechin, flavonoids like alpinumisoflavone, and methoxy licoflavanone show an apoptosis effect. The phytoconstituents including curcumin, rutin, allicin, epigallocatechin gallate, β-carotene, thymol, quercetin, rosmarinic acid, and coumarin treat cancer by antioxidant mechanism (Singh et al. 2016).

C. Gong et al. developed curcumin polymeric micelles for antiangiogenesis and anticancer purposes. Curcumin polymeric micelles are prepared by solid dispersion method using monomethyl poly(ethylene glycol)-poly(ε-caprolactone) (MPEG-PCL) as a copolymer. MPEG-PCL is a di-block copolymer with an atactic structure formed by ring-opening polymerization between poly(ethylene glycol) methyl ether (MPEG) and ε-caprolactone and different ratios of the copolymer used for the preparation of polymeric micelles. Using the MTT test, curcumin micelles and free curcumin were studied for cytotoxicity and apoptosis. The percentage cell viability of curcumin micelles is less than free curcumin and the apoptosis rate of curcumin micelles is higher than free curcumin showing that polymeric micelles of curcumin improved cytotoxic activity. In vitro antiangiogenic activity was studied by the MTT method. Here inhibitory activity of Cur micelles inside endothelial cells is greater than free curcumin. The conclusion found that curcumin micelles have a more suppressive effect on tumor growth and also enhance retention time and plasma, survival rate, and antiangiogenesis effect (Gong et al. 2013). L. Liu et al. also develop curcumin-loaded polymeric micelles using MPEG-PCL as a biodegradable copolymer for inhabiting breast tumor. Surgery is another option for breast tumor but the chances of recurrence and metastasis are high. To avoid recurrence and metastasis chemotherapy is a better option. Chemotherapy shows side effects like myelosuppression and immunosuppression. The nanodelivery of curcumin-loaded polymeric micelles promises delivery with high drug accumulation at the target site. The neutral surface charge of curcumin micelles provides stability to the formulation. Poly(ethylene glycol) forms the core and poly(ε-caprolactone) forms the shell. These copolymers cause less aggregation and have a limited affinity toward micelles. A cytotoxicity study was performed on 4T1 cells placed on 96-well plates and then incubated for 24 h. Curcumin polymeric micelles have high cell uptake, with high cytotoxicity observed during the experiment. Apoptosis was performed by TUNEL

Table 9.2 Phytoconstituent shows anticancer effect delivered through polymeric micelles

Drug	Copolymer	Type of tumor	Size (nm)	Key observation and outcome	Cell line/animal model	Ref
Curcumin (CUR) (polyphenol)	MPEG-PCL	Colorectal carcinoma	27.1	CUR micelles further converted into curcumin thermosensitive hydrogel with slow-release rate with high cytotoxicity. Also increases half-life and plasma concentration of CUR	CT26 cell line/ BALB mice	Zhang et al. (2015)
Curcumin (CUR) (polyphenol)	MPEG-PCL	Breast cancer and pulmonary metastasis	28.1	CUR containing biodegradable polymeric micelles with sustained release, enhanced survival rate, inhabits breast cancer and pulmonary metastasis	L929 and 4T1 cell culture/BALB mice	Liu et al. (2013)
Icariside II (flavonoid)	mPEG-b-PLA-Phe (Boc)	Liver cancer	20	Sustained release over 10 days with DL is 10% and entrapment efficiency is 98%. Potent anticancer efficacy observed in cell line study and biopharmaceutical parameter	SMMC-7721 cell line/nude mice model	Sun et al. (2022)
Quercetin (polyphenol)	Pluronics (P123 and P407)	Breast and ovarian cancer	24.83	Cell cycle arrest at G2/M phase, microtubule depolymerization, increase antioxidant activity and in vitro cytotoxicity. Also decrease MDR	SKOV-3 (ovarian) and MDA-MB-231 (breast)	Patra et al. (2018)
Okra (flavonoid)	Poloxamer 40/7	Lung, breast, liver, colorectal, and cervical cancer	190 nm	Inhibition of VEGF, chemoprotective effect, high intracellular penetration in cells with inhibition of P-gp efflux	MCF-7 (breast) HepG2 (liver), and HeLa (cervical)	Chaemsawang et al. (2019)
Naringin (flavonoid)	Pluronic F68	Colon cancer	74.80	Inhibition of proliferation of cancer cell and cytoprotective activity for treatment of ulcer and decreased cell viability without damaging microflora	HepG2, MCF-7 and Caco-2/MSD rats	Mohamed et al. (2018)

(continued)

Table 9.2 (continued)

Drug	Copolymer	Type of tumor	Size (nm)	Key observation and outcome	Cell line/animal model	Ref
Chalcone (flavonoid)	mPEG-PLA	Breast cancer	20.8	Release rate increase in acidic condition, selective inhibition breast cancer cell growth	MDA-MB-231 and MCF-7 cells	Kim et al. (2019)
Apigenin (flavonoid)	Pluronic F127, pluronic F68, and pluronic P123	Breast cancer and liver cancer	16.9	Inhibits free radicles production and also invasive mechanism. Up to 84% drug release in 36 h, high accumulation at target site due to high EPR	MCF-7 and HepG2 cells	Zhai et al. (2013)
Alfa-hederin (triterpenoid)	PCL-b-P (OEGMA-co-RGD)	Colorectal cancer	108	High apoptosis rate, enhanced activity by decreasing protein absorption. High drug accumulation and arrest cell cycle in S phase	HCT8, HCT116 and LoVo cells	Sun et al. (2016)
Quercetin and doxorubicin	PMECL-b-PBnCL	Liver cancer	29.2	Synergistic action, chemo sensitizing and cardioprotective effects, thermoresponsive behavior of polymer and protect to normal cells	H9c2 (rat heart) and HepG2 cells	Soltantabar et al. (2020) and Handa et al. (2021)

staining assay on 4T1 cells using immunofluorescence. It is observed that curcumin polymeric micelles have a more apoptosis induction effect than free curcumin (Liu et al. 2013).

Berberine is a quaternary derivative of isoquinoline alkaloid obtained from *Berberis vulgar* (European barberry) and belongs to the Berberidaceae family. Various advanced techniques are used for the extraction of berberine showing potential anticancer properties. Berberine has poor water solubility with low bioavailability, tissue uptake, and distribution (Khan et al. 2022). Shen et al. reported the preparation of berberine-containing polymeric phospholipid micelles as a promising strategy to enhance the delivery of berberine to target cancer. The solubility of berberine is enhanced by mixing phospho-ethanolamine-N-[methoxy (poly-ethyleneglycol)-2000] (PEG-PE) with d-α-tocopheryl polyethylene glycol 1000 succinate (TPGS) in ratio 3:1, helps to solubilize berberine in the hydrophobic core of micelles. PEG–PE conjugates formed by polyethylene glycol (PEG) and diacyl-lipids help to achieve a size range between 10 and 100 nm. TPGS is a pegylated derivative of vitamin E that provide increased solubility and absorption rate of berberine. TPGS also inhabits Pgp efflux pumps and helps decrease multidrug-resistant to various chemotherapeutic drugs including paclitaxel, vinblastine, doxorubicin, and gemcitabine. These increase up to 5 times improvement in oral bioavailability of berberine. Uptake study results show berberine lipopolymeric micelles have 18 times high in vitro cellular uptake than free berberine and a cytotoxicity study on cancer cell spheroid models like PC3 shows that free berberine kills only 20% of cancer cells, while berberine lipopolymeric micelles kill>60% of PC3 cells in 48 h (Shen et al. 2016).

An effective anticancer activity is shown by Paclitaxel (Taxol®) obtained from the bark of the *Taxus Brevifolia* (Pacific yew tree) belonging to family Taxaceae. The paclitaxel has tetracyclic diterpenoid that is sparingly soluble (0.3 mg/L at 37 °C) in water. To enhance the solubility of paclitaxel use Cremophor EL and dehydrated ethanol. They are commercially available in the market under the brand name Taxol®. The use of Cremophor EL shows hypersensitivity reactions and side effects in patients. Also, the organic medium used for the preparation of formulation leached a carcinogenic compound from the infusion bag used in the hospital. W.Y. Seow et al. reported for targeted delivery of paclitaxel with the help of multifunctional polymeric micelles contain poly (N-isopropylacrylamide-co-N,N-dimethylacrylamide-co-undecenoic acid). These help encapsulation of paclitaxel in core conjugation of cholesterol to carboxylic group of polymer and attachment of folate to amine group within the hydrophilic chain segment as a targeting moiety as well as nonimmunogenic property. This formulation is temperature and pH sensitive, mostly in cancer cells that produce an acidic environment that helps the rapid release of a drug. Dropping the lower critical solution temperature (LCST) of micelles below 37 °C destruction of the core and drug release occurs. In the release study at pH 7.4 only 8% of total drug release occurs, while at pH 5.0 about 21% of its total drug content occurs in the first 4 h and 44% by the first 24 h. The paclitaxel encapsulated polymeric micelles coated with folate start precipitation and deformation at a temperature below 37 °C and maximum release occurs. In a condition like

LCST of micelles above 37 °C at pH 7.4 drug is encapsulated in polymer and does not show any release property. The cell viability of paclitaxel PM conjugated with folate increases with an increase in free folate concentration, these suggest that free folate shows competitive binding to paclitaxel PM conjugated with folate. For increasing in vitro therapeutic efficiency of paclitaxel PM conjugated with folate avoid a folate-contain diet (Seow et al. 2007). For enhancing the chemotherapeutic efficacy and producing synergistic action the combination of two drugs is used in nanoformulation. Paclitaxel and cisplatin are used by X Wan et al. to treat ovarian and breast cancer. In that cisplatin are delivered together in poly(2-oxazoline) polymeric micelles. Paclitaxel work as anti-microtubule agent, they bind to the microtubules and prevent them from disassembly. Also At high concentration of paclitaxel arrest cell growth at G2/M phase. Cisplatin is the most powerful member of the platinum family. They form an adduct with DNA and inhibit protein synthesis. They show dose-limiting toxicity and rapid development of MDR. Both cisplatin and paclitaxel have low solubility in water. The poly(2-oxazoline) copolymer containing micelles enhances the solubility of the drug up to 100,000 times with 4–100 times high drug loading of paclitaxel micelle and slow down release rate (Wan et al. 2019). B. Cote et al. describe the use of resveratrol and quercetin polymeric micelles to decrease cardiotoxicity induced by doxorubicin. Natural phytoconstituent resveratrol (polyphenols) and quercetin (flavonoid) were obtained from grapes. They prevent the heart from myocardial damage and show cardioprotective effects. The mechanism behind cardioprotective effects is free radical scavenging and antioxidant effect. Resveratrol behaves as a chemotherapeutics and chemosensitizer. The anticancer drug quercetin exhibits cell cycle arrest at G0/G1 phase with significant antioxidant action through reactive oxygen species (ROS) generation. Both resveratrol and quercetin have a low level of oral bioavailability and aqueous solubility. Most of the antibiotics belong to the anthracycline class effective for cancer treatment. That doxorubicin (DOX) hydrochloride (Adriamycin for Injection, USP) is used to treat various cancer like breast, ovarian, Wilms' tumor, and neuroblastoma. The side effect of DOX formulation shows cardiotoxic side effects at a dose greater than 15 mg/kg, with an unknown mechanism. Mostly adriamycin plays an important role in the generation of free radicles and lipid peroxidation, these are the main reason behind the generation of cardiotoxicity. Triblock copolymer is formed by one polypropylene oxide (PPO) chain and two polyethylene oxides (PEO) known as Pluronics®. The resveratrol and quercetin (1:1) were solubilized using Pluronics® by solvent casting method formation. Polymeric micelles co-administered with DOX by enhancing anticancer effect as well as cardioprotective action (Cote et al. 2015).

9.8.2 Antibacterial Agent

Antibacterial agents are compounds that either kill bacteria or halt their development. The main drawback of antibacterial drugs is they show resistance produced due to consumption of excessive antibiotics beyond the limit. They produce aggressive strains that do not respond to rational treatment. According to a WHO report, antibiotic resistance results in 700,000 deaths worldwide every year. The antibiotic resistance crisis attribute huge loss of about $55–70 billion per annum in the USA and €1.5 billion annually in the UK. Other reasons like low water solubility and oral bioavailability, stability problems, toxicity, low patient compliance create a major challenge in antibiotic therapy. The polymeric micelles play a key role as a nanocarrier to overcome bacterial resistance. Curcumin-loaded polymeric micelles have a size range between 90 and 95 nm and contain silver-coated amphiphilic di-block copolymers made up of poly (ε-caprolactone) and poly (aspartic acid). Biofilm-associated bacteria contain *Pseudomonas aeruginosa* and *Staphylococcus aureus* show synergistic activity against both gram-positive and gram-negative bacteria. The curcumin-loaded polymeric micelles show high biocompatibility with RBC and decrease the rate of drug release in the absence of lipase enzyme. When incubated with P. lipase, the release rate increases; approximately 95% of the release occurs within 48 h (Eleraky et al. 2020). The biodegradable antimicrobial polycarbonates are important in polymeric micelles for the delivery of the antibacterial agent. The triblock polymer of poly(ethylene oxide) as PEO, poly (ε-caprolactone) as PCL, poly[(2-tert-butylaminoethyl) methacrylate] as PTA self-assembled by ring-opening polymerization reaction give rise to (PEO-b-PCL-b-PTA). In that PEO block gives rise to biocompatibility and stability. PTA block shows antibacterial activity by bacterial lysis with divalent cation exchange mechanism. PCL block works as a biodegradable hydrophobic core in an aqueous solution (Ding et al. 2019).

F. Huang et al. designed the polymeric micelles containing silver nanoparticles with encapsulation of curcumin to produce synergistic antibacterial action. The rapid development of MDR is a global issue. In such a way antibacterial produces resistance within a very short period. Silver is the best antibacterial agent with less resistance and low toxicity. They enter bacterial cytoplasm and denature the protein by interfering with DNA synthesis. Silver is used for combination treatment with other antibiotics including penicillin, kanamycin, and vancomycin helps significantly decrease the MDR. The polymeric micelles contain block copolymer to prevent aggregation of silver nanoparticles and enhance their stability. Curcumin is used for its antibacterial, antifungal, antioxidant, anticancer, and other multiple activities. Curcumin produces a synergistic effect with antibiotics. Curcumin shows poor bioavailability and less water solubility, these can be overcome by preparing polymeric micelles. The composition of curcumin and silver nanoparticles in polymeric micelles prepared by using di-block copolymer formed by poly(-ε-caprolactone) as PCL and poly(aspartic acid) as Asp comes together to form (PCL-b-Asp) di-block copolymer biodegradable in nature. In polymeric micelles silver nanoparticles decorated on the shell and curcumin encapsulate in a hydrophobic core. The PCL-b-Asp di-block copolymer shows electrostatic interaction by absorbing Ag + ions on poly(aspartic acid). Antibacterial activity is determined by performing an antibacterial test on *P. aeruginosa* (Gram-negative bacteria) and *S. aureus* (Gram-positive bacteria), by developing a colony on an agar plate and then transferring it into culture media incubate overnight at 37 °C. The result obtained is evaluated by measuring optical density. The test result shows that silver-curcumin contains polymeric micelles that have strong antibacterial activity on both *P. aeruginosa and S. aureus* than only silver-decorated polymeric micelles. The release of curcumin micelles increased in the presence of lipase about 95% release over 48 h. Hemolysis study of silver-curcumin contains polymeric micelles good biocompatibility with RBCs (Huang et al. 2017).

R. Bakr et al. developed a polymeric micelles formulation containing mentha showing anti-Helicobacter activity. Mentha species contain various phytoconstituents and various therapeutic applications. Different species of mentha like *M. suaveolens, M. sylvestris, M. piperita, M. longifolia, and M. viridis* belong to family Lamiaceae (Labiatae). They contain various types of volatile oil, flavonoids, tannins, and phenolic oil with various therapeutic effects including antiviral, antibacterial, antioxidant, and antiobesity effects. To study functional genomics, it is necessary to understand metabolite levels using plant metabolomics. Various mentha species are used for treating peptic ulcers caused by *Helicobacter pylori* bacteria. These are flagella-shaped, gram-negative bacteria also cause gastric cancer. Most antibiotics develop MDR against rational therapy. Antibiotics used to treat ulcers cause resistance; this issue is solved by using natural phytochemicals. The preparation of the nano formulation is needed due to the limited solubility and bioavailability of mentha. The polymeric micelles were prepared by using Pluronic® F127 as a polymer. The Pluronic® F127 has a biodegradable and biocompatible character with enhanced drug encapsulation and increases the therapeutic potential of mentha species. That disc diffusion method helps to determine the activity of different Mentha species against the *H. pylori* strain. Metabolomics analysis was performed using a mass spectrometer. The minimum inhibitory concentration (MIC) and minimum bactericidal concentrations (MBC) assist in selecting mentha species with anti-Helicobacter properties. The result obtained from cytotoxic activities shows *M. viridis* expose excellent anti-Helicobacter activity with a maximum zone of inhibition (14 ± 1.5 mm). After that *M. piperita* shows good anti-Helicobacter activity delivered with polymeric micelles (Bakr et al. 2021).

G. D. Kumar et al. used pelargonic acid (PA) obtained from tomatoes as an antimicrobial agent. It commonly acts as an antifungal agent approved by Generally Recognized as Safe (GRAS) status, other use like in food additives and sanitizer purposes. Quillaja contains high amount of saponin obtained from the bark of *Quillaja saponaria* used as micelle formulation for antimicrobial activity. A variety of fatty acids including oleic acid, capric acid, and palmitoleic acid derived from plants show good antibacterial activity. The low solubility of fatty acid in oil increased by adding a surfactant to overcome the solubility problem. PA micelles were prepared using surfactants including Tween 80, Triton X100, and Sodium Dodecyl Sulfate (SDS) (Dev Kumar et al. 2020).

9.8.3 Antidiabetic Treatment

Diabetes mellitus (DM) is the third leading disease after a heart attack and cancer. DM is caused due to insufficient insulin secretion or insulin resistance or both leading to an increased concentration of glucose in the blood (hyperglycemia). The survey found that about 69.1 million cases of diabetes in India and 415 million cases of diabetes in the world increase every day. It will forecast that there would be 140 million cases of diabetes in 2040. DM is classified into two types in that Type 1 diabetes (insulin-dependent) causes a lack of insulin production due to the destruction of β islets cells present in the pancreas. Type 2 diabetes (non-insulin-dependent) is due to ineffective use of insulin and the reason behind is a sedentary lifestyle and carbohydrate metabolism problems. In India, about 8.7% diabetic population is between the age 20–70. There are additional methods to maintain the sugar level in the body, including a balanced diet, exercise, yoga, and weight management, and these help minimize the use of oral hypoglycemic tablets and insulin injections. Herbal medicines are the drug of choice for a diabetic patient showing excellent therapeutic action. About 300 phytoconstituents extracted from 100 herbs establish antidiabetic action with different mechanistic pathways. The main aim is to develop pharmacologically active phytoconstituent for diabetes patients with novel delivery in a cost-effective manner (Bharti et al. 2018). Y. Fu et al. report the use of PEGylated micelles in antidiabetic treatments to enhance targeting delivery and cellular uptake. The amphiphilic block copolymer is self-assembled with enhanced solubility of poorly water-soluble phytoconstituent. Cationic polymeric micelles efficiently deliver insulin-producing electrostatic interaction between insulin and cations on the PEI-PEG layer with excellent glucose regulation. Glucose-responsive polymeric micelles are novel therapy to deliver the antidiabetic drug that decreases the level of insulin under 100 mg/dL with rapid release of insulin from polymeric nanovesicles (Fu et al. 2021). M.U. Akbar et al. described the use of curcumin polymeric micelles for antidiabetic and wound healing effects using chitosan, alginate, maltodextrin, and pluronic. In the daily use of plastic containers for food and beverage, synthetic plastic bags contain bisphenol A (BPA) to enhance the durability of that material. Various environmental conditions and changes in pH cause the release of BPA. According to a literature survey if daily intake of BSA is greater than 0.05 mg/kg body weight it causes harmful effect on insulin secretion and pancreatic glands. Curcumin-loaded mixed polymeric micelles are used to reduce the toxicity of BPA and show antidiabetic and wound healing activity. Mixed polymeric micelles containing curcumin prepared by thin-film hydration method give thin layer of curcumin-pluronic copolymer matrix. Add different polymer in appropriate amount to obtain formulation. Curcumin-loaded polymeric micelles show decrease in glucose level with increase in red blood cells and excellent wound healing response due to antioxidant, antibacterial, and anti-inflammatory properties of curcumin (Akbar et al. 2018a). The use of poloxamers or pluronic are amphiphilic, non-ionic surfactants that show the phenomenon of reverse thermal gelation, in that conversion from liquid to gel with a temperature change. Pluronics is FDA approved excipient mentioned in the US and British Pharmacopoeias. Two grades of pluronic, PF127 and P123 made a triblock structure. Micelles containing pluronic solutions show gelation properties determined by altering temperature with the movement of the magnetic bar. The rpm speed of the magnetic bar is inversely proportional to the gelation process. The antidiabetic activity of curcumin polymeric micelles is due to inhabiting α-amylase activity (Akbar et al. 2018b).

P. Mukhopadhyay et al. describe the use of quercetin as a hypoglycemic agent. The multiple medicines prescribed to older patients may produce complication in anti-diabetic treatment and high chances to display adverse drug reaction. Most secondary metabolites and bioactive drugs isolated from plants are used for antidiabetic purposes. These help to reduce hyperglycemic episodes and improve the life of the diabetic patient. Quercetin is a flavonoid, easily found in green vegetables and fruit. They have been used for multiple biological activities antidiabetic, anti-ulcer effects, and anticancer activity. The problem related to low bioavailability and permeability, short half-life, and first-pass metabolism is overcome by using polymeric nanocarriers like nanoparticles, polymeric micelles, and hydrogels. Quercetin maintains the glucose level by enhancing glucose uptake. It stimulates the glucose transporter 4 (GLUT 4) receptors as well as insulin receptors present on adipose tissue and skeletal muscles of the body. Another mechanism is a decrease in hyperglycemic effects due to enhanced glucokinase activity in the liver leads to high storage of glucose. Various study shows that quercetin helps to regenerate beta cells of islets of Langerhans by stimulating the ductal stem cells. In the small intestine glucose transporter 2 (GLUT 2) receptors present for enhancing glucose absorption. Quercetin decrease GLUT 2 expression, that leads to decrease the glucose absorption in gastrointestinal tract. The function of polymeric micelles for the delivery of quercetin is to protect from endolysosomal damage and enhanced oral drug delivery. Various polymers encapsulate quercetin and easily transport it to the colon by avoiding enzymatic degradation (Mukhopadhyay and Prajapati 2015).

9.8.4 Antifungal Treatment

The various phytochemicals and standaedized plant extract work against opportunistic fungal pathogens. These phytoconstituents have a potential activity to treat fungal activity caused due to *Candida, Aspergillus*, and *Cryptococcus* spp. These mostly affect skin and other body parts like skin, scalp, vagina, mouth, armpits, feet, and fingers. The warm and wet parts of the body include the part between two fingers, between the thighs, and wet armpits due to sweat. Fungal infections may be categorized according to the site of infection, route of acquisition and type of virulence, most common fungal diseases include dermatophytosis, Ringworm (*Tinea corporis*), Candidiasis or yeast infection, Jock itch *(Tinea cruris), Tinea pedis, Tinea faciei, Tinea barbae* belongs Trichophyton species. In the current approach for the treatment of both systemic and topical fungal diseases, various antifungal formulations are available in the market including azole drugs, polyene, allylamines, echinocandins, and morpholine. Most of them show side effects like erythema, burning sensation, redness of the skin, stinging, and some allergic reactions. Toxicity and development of resistance and uneven distribution of plasma profile is a limitation of an antifungal agent which is overcome by nanocarriers. Polymeric delivery systems mostly use biodegradable polymers for targeted activity. The chitosan-containing nanoparticles help to reduce toxicity profile of the antifungal drug and excellent antifungal activity against *Fusarium solani, A. niger, and C. albicans (*Priya et al. 2017*)*.

N. Kaur et al. described the use of various plant extracts as an antifungal. The antimycotic drug is also known as antifungal drug. The natural plants include *Bucida buceras, Vangueria infausta,* and *Olinia ventosa*. Various essential oil including clove, peppermint, cinnamon, citronella, camphor behaves as antifungal agent. The synthesis process affects the activity of herbal antifungal formulation due to high temperature. The epimerization process disrupt the structural integrity of phytoconstituent, that create significant problem in herbal extract. The stability of phytoconstituent is enhanced using a polymeric delivery system. Polymer shows crosslinking phenomenon, these help phytoconstituent for an effective and controlled delivery system. The copolymer binds with crosslinkers and gives better stability (Kaur et al. 2021b). T. Suwan et al. described the use of the herbal plant *Psidium guajava* aqueous extract (PE) delivered through polymeric micelles with silver nanoparticles shows synergistic antifungal activity. The plant extract obtained from leaves of *Psidium guajava* belongs to family Myrtaceous. The common *Psidium guajava* is commonly known as guava, mainly found in Asia, European countries, South America, and Africa. The main chemical constituent present in *P. guajava* is gallic acid, kaempferol, catechin, naringenin, rutin, and epicatechin. The extract shows excellent biological therapeutic effects including antimicrobial activity, hepatoprotective, anti-hyperlipidemic effect, anti-inflammatory activity as well as antidiabetic effect. *P. guajava* acts as a reducing agent and helps to synthesize silver nanoparticles. These synthesis processes increase the size of silver nanoparticles due to aggregation. Polymeric micelles help to obtain small-sized silver nanoparticles. In literature Poloxamer 407 (F127) triblock copolymer used to stabilize silver nanoparticles with *Psidium guajava* extract work as a reducing agent. A Kirby-Bauer method used to inspect antifungal activity of silver polymeric micelles formulation against *C. albicans*. (Suwan et al. 2019).

9.8.5 Antimalarial Treatment

Malaria is an acute febrile parasitic disease caused by a single-celled parasite belonging to the genus Plasmodium and transmitted through female Anopheles mosquitoes. Majorly four *Plasmodium* species spread malaria including *Plasmodium malariae, Plasmodium falciparum, Plasmodium vivax,* and *Plasmodium ovale*. In that *P. falciparum* is the most dangerous species found in Africa, America, and Asia. According to a WHO global survey report, there will be 241 million cases of malaria and 627,000 deaths due to malaria in 2020. The data obtained from the WHO global survey reported that in 2020 there would be an additional 69,000 deaths and 14 million cases as compared to 2019 (World Malaria Report 2021). Various phytoconstituents show positive results against *Plasmodium* parasites. Since ancient times, medicinal plants have been used as effective antimalarial drugs. The first antimalarial drug is quinine extracted from the bark of *Cinchona calisaya* belongs to the family *Rubiaceae*, having potent antimalerial action. In 1632 first infusion of quinine has been given for the treatment of human malaria and first case of resistance is observed in 1910. In 1972 Chinese scientists discovered new phytochemical called artemisinin isolated from *Artemisia annua* (sweet wormwood) belonging to the family Asteraceae. Artemisinin act as a novel natural product and it is used with a combination of another malaria medicines to enhance half-life and efficacy. Artemisinin-based combination therapy (ACT) is a promising statergy that is currently employed to overcome drug resistance (Erhirhie et al. 2021).

B. Zhai et al. described the use of *Bixa orellana* as antimalarial. The plant extracts including ishwarane, 5-tocotrienol, sitosterol, and stigmasterol show antimalarial activity. In that stigmasterol is potent antimalarial compound and shows modest activity against3D7 and K1 malerial strains (Zhai et al. 2014). D. Bhadra et al. use copolymeric dendritic micelles for delivery of artemether. Copolymeric micelles contain PEG conjugation at outer shell for avoiding reticuloendothelial system (RES) barrier and core formed using block copolymers including poly(ortho ethers) (POE), PLGA (polylactide co-glycolide), poly(butylene terephthalate) (PBT), poly-l-lysine (dendrimers), and PCL as (poly-caprolactone). The negative side of artemisinin is poor bioavailability and shorter half-life up to 5 h. Here the use of amphiphilic peptide-based AB-dendritic copolymeric core is used for enhancing the solubilization of artemether and increases half-life. The challenge like resistance produced by *P. falciparum* strains and chloroquine-sensitivity is overcome by using a derivative obtained from artemisinin (Bhadra et al. 2005).

9.9 Future Prospective

Globally the demand of natural product is increased per day, but poor solubility and bioavailability limit their application. Phytochemical-based nanoformulation is promising approach to overcome solubility and bioavailability parameter. The formulation strategy is based on the treatment of various chronic diseases with targeted delivery. The poor solubility of phytoconstituents creates a challenge regarding the delivery of drugs to the targeted site. The role of nanocarriers is to enhance solubility, bioavailability and stability in phytoconstituent. The multifunctional property of a single phytoconstituent including antioxidant, antibacterial, antiviral, and much more helps to cure multiple diseases at the same time by a single formulation. The nanocarriers containing phytoconstituent will change the approach toward the modern delivery of phytoconstituent (Handa et al. 2020, 2022).

In the future, the availability of modern extraction techniques increases the number of phytoconstituent with potent therapeutic action. The challenge is the development of a nanodelivery system. As compared to other nano-drug delivery systems, polymeric micelles are an effective delivery system with good stability. Polymeric micelles are excellent drug carriers for the delivery of poorly water-soluble phytoconstituent. The hydrophobic core of amphiphilic block copolymer micelles loaded with poorly water soluble phytoconstituents that enhance solubility, bioavailability and therapeutic efficacy of drug. Polymeric micelles possess good stability in blood. The micelles loaded with poorly water soluble anticancer drugs The targeting ability of polymeric micelles increased by attaching various antibodies, carrier ligands and protein on the surface. Also, stimuli-sensitive including pH and temperature sensitive polymeric micelles enhance targeting ability with less damage to other cells. In future major challenge for polymeric micelles is to enhance the drug efficacy (Kulthe et al. 2012; Varma et al. 2020).

9.10 Conclusion

The herbal plants and their phytoconstituent play promising therapeutic role for treating various diseases. The extraction of phytoconstituent from herbs gives flavonoid, alkaloid, glycoside, tannins, polyphenol, carbohydrates, and proteins. These phytoconstituents had medicinal properties with good therapeutic action. Various phytoconstituent is avoided in formulation due to solubility, absorption, degradation, short half-life with high clearance rate, and toxicity issue. These problems are overcome by developing a nanocarrier delivery system. Various nanocarriers dendrimers, organic nanoparticles, nanoemulsions, emulsomes, liposomes, phytosomes, and nanocrystals increase solubility and bioavailability of phytoconstituent, for delivery of poorly water-soluble phytoconstituent polymeric micelles helps in enhanced delivery. The graft and block copolymers play an important role for enhance solubility. The stability of polymeric micelles is more than liposomes. The delivery of phytoconstituent like curcumin, berberine, and paclitaxel through polymeric micelles shows significant anticancer activity. They also decrease MDR and increase intracellular uptake in cancer treatment. Phytoconstituent showing antibacterial, antifungal, antidiabetic, and antimalarial activity is improved by polymeric micelles. The combination of two drugs like Quercetin and Doxorubicin containing polymeric micelles shows synergistic effects with minimal adverse effect of doxorubicin. The polymeric micelles are first choice formulation for poorly water-soluble drugs and to enhance targeting ability.

Acknowledgment The authors acknowledge the Department of Pharmaceuticals, Ministry of Chemical and Fertilizers, Govt. of India for support. The NIPER-R communication number for the review article is NIPER-R/Communication/394.

Conflict of Interest The authors declare no conflict of interest among themselves.

References

Aguilar ZP (2013) Types of nanomaterials and corresponding methods of synthesis. In: Nanomaterials for medical applications. Elsevier, Amsterdam, pp 33–82

Akbar MU, Zia KM, Akash MS, Nazir A, Zuber M, Ibrahim M (2018a) In-vivo anti-diabetic and wound healing potential of chitosan/alginate/maltodextrin/pluronic-based mixed polymeric micelles: curcumin therapeutic potential. Int J Biol Macromol 120:2418–2430

Akbar MU, Zia KM, Nazir A, Iqbal J, Ejaz SA, Akash MS (2018b) Pluronic-based mixed polymeric micelles enhance the therapeutic potential of curcumin. AAPS PharmSciTech 19(6):2719–2739

Alamgir AN (2018) Phytoconstituents—active and inert constituents, metabolic pathways, chemistry and application of phytoconstituents, primary metabolic products, and bioactive compounds of primary metabolic origin. In: Therapeutic use of medicinal plants and their extracts, vol 2. Springer, Cham, pp 25–164

Aliabadi HM, Lavasanifar A (2006) Polymeric micelles for drug delivery. Expert Opin Drug Deliv 3(1):139–162

Bader H, Ringsdorf H, Schmidt B (1984) Water soluble polymers in medicine. Angew Makromol Chem 123(1):457–485

Bakr RO, Tawfike A, El-Gizawy HA, Tawfik N, Abdelmohsen UR, Abdelwahab MF, Alshareef WA, Fayez SM, El-Mancy SM, El-Fishawy AM, Abdelkawy MA (2021) The metabolomic analysis of five Mentha species: cytotoxicity, anti-helicobacter assessment, and the development of polymeric micelles for enhancing the anti-helicobacter activity. RSC Adv 11(13):7318–7330

Bhadra D, Bhadra S, Jain NK (2005) Pegylated lysine based copolymeric dendritic micelles for solubilization and delivery of artemether. J Pharm Pharm Sci 8(3):467–482

Bharti SK, Krishnan S, Kumar A, Kumar A (2018) Antidiabetic phytoconstituents and their mode of action on metabolic pathways. Ther Adv Endocrinol Metab 9(3):81–100

Biswal DR, Singh RP (2004) Characterisation of carboxymethyl cellulose and polyacrylamide graft copolymer. Carbohydr Polym 57(4):379–387

Chaemsawang W, Prasongchean W, Papadopoulos KI, Ritthidej G, Sukrong S, Wattanaarsakit P (2019) The effect of okra (Abelmoschus esculentus (L.) Moench) seed extract on human cancer cell lines delivered in its native form and loaded in polymeric micelles. Int J Biomater 2019:1

Cote B, Carlson LJ, Rao DA, Alani AW (2015) Combinatorial resveratrol and quercetin polymeric micelles mitigate doxorubicin induced cardiotoxicity in vitro and in vivo. J Control Release 213: 128–133

De Leo V, Milano F, Mancini E, Comparelli R, Giotta L, Nacci A, Longobardi F, Garbetta A, Agostiano A, Catucci L (2018) Encapsulation of curcumin-loaded liposomes for colonic drug delivery in a pH-responsive polymer cluster using a pH-driven and organic solvent-free process. Molecules 23(4):739

Dev Kumar G, Mis Solval K, Mishra A, Macarisin D (2020) Antimicrobial efficacy of pelargonic acid micelles against salmonella varies by surfactant, serotype and stress response. Sci Rep 10(1):1–13

Ding X, Wang A, Tong W, Xu FJ (2019) Biodegradable antibacterial polymeric nanosystems: a new hope to cope with multidrug-resistant bacteria. Small 15(20):1900999

Eleraky NE, Allam A, Hassan SB, Omar MM (2020) Nanomedicine fight against antibacterial resistance: an overview of the recent pharmaceutical innovations. Pharmaceutics 12(2):142

Erhirhie EO, Ikegbune C, Okeke AI, Onwuzuligbo CC, Madubuogwu NU, Chukwudulue UM, Okonkwo OB (2021) Antimalarial herbal drugs: a review of their interactions with conventional antimalarial drugs. Clin Phytosci 7(1):1–10

Feng C, Li Y, Yang D, Hu J, Zhang X, Huang X (2011) Well-defined graft copolymers: from controlled synthesis to multipurpose applications. Chem Soc Rev 40(3):1282–1295

Fu Y, Ding Y, Zhang L, Zhang Y, Liu J, Yu P (2021) Poly ethylene glycol (PEG)-related controllable and sustainable antidiabetic drug delivery systems. Eur J Med Chem 217:113372

Gaucher G, Dufresne MH, Sant VP, Kang N, Maysinger D, Leroux JC (2005) Block copolymer micelles: preparation, characterization and application in drug delivery. J Control Release 109(1–3):169–188

Gelderblom H, Verweij J, Nooter K, Sparreboom A (2001) Cremophor EL: the drawbacks and advantages of vehicle selection for drug formulation. Eur J Cancer 37(13):1590–1598

Gokhale JP, Mahajan HS, Surana SJ (2019) Quercetin loaded nanoemulsion-based gel for rheumatoid arthritis: in vivo and in vitro studies. Biomed Pharmacother 112:108622

Gong C, Deng S, Wu Q, Xiang M, Wei X, Li L, Gao X, Wang B, Sun L, Chen Y, Li Y (2013) Improving antiangiogenesis and anti-tumor activity of curcumin by biodegradable polymeric micelles. Biomaterials 34(4):1413–1432

Handa M, Ujjwal RR, Vasdev N, Flora SJ, Shukla R (2020) Optimization of surfactant-and cosurfactant-aided pine oil nanoemulsions by isothermal low-energy methods for anticholinesterase activity. ACS Omega 6(1):559–568

Handa M, Beg S, Shukla R, Barkat MA, Choudhry H, Singh. (2021) KK advances in lipid-engineered multifunctional nanophytomedicines for cancer targeting. J Control Release 10(340):48–59

Handa M, Kumar K, Garabadu D, Kushawaha PK, Shukla R (2022) Bilayer fixed-dose combination tablet for curcumin microparticles and piroxicam and in vitro evaluation. Ther Deliv

Harwansh RK, Deshmukh R, Rahman MA (2019) Nanoemulsion: promising nanocarrier system for delivery of herbal bioactives. J Drug Deliv Sci Technol 51:224–233

Hasegawa H, Hashimoto T, Hyde ST (1996) Microdomain structures with hyperbolic interfaces in block and graft copolymer systems. Polymer 37(17):3825–3833

Huang F, Gao Y, Zhang Y, Cheng T, Ou H, Yang L, Liu J, Shi L, Liu J (2017) Silver-decorated polymeric micelles combined with curcumin for enhanced antibacterial activity. ACS Appl Mater Interfaces 9(20):16880–16889

Jiao J (2008) Polyoxyethylated nonionic surfactants and their applications in topical ocular drug delivery. Adv Drug Deliv Rev 60(15):1663–1673

Jones MC, Leroux JC (1999) Polymeric micelles—a new generation of colloidal drug carriers. Eur J Pharm Biopharm 48(2):101–111

Karar MG, Kuhnert N (2017) Herbal drugs from Sudan: traditional uses and phytoconstituents. Pharmacogn Rev 11(22):83

Kaur J, Mishra V, Singh SK, Gulati M, Kapoor B, Chellappan DK, Gupta G, Dureja H, Anand K, Dua K, Khatik GL (2021a) Harnessing amphiphilic polymeric micelles for diagnostic and therapeutic applications: breakthroughs and bottlenecks. J Control Release 334:64–95

Kaur N, Bains A, Kaushik R, Dhull SB, Melinda F, Chawla P (2021b) A review on antifungal efficiency of plant extracts entrenched polysaccharide-based nanohydrogels. Nutrients 13(6):2055

Kawabata Y, Wada K, Nakatani M, Yamada S, Onoue S (2011) Formulation design for poorly water-soluble drugs based on biopharmaceutics classification system: basic approaches and practical applications. Int J Pharm 420(1):1–10

Khan S, Hussain A, Attar F, Bloukh SH, Edis Z, Sharifi M, Balali E, Nemati F, Derakhshankhah H, Zeinabad HA, Nabi F (2022) A review of the berberine natural polysaccharide nanostructures as potential anticancer and antibacterial agents. Biomed Pharmacother 146:112531

Kim YJ, Lee KP, Lee DY, Kim YT, Koh D, Lim Y, Yoon MS (2019) Anticancer activity of a new chalcone derivative-loaded polymeric micelle. Macromol Res 27(1):48–54

Kulthe SS, Choudhari YM, Inamdar NN, Mourya V (2012) Polymeric micelles: authoritative aspects for drug delivery. Des Monomers Polym 15(5):465–521

Lawrence MJ (1994) Surfactant systems: their use in drug delivery. Chem Soc Rev 23(6):417–424

Li C, Tho CC, Galaktionova D, Chen X, Král P, Mirsaidov U (2019) Dynamics of amphiphilic block copolymers in an aqueous solution: direct imaging of micelle formation and nanoparticle encapsulation. Nanoscale 11(5):2299–2305

Liu L, Sun L, Wu Q, Guo W, Li L, Chen Y, Li Y, Gong C, Qian Z, Wei Y (2013) Curcumin loaded polymeric micelles inhibit breast tumor growth and spontaneous pulmonary metastasis. Int J Pharm 443(1–2):175–182

Lu M, Qiu Q, Luo X, Liu X, Sun J, Wang C, Lin X, Deng Y, Song Y (2019) Phyto-phospholipid complexes (phytosomes): a novel strategy to improve the bioavailability of active constituents. Asian J Pharm Sci 14(3):265–274

Mohamed EA, Hashim II, Yusif RM, Shaaban AA, El-Sheakh AR, Hamed MF, Badria FA (2018) Polymeric micelles for potentiated antiulcer and anticancer activities of naringin. Int J Nanomedicine 13:1009

Mukhopadhyay P, Prajapati AK (2015) Quercetin in anti-diabetic research and strategies for improved quercetin bioavailability using polymer-based carriers—a review. RSC Adv 5(118): 97547–97562

Ng PQ, Ling LS, Chellian J, Madheswaran T, Panneerselvam J, Kunnath AP, Gupta G, Satija S, Mehta M, Hansbro PM, Collet T (2020) Applications of nanocarriers as drug delivery vehicles for active phytoconstituents. Curr Pharm Des 26(36):4580–4590

Nishiyama N, Kato Y, Sugiyama Y, Kataoka K (2001) Cisplatin-loaded polymer-metal complex micelle with time-modulated decaying property as a novel drug delivery system. Pharm Res 18(7):1035–1041

Pal SK, Shukla Y (2003) Herbal medicine: current status and the future. Asian Pac J Cancer Prev 4(4):281–288

Parasuraman S, Thing GS, Dhanaraj SA (2014) Polyherbal formulation: concept of Ayurveda. Pharmacogn Rev 8(16):73–80

Patra A, Satpathy S, Shenoy AK, Bush JA, Kazi M, Hussain MD (2018) Formulation and evaluation of mixed polymeric micelles of quercetin for treatment of breast, ovarian, and multidrug resistant cancers. Int J Nanomedicine 13:2869

Priya LB, Baskaran R, Padma VV (2017) Phytonanoconjugates in oral medicine. micro and nano technologies, nanostructures for oral medicine. Elsevier, Amsterdam, pp 639–668

Roy D, Semsarilar M, Guthrie JT, Perrier S (2009) Cellulose modification by polymer grafting: a review. Chem Soc Rev 38(7):2046–2064

Seow WY, Xue JM, Yang YY (2007) Targeted and intracellular delivery of paclitaxel using multi-functional polymeric micelles. Biomaterials 28(9):1730–1740

Shen R, Kim JJ, Yao M, Elbayoumi TA (2016) Development and evaluation of vitamin E D-α-tocopheryl polyethylene glycol 1000 succinate-mixed polymeric phospholipid micelles of berberine as an anticancer nanopharmaceutical. Int J Nanomedicine 11:1687

Singh S, Sharma B, Kanwar SS, Kumar A (2016) Lead phytochemicals for anticancer drug development. Front Plant Sci 7:1667

Singh A, Ujjwal RR, Kumar A, Verma RK, Shukla R (2022) Formulation and optimization of silymarin-encapsulated binary micelles for enhanced amyloid disaggregation activity. Drug Dev Ind Pharm 47(11):1775–1785

Soltantabar P, Calubaquib EL, Mostafavi E, Biewer MC, Stefan MC (2020) Enhancement of loading efficiency by coloading of doxorubicin and quercetin in thermoresponsive polymeric micelles. Biomacromolecules 21(4):1427–1436

Sun J, Liu T, Xu J (2016) Improving the anticancer activity of α-hederin by physically encapsulating it with targeted micelles assembled from amphiphilic block copolymers. J Drug Deliv Sci Technol 35:252–259

Sun Z, Li Y, Gu X, Wang L, Yu B, Yang X, Xu H (2022) Development of Icariside II loaded polymeric micelles and evaluation of anticancer activity in vitro and in vivo. J Drug Deliv Sci Technol 75:103652. https://linkinghub.elsevier.com/retrieve/pii/S1773224722005639

Suwan T, Khongkhunthian S, Okonogi S (2019) Antifungal activity of polymeric micelles of silver nanoparticles prepared from Psidium guajava aqueous extract. Drug Discov Ther 13(2):62–69

Tamboli V, Mishra GP, Mitra AK (2013) Novel pentablock copolymer (PLA–PCL–PEG–PCL–PLA)-based nanoparticles for controlled drug delivery: effect of copolymer compositions on the crystallinity of copolymers and in vitro drug release profile from nanoparticles. Colloid Polym Sci 291(5):1235–1245

Tsai YM, Chien CF, Lin LC, Tsai TH (2011) Curcumin and its nano-formulation: the kinetics of tissue distribution and blood–brain barrier penetration. Int J Pharm 416(1):331–338

Varma LT, Singh N, Gorain B, Choudhury H, Tambuwala MM, Kesharwani P, Shukla R (2020) Recent advances in self-assembled nanoparticles for drug delivery. Curr Drug Deliv 17(4): 279–291

Wan X, Beaudoin JJ, Vinod N, Min Y, Makita N, Bludau H, Jordan R, Wang A, Sokolsky M, Kabanov AV (2019) Co-delivery of paclitaxel and cisplatin in poly (2-oxazoline) polymeric micelles: implications for drug loading, release, pharmacokinetics and outcome of ovarian and breast cancer treatments. Biomaterials 192:1–14

WHO (n.d.) Cancer. https://www.who.int/health-topics/cancer#tab=tab_1

World Malaria Report (2021). https://www.who.int/teams/global-malaria-programme/reports/world-malaria-report-2021

Zhai Y, Guo S, Liu C, Yang C, Dou J, Li L, Zhai G (2013) Preparation and in vitro evaluation of apigenin-loaded polymeric micelles. Colloids Surf A Physicochem Eng Asp 429:24–30

Zhai B, Clark J, Ling T, Connelly M, Medina-Bolivar F, Rivas F (2014) Antimalarial evaluation of the chemical constituents of hairy root culture of Bixa orellana L. Molecules 19(1):756–766

Zhang W, Cui T, Liu L, Wu Q, Sun L, Li L, Wang N, Gong C (2015) Improving anti-tumor activity of curcumin by polymeric micelles in thermosensitive hydrogel system in colorectal peritoneal carcinomatosis model. J Biomed Nanotechnol 11(7):1173–1182

Zhang Y, Long Y, Yu S, Li D, Yang M, Guan Y, Zhang D, Wan J, Liu S, Shi A, Li N (2021) Natural volatile oils derived from herbal medicines: a promising therapy way for treating depressive disorder. Pharmacol Res 164:105376

Diagnostic Applications of Surface-Engineered Polymeric Micelles

10

Jaskiran Kaur, Monica Gulati, Kamal Dua, Leander Corrie, Ankit Awasthi, and Sachin Kumar Singh

Abstract

In the past few years, the flexible chemistry of amphiphilic block copolymers (ABCs) has made them one of the unique drug delivery vehicles for biomedical applications. Most of these have been extensively explored in oncology. Their self-assembly property is one of the remarkable properties that lead to the production of polymeric micelles (PMs) above critical micelle concentration (CMC) upon addition in water. PMs have been widely explored in drug delivery-based systems but very limited work is available on their use as a diagnostic agent. Their unique-physicochemical property mostly ease of functionalization is one of the most exploited properties utilized in the diagnosis of various diseases. With agreement to this, the present chapter emphasizes the role of PMs as diagnostic agents using functionalization properties with relevant case studies.

Keywords

Copolymers · Micelles · Cancer · Theranostic · Theragnosis · Sensor

---

J. Kaur · M. Gulati · L. Corrie · A. Awasthi · S. K. Singh (✉)
School of Pharmaceutical Sciences, Lovely Professional University, Phagwara, Punjab, India
e-mail: sachin.16030@lpu.co.in

K. Dua
Discipline of Pharmacy, Graduate School of Health, University of Technology Sydney, Ultimo, NSW, Australia
e-mail: Kamal.Dua@uts.edu.au

S. K. Singh et al. (eds.), *Polymeric Micelles: Principles, Perspectives and Practices*,
https://doi.org/10.1007/978-981-99-0361-0_10

10.1 Introduction

Various imaging techniques used in recent times are non-invasive and operate at morphological, functional, and molecular levels to identify the diseased site (Kaur et al. 2021). In such techniques, a contrast agent is used to get high-resolution images of the diagnosed areas (Kaur et al. 2022a). In addition, several contrast agents have high relaxivity and durability used to identify the pathological condition in cells and tissue (Wahsner et al. 2019). These agents have the potential to absorb signals more strongly than neighbour cells without any adverse effects. However, the major challenge observed is their systemic distribution which limits the attainment of optimum local concentration of these agents at the site of interest (Kaur et al. 2021).

In the past few years, PMs based on ABCs have gained extensive attention in the diagnosis of various diseases mostly cancer owing to their flexible chemistry and self-assembly property (Kaur et al. 2022a). These ABCs has been widely utilized in the formation of PMs owing to their low critical micelle concentration over surfactant-based PMs (Kaur et al. 2022b). As a consequence, ABCs provide good thermodynamic stability to the PMs.

PMs have the potential in restricting the distribution of diagnostic agents at the site of interest using three main approaches. This mainly includes surface-engineering with the help of targeting ligands, covalent or non-covalent conjugation with the diagnostic moieties, and incorporation of stimuli-responsive moieties in the copolymer network for micellar disassembly in response to a pathological condition (Kaur et al. 2021). Therefore, this approach has the potential to offer high-quality images and can improve the existing diagnostic approach by simplifying the task of detection, monitoring, and visual inspection (Xu et al. 2020). With this agreement, the main objective of this work is to highlight the role of PMs as an effective diagnostic nanocarrier with the help of recent case studies.

10.2 Diagnostic Applications

The surface-functionalization property of PMs imparts magnetic resonance as well as certain, fluorescent and optical properties to the PMs that have led to their role in the clinical diagnosis of several diseases (Kaur et al. 2021). In addition, such property makes PMs a suitable nanocarrier for mapping drug therapy loaded with contrast/imaging agents along with imaging at the target site. These PMs can be further designed with several targeting moieties for site-specificity to perform effective imaging of the molecular changes in a cell (Kaur et al. 2022c). To achieve this quality, multiple stimuli-responsive moieties can be imparted in copolymer networks for their use as imaging probes (Guragain et al. 2015). All such properties of PMs have led to their use in the diagnosis of various diseases. Various studies related to diagnostic applications of PMs are discussed below.

10.2.1 Diagnosis of Tumour

In the case of MRI, Feiner-Gracia et al. developed spectrally active PEG-dendron PMs entrapped with 7-diethylamino-3-carboxy coumarin as a responsive dye for real-time imaging and evaluation of drug delivery carrier in tumour microenvironment. The results of this study revealed leaky vasculature in the HeLa cells along with heterogeneity. In addition, the study indicated free diffusion of developed PMs in endothelial cellular matrix of cancer cells. This developed micellar system indicated potential for recapitulating the tumour 3D microenvironment (Feiner-Gracia et al. 2021). Sun et al. developed fluorescent PMs based on perylene diimide-block-poly(D,L lactide)-b-poly(ethyl ethylene phosphate) ABC loaded with camptothecin for the development of image guided therapy of cancer. The result of imaging study revealed high intensity of yellow fluorescence in the nuclei of HeLa cells indicating effective internalization in tumour cells. Thus the study indicated successful delivery of the loaded camptothecin in the nucleus region. Furthermore, PMs showed strong and uniformly distributed fluorescent signals in HeLa cells indicating successful bioimaging of the cancer cells. The developed formulation resulted in excellent optical imaging of the cancer cells (Sun et al. 2016).

Similarly, Yang et al. reported strong fluorescence intensity and photostability of PMs made of poly (fluorene-alt-(4, 7-bis (hexylthien)-2, 1, 3-benzothiadiazole))-graft- polycaprolactone-block-poly [oligo (ethylene glycol) methyl ether methacrylate] copolymer upon fluorescence imaging of cancer cells (Yang et al. 2016). Vinh et al. reported strong and specific tumour contrast enhancement and reduced tumour size in rats with hepatic cancer upon single administration of PMs loaded with gadolinium and oxaliplatin. The PMs were designed for the simultaneous imaging and treatment of cancer. The results of MR images revealed higher reduction in tumour growth ($p < 0.05$) upon treatment with micelles in comparison to oxaliplatin treated rats. Also, PMs increased the survival rate by 1.0-folds in comparison to oxaliplatin treated rats, respectively (Vinh et al. 2015).

Kim et al. reported enhancement in T1 MR imaging of tumour site using PMs consisting pH-responsive moieties in copolymer network, i.e., poly-L-histine and poly-L-lactic acid and loaded with gadolinium. The developed micellar system indicated potential for the imaging of small tumour cells via responsive-disassembly phenomenon owing to protonation of moieties incorporated in PMs (Kim et al. 2014).

Several studies claimed the use of mPEG-Lys3-CA4-based PMs for magnetic resonance and fluorescence imaging (dual mode) by loading superparamagnetic iron oxide nanoparticles and the Nile red as a lipophilic moiety in copolymer assembly for targeted cancer therapy (Li et al. 2018).

In addition to this, PMs have been utilized as a cancer theranostic agent for simultaneous drug delivery and diagnosis. For instance, Cao et al. claimed that PEG-PO-based PMs conjugated with paclitaxel drug along with cypate as a contrast agent resulted in image-assisted chemo-phototherapy against liver carcinoma cells (Li et al. 2019). Similarly, mPEG-PG-b-PCL and PEG-b-PCL-based complex PMs loaded with DOTA (Gd) showed higher MR contrast in keratin-forming tumour cells

Table 10.1 PMs in cancer diagnosis/theranostic

ABCs	Diagnostic moiety/drug	Outcome	References
mPEG-PCL-g-PEI-FITC	SPION/ DOX	Novel bifunctional nanoprobe for MR and optical imaging of breast cancer cells	Guo et al. (2015)
PEG-b-PCL (with glucose analogue)	SPION	27-fold higher cellular uptake of developed PMs in prostate cancer cells than PMs without glucose analogue Exhibited high transverse relativities (MRI contrast agent)	Theerasilp et al. (2017)
DSPE-b-PEG2000-DTPA	Technetium-99 m	Good target effect on malignant cells for confocal fluorescence imaging in liver cancer cells	Oda et al. (2020)
HA-C16	SPION/ DTX	Dual tumour-targeted drug delivery with MR imaging Photothermal therapy and chemotherapy resulted in a synergistic therapeutic effect	Zheng et al. (2018)
H40-poly(L-Glu-Hyd-dox)-b-PEG functionalized with cRGD peptide	^{64}Cu-labelling/ DOX	pH-controlled release with PET imaging of glioma cells	Xiao et al. (2012)

(Cao et al. 2017). The use of PMs in cancer diagnosis/theranostic is presented in Table 10.1.

Xiao et al. developed multifunctional PMs composed of H40-poly(L-glutamate-hydrazone)-b-poly(ethylene glycol) conjugated with DOX via hydrazine bond for site-specific release in the brain. These PMs were further functionalized using cRGD peptide at the surface of the corona for active targeting of glioma. Furthermore, these PMs were loaded with 1,4,7-triazacyclononane-N, N′, N″-triacetic acid for PET imaging of glioma. The developed multifunctional PMs showed higher fluorescence intensity of DOX entrapped in the core inside the tumour cells after PET scanning over simple PMs. The results showed higher internalization of the multifunctional PMs post 4 h of injection in mice bearing glioma cells (Xiao et al. 2012).

Chen et al. claimed effective targeting and imaging of glioma cells using poly (ethylene glycol) based se2-fluorescent functionalized PMs along with cRGD peptide as targeting ligand at corona surface to target integrin proteins expressed by the glioma cells. The developed PMs showed higher internalization in glioma cells as a measure of blue fluorescence emitted by the cells upon targeting integrin receptors via endocytosis pathway after 6 h of incubation ($p < 0.05$) over non-functionalized PMs, respectively (Chen et al. 2015).

10.2.2 Diagnosis of Glucose

For the early detection of blood glucose levels, PMs as a nanocarrier owing to their functional properties could be employed (Jang et al. 2020). Such properties mainly include a high surface-to-volume ratio, low toxicity, biocompatibility, prolonged residence time and tailor-made recognition sites for the target analytes (Shukla et al. 2016; Ngamdee et al. 2011). The fluorescence quenching property of PMs can be used for fluorescing sensing of glucose using phenylboronic acid as glucose sensitive moiety and alizarin red as a dye. Overall, such sensing of glucose using PMs is based on dye displacement strategy via self-regulated, i.e., switch on-off approach in the presence of high glucose. This approach has resulted in a 13-fold increase in fluorescence intensity of the developed PMs in response to high glucose (Ngamdee et al. 2011).

In addition to this, PMs can be developed as a biosensor for glucose detection using the catalytic activity of the glucose oxidase enzyme. The nanosized PMs formed from conducting/non-conducting polymers have been used for the rapid adsorption of analytes owing to their high surface area and effective surface functional groups (Anusha et al. 2015). Their coating over carbon electrodes (cathodic potential) can provide electrical properties along with excellent catalytic efficiencies and higher selectivities without the interference of uric acid or L-ascorbic acid (Kaur et al. 2022a).

In one of the studies, polymaleimidostyrene-polystyrene based PMs were designed for amperometric sensing of glucose. The developed PMs showed pH-dependent glucose sensitivity, i.e., higher current response of glucose generated in phosphate buffer pH 6.5. PMs also indicated prolonged stability even after storing the formulation for months (Wang et al. 2008). In another study, a fluorescent sensor based on PMs composed of CTAB loaded with alizarin/boronic acid was designed for fluorescing sensing glucose. The developed PMs showed 14-fold higher calibration sensitivity than the buffer solution (Ngamdee et al. 2011).

In the next study, fluorescent PMs were developed using tetraphenylethylenes, styrene, 4-vinylpyridine, and methylated 4-vinylpyridine moieties for glucose detection. The developed PMs were able to detect the lowest amount of glucose present in the sample owing to hydrogen-peroxide-induced fluorescence quenching in the presence of glucose oxidase and potassium iodide. In addition, PMs exhibited higher specificity and selectivity for glucose even in the presence of proteins (Shen et al. 2012).

10.2.3 Diagnosis of Neuronal Diseases

Despite having good potential, PMs have been relatively less explored for the imaging of brain diseases. This could be because of the higher complexity of the brain or neuronal network. Some of the studies regarding the same are discussed in the subsequent sections.

Shiraishi et al. claimed effective MR imaging of cerebral ischemia-reperfusion injury in rat transient middle cerebral artery occlusion-reperfusion model using PMs composed of poly (ethylene glycol)-b-poly(L-lysine-DOTA-gadolinium). The developed PMs showed clear/stark contrast images in 15.5 ± 10.3% of the ischemic hemisphere after reperfusion within 30 min of post-intravenous injection in rats than classic gadolinium chelate contrast agents followed by T1-hyperintense area of the developed PMs in the striatum and cerebral cortex (Shiraishi et al. 2017).

In the next study, Wu et al. reported MR imaging of neuroglioma in mice using PMs based on monomethoxy-poly(ethylene glycol)-S-S-hexadecyl and SPIONs. The results of in vivo study revealed shorter T2 relaxation time with higher contrast of the tumour. The developed PMs increased the transverse relaxivity by 1.83-folds via glutathione-sensitive aggregation of PMs in tumour-bearing mice over non-sensitive PMs upon intravenous injection. In addition, PMs showed more robustness in T2 enhancement at the tumour site (Wu et al. 2021).

10.3 Conclusion

The incidence of diseases across the globe is increasing year by year due to ageing, genetic and environmental factors. The effective management of diseases requires novel drug delivery systems that can offer either simultaneous mapping/monitoring/imaging and drug delivery or both. The nanosized PMs over other nanocarriers offer relatively more stability, and monodispersity, and are remunerative. In addition, the surface of PMs can be modified or stimuli-sensitized using the functional properties of ABCs for targeting and imaging. Furthermore, the high surface-to-volume ratios enable their use as biosensors for the detection of specific analytes in biological samples. Despite having multifaceted potential, PMs have not been explored to that extent for imaging purpose. This could be because of the lack of targeting ligands required for the targeting of desired diseased sites. Mostly it has been explored in active targeting of tumours but the data is extremely limited in terms of diagnosis of brain diseases (lack of ligand-mediated endocytosis). These limitations will provide an opportunity for the researchers to design smart PMs with less complexity for non-invasive imaging modalities.

References

Anusha JR, Raj CJ, Cho BB, Fleming AT, Yu KH, Kim BC (2015) Amperometric glucose biosensor based on glucose oxidase immobilized over chitosan nanoparticles from gladius of Uroteuthis duvauceli. Sens Actuators B 215:536–543. https://doi.org/10.1016/j.snb.2015.03.110

Cao Y, Liu M, Zhang K, Zu G, Kuang Y, Tong X et al (2017) Poly(glycerol) used for constructing mixed polymeric micelles as T1 MRI contrast agent for tumor-targeted imaging. Biomacromolecules 18(1):150–158

Chen G, Wang L, Cordie T, Vokoun C, Eliceiri KW, Gong S (2015) Multi-functional self-fluorescent unimolecular micelles for tumor-targeted drug delivery and bioimaging. Biomaterials 47:41–50. https://doi.org/10.1016/j.biomaterials.2015.01.006

Feiner-Gracia N, Glinkowska Mares A, Buzhor M, Rodriguez-Trujillo R, Samitier Marti J, Amir RJ et al (2021) Real-time ratiometric imaging of micelles assembly state in a microfluidic cancer-on-a-chip. ACS Appl Bio Mater 4(1):669–681

Guo Q, Kuang L, Cao H, Li W, Wei J (2015) Self-assembled mPEG-PCL-g-PEI micelles for multifunctional nanoprobes of doxorubicin delivery and magnetic resonance imaging and optical imaging. Colloids Surf B Biointerfaces 136:687–693. https://doi.org/10.1016/j.colsurfb.2015.10.013

Guragain S, Bastakoti BP, Malgras V, Nakashima K, Yamauchi Y (2015) Multi-stimuli-responsive polymeric materials. Chem A Eur J 21(38):13164–13174

Jang YJ, Kim B, Roh E, Kim H, Lee SH (2020) Micellization-induced amplified fluorescence response for highly sensitive detection of heparin in serum. Sci Rep 10(1):1–9. https://doi.org/10.1038/s41598-020-66360-8

Kaur J, Mishra V, Singh SK, Gulati M, Kapoor B, Chellappan DK et al (2021) Harnessing amphiphilic polymeric micelles for diagnostic and therapeutic applications: breakthroughs and bottlenecks. J Control Release 334:64–95. https://doi.org/10.1016/j.jconrel.2021.04.014

Kaur J, Gulati M, Zacconi F, Dureja H, Loebenberg R, Ansari MS et al (2022a) Biomedical applications of polymeric micelles in the treatment of diabetes mellitus: current success and future approaches. Expert Opin Drug Deliv 19(7):771–723. https://doi.org/10.1080/17425247.2022.2087629

Kaur J, Gulati M, Jha NK, Disouza J, Patravale V, Dua K et al (2022b) Recent advances in developing polymeric micelles for treating cancer: breakthroughs and bottlenecks in their clinical translation. Drug Discov Today 27:1495. https://doi.org/10.1016/j.drudis.2022.02.005

Kaur J, Gulati M, Kapoor B, Jha NK, Gupta PK, Gupta G et al (2022c) Advances in designing of polymeric micelles for biomedical application in brain related diseases. Chem Biol Interact 361:109960. https://doi.org/10.1016/j.cbi.2022.109960

Kim KS, Park W, Hu J, Bae YH, Na K (2014) A cancer-recognizable MRI contrast agents using pH-responsive polymeric micelle. Biomaterials 35(1):337–343. https://doi.org/10.1016/j.biomaterials.2013.10.004

Li WJ, Wang Y, Liu Y, Wu T, Cai WL, Shuai XT et al (2018) Preliminary study of MR and fluorescence dual-mode imaging: combined macrophage-targeted and superparamagnetic polymeric micelles. Int J Med Sci 15(2):129–141

Li X, Zhang C, Zheng Q, Shi X (2019) ROS-responsive targeting micelles for optical imaging-guided chemo-phototherapy of cancer. Colloids Surf B Biointerfaces 179:218–225. https://doi.org/10.1016/j.colsurfb.2019.04.005

Ngamdee K, Noipa T, Martwiset S, Tuntulani T, Ngeontae W (2011) Enhancement of sensitivity of glucose sensors from alizarin-boronic acid adducts in aqueous micelles. Sens Actuators B 160(1):129–138. https://doi.org/10.1016/j.snb.2011.07.023

Oda CMR, de Oliveira Silva J, Fernandes RS, Braga AV, de Machado RM, de Coelho MM et al (2020) Encapsulating paclitaxel in polymeric nanomicelles increases antitumor activity and prevents peripheral neuropathy. Biomed Pharmacother 132:110864

Shen X, Shi Y, Peng B, Li K, Xiang J, Zhang G et al (2012) Fluorescent polymeric micelles with tetraphenylethylene moieties and their application for the selective detection of glucose. Macromol Biosci 12(11):1583–1590

Shiraishi K, Wang Z, Kokuryo D, Aoki I, Yokoyama M (2017) A polymeric micelle magnetic resonance imaging (MRI) contrast agent reveals blood–brain barrier (BBB) permeability for macromolecules in cerebral ischemia-reperfusion injury. J Control Release 253:165–171. https://doi.org/10.1016/j.jconrel.2017.03.020

Shukla SK, Govender PP, Tiwari A (2016) Polymeric micellar structures for biosensor technology. In: Advances in biomembranes and lipid self-assembly, vol 24, 1st edn. Elsevier, Amsterdam, pp 143–161. https://doi.org/10.1016/bs.abl.2016.04.005

Sun M, Yin W, Dong X, Yang W, Zhao Y, Yin M (2016) Fluorescent supramolecular micelles for imaging-guided cancer therapy. Nanoscale 8(9):5302–5312

Theerasilp M, Sunintaboon P, Sungkarat W, Nasongkla N (2017) Glucose-installed, SPIO-loaded PEG-b-PCL micelles as MR contrast agents to target prostate cancer cells. Appl Nanosci 7(8): 711–721

Vinh NQ, Naka S, Cabral H, Murayama H, Kaida S, Kataoka K et al (2015) MRI-detectable polymeric micelles incorporating platinum anticancer drugs enhance survival in an advanced hepatocellular carcinoma model. Int J Nanomedicine 10:4137–4147

Wahsner J, Gale EM, Rodríguez-Rodríguez A, Caravan P (2019) Chemistry of MRI contrast agents: current challenges and new frontiers. Chem Rev 119(2):957–1057

Wang X, Watanabe H, Uchiyama S (2008) Amperometric l-ascorbic acid biosensors equipped with enzyme micelle membrane. Talanta 74(5):1681–1685

Wu B, Deng K, Lu ST, Zhang CJ, Ao YW, Wang H et al (2021) Reduction-active Fe3O4-loaded micelles with aggregation- enhanced MRI contrast for differential diagnosis of Neroglioma. Biomaterials 268:120531

Xiao Y, Hong H, Javadi A, Engle JW, Xu W, Yang Y et al (2012) Multifunctional unimolecular micelles for cancer-targeted drug delivery and positron emission tomography imaging. Biomaterials 33(11):3071–3082. https://doi.org/10.1016/j.biomaterials.2011.12.030

Xu H, Ma B, Jiang J, Xiao S, Peng R, Zhuang W et al (2020) Integrated prodrug micelles with two-photon bioimaging and pH-triggered drug delivery for cancer theranostics. Regen Biomater 7(2):171–180

Yang C, Huang S, Wang X, Wang M (2016) Theranostic unimolecular micelles of highly fluorescent conjugated polymer bottlebrushes for far red/near infrared bioimaging and efficient anticancer drug delivery. Polym Chem 7(48):7455–7468

Zheng S, Han J, Jin Z, Kim CS, Park S, Kim K et al (2018) Dual tumor-targeted multifunctional magnetic hyaluronic acid micelles for enhanced MR imaging and combined photothermal-chemotherapy. Colloids Surf B Biointerfaces 164:424–435. https://doi.org/10.1016/j.colsurfb.2018.02.005

Ligand Conjugated Polymeric Micelles for Targeted Delivery of Drug Payloads in Cancer Therapy

11

Simran Deep Kaur, Sachin Kumar Singh, Dinesh Kumar Chellappan, Alaa A. Aljabali, Kamal Dua, and Deepak N. Kapoor

Abstract

Therapeutics can be delivered to specific areas in the body using ligand conjugated polymeric micelles. Both passive and active targeting methods can be used to deliver polymeric micelles (PMs) to tumor locations. Polymeric micelles are useful for encapsulating hydrophobic anticancer medicines that are poorly water soluble. Additionally, the outer shell of PMs can be used to target the anticancer medicine to tumors by active mechanisms. Small size, stability, sustained and controlled release of therapeutics, and clinical aspects of malignancies make it possible for polymeric micelles to deliver drugs to tumor locations by virtue of their intrinsic properties. Immuno-fragmentation, cRGD, epidermal growth factors, folate, and transferrin are some of the ligands that are conjugated with polymeric micelles for active targeting. Micelles can also be used to target tumors, which is a prospective target for anticancer medications. In experimental

S. D. Kaur · D. N. Kapoor (✉)
School of Pharmaceutical Sciences, Shoolini University of Biotechnology and Management Sciences, Solan, India
e-mail: deepaknandkishore.558@shooliniuniversity.com

S. K. Singh
School of Pharmaceutical Sciences, Lovely Professional University, Phagwara, Punjab, India

D. K. Chellappan
Department of Life Sciences, School of Pharmacy, International Medical University, Kuala Lumpur, Malaysia

A. A. Aljabali
Department of Pharmaceutical Sciences, Faculty of Pharmacy, Yarmouk University, Irbid, Jordan

K. Dua
Discipline of Pharmacy, University of Technology Sydney, Sydney, NSW, Australia
e-mail: Kamal.Dua@uts.edu.au

S. K. Singh et al. (eds.), *Polymeric Micelles: Principles, Perspectives and Practices*,
https://doi.org/10.1007/978-981-99-0361-0_11

and clinical investigations, different anticancer drugs have been successfully delivered by PMs. An overview of the most recent knowledge on the use of PMs to target anticancer medications to tumor sites is provided in this chapter.

Keywords

Polymeric micelles · Immuno-fragmentation · Ligands · cRGD · Conjugation · Epidermal growth factors

11.1 Introduction

Poor water solubility is quickly becoming the primary impediment to the clinical implementation of many prospective novel medications owing to the lipophilic groups essential for receptor recognition and membrane permeability (Da Silva et al. 2020). Because of the non-selective targeting, many traditional chemotherapeutic treatments are not in use despite their long-term benefits. Conventional drug delivery methods for anticancer chemotherapeutics provide a variety of distinct challenges, including inadequate selectivity, significant toxicity, and the formation of drug resistance (Senapati et al. 2018). As a result, many anticancer medications lose some of their effectiveness. To maximize the therapeutic benefit of a treatment, a nanocarrier drug delivery system is developed (Din et al. 2017). Because of large surface area, nanocarriers may modify the fundamental characteristics and bioactivity of pharmaceuticals. Incorporating nanocarriers into drug delivery systems has the potential to improve drug release, solubility, stability, toxicity, localization, pharmacokinetics, and biodistribution (Edis et al. 2021). The pathophysiology of tumor microenvironment has been exploited by nanocarrier-based platforms to transport anticancer medications effectively into tumors, greatly enhancing therapy results for various kinds of tumors (Fernandes et al. 2018). Furthermore, rising resistance to anticancer medications necessitates higher dosages, which in turn increases the treatment's toxicity. Thus, in order to minimize toxicity, a targeted delivery mechanism for the medicine at the target tissue is strongly recommended (Din et al. 2017). To target the receptors that are highly expressed on the surface of tumor cells, nanocarrier systems were coated with specific ligands (Lei et al. 2022). An anticancer therapeutics may be delivered to a tumor by using polymeric micelles (PMs). Polymeric micelles are a novel kind of colloidal delivery system that has attracted a lot of attention in recent years due to their potential to serve as a versatile and effective drug carrier (Majumder et al. 2020). An increased therapeutic index may be attained with better pharmacological activity at the therapeutic sites by enhancing drug delivery to the targeted sites and minimizing delivery to undesirable locations (Wen et al. 2015; Anselmo and Mitragotri 2014).

Kataoka's team created doxorubicin-encapsulated polymeric micelles in the 1990s as a cancer drug delivery system (Kataoka et al. 1999). Preclinical as well as clinical research has found that they can be utilized to deliver anticancer therapeutics. There have been reports of targeting tumors using nanocarrier drug delivery systems like liposomes and lipid-based DDSs. However, these have been hampered by drug resistance and non-targeting (Edis et al. 2021). Targeted drug delivery methods are sorely needed because of the intricacy of tumor cells and medication resistance. For the first time, PMs have been shown to be the only DDS capable of eliminating multidrug resistance by absolute targeting regardless of the method used (Hari et al. 2022). Polymeric micelles have been demonstrated to be effective in both preclinical and clinical trials; therefore, it is essential that excipients be both rapidly biodegradable and biologically compatible (Hwang et al. 2020). One of the earliest reported self-assembled polymers nanocarrier drug delivery methods was polymeric micelles. Polymeric micelles are made up of an outer and inner core each with a distinct function. The outer layer regulates the pharmacokinetic (PK) action in vivo, whereas the inner core manages drug loading, stability, and release (Parra et al. 2021). The polar area confronts the micelle's outer surface, whereas the non-polar region forms its inner core. Micelles encapsulate both hydrophilic and hydrophobic therapeutics. The micelles enclose the hydrophobic therapeutics in their inner cores and the hydrophilic drugs in their outer cores (Ekladious et al. 2019). It consists of amphiphilic block copolymers which are self-assembled in an aqueous solvent. These amphiphilic block copolymers used in preparation of polymeric micelles are diblock, triblock copolymers, and graft polymers (Ekladious et al. 2019). In an aqueous solution, micelles develop when block copolymer concentration is higher than the critical micelle concentration threshold (CMC) (Fluksman and Benny 2019). Polymeric micelles have prodigious potential for treating cancer because of their enhanced biopharmaceutical and pharmacokinetic attributes, including their small size, propensity to dissolve hydrophobic therapeutics, and ligand-mediated active targeting, which enables site-specific delivery of pharmaceuticals (Yu et al. 2019).

11.2 Active Targeting of Tumors

Most anticancer drugs have a low therapeutic index, thus developing targeted drug delivery methods to treat different types of cancer is an essential need. It is feasible to transport medications to target cells in vivo by actively targeting tumor cells that include targeting moieties and active therapeutics (Veselov et al. 2022). There have been several conjugating ligands, for instance, proteins, oligopeptide conjugated ligands, vitamins, and small targeted molecules which can be attached to a surface of nanoparticle drug delivery systems (Salahpour Anarjan 2019). Molecular recognition and the use of appropriate ligands that can identify its biomarker on the target site are essential for achieving active targeting. Oncogenic cells usually have receptors that are overexpressed from those found in healthy cells and these receptors are overexpressed in cancerous cells to facilitate the nourishment required for metabolism. The chemotherapeutic drugs are derivatized with a specific ligand that recognizes specific receptors on target tumor cells. When this ligand–chemotherapeutic drug delivery system is administered to patients they bind to the specific receptors of targeted tumor cells and deliver therapeutics to these specific tumor cells

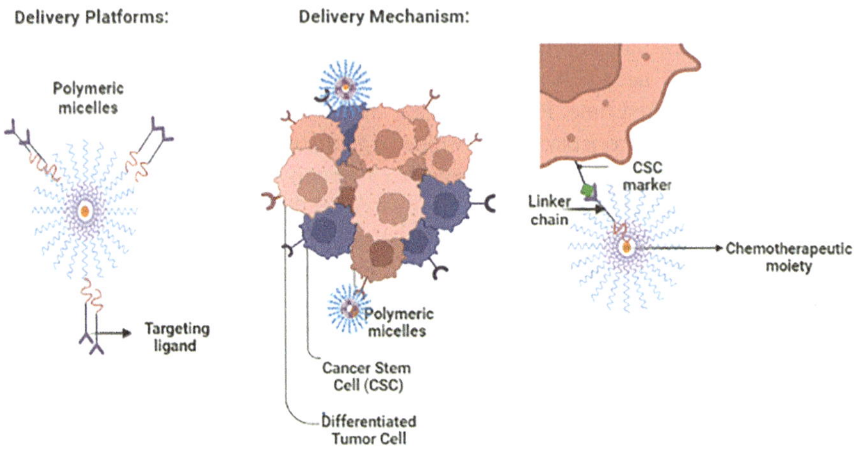

Fig. 11.1 Mechanism of active targeting

(Large et al. 2019). Chemotherapeutic drugs can be transported to specific cancer cells by developing a targeted nanocarrier drug delivery system that bypasses the activity of multidrug resistance transporters (Sutradhar and Amin 2014). Due to this therapeutic effectiveness of drugs is decreased and these drugs are unable to take place of more traditional chemotherapy options in clinical practice today. Targeting nanoparticles directly toward tumor cells may be an effective way to circumvent this problem, increasing the concentration of drug at targeted site and also enhancing the bioavailability of drug (Bharali and Mousa 2010). The mechanism of active targeting is illustrated in Fig. 11.1.

11.3 Various Ligands for Active Targeting

Many ligands have been developed and several of them have been examined for their potential role in the active targeting of nanoparticles. Therapeutic efficacy and uptake of drugs encapsulated in nanoparticles are enhanced by attaching ligands to the surface of target tumor cells (Ahmad et al. 2019). For active targeting ligands including nucleic acids, proteins, peptides, small molecules, and polysaccharides have been used (Gu et al. 2021). There are two methods by which these ligands may be attached to nanoparticles. After NPSs are formed, they may be chemically coupled or physically adsorbed on top of surface of nanoparticles, or they can be connected to polymers prior to the preparation of nanoparticles (Heuer-Jungemann et al. 2019). The several types of ligands conjugated with polymeric micelles to deliver therapeutics to targeted sites are described below.

11.3.1 Antibodies Targeted Polymeric Micelles

Antibody-targeted drug delivery systems preferentially transport anticancer medications to cancer cells rather than the normal cells of the body. Recently they have received a lot of attention in the realm of cancer treatment (Kadkhoda et al. 2021). It is feasible to selectively kill tumor cells while retaining an appropriate cytotoxic activity by using particular or related antigens to target tumors. Additionally, therapeutic antibodies aimed at activating the patient's antitumor immune response are also being produced. Even though immunotherapy is a promising treatment option, further research is needed to improve clinical outcomes and overcome therapeutic resistance (van Elsas et al. 2020). In cancer immunotherapy, the IgG isotype is the most usually employed, although there are several other isotypes as well, such as IgD, IgE, and IgM. Each antibody consists of three parts: two variable fragments that are responsible for binding antigen also known as antigen-binding fragments (Fabs) and one variable known as "light chain" (Fc). Antibody specificity is determined by complementarity determining regions (CDRs) in the antibody's antigen-binding fragment, whereas the Fc domain links antibodies to immune effector mechanisms through fragment receptors on cells such as neutrophils, NK cells, DCs, monocytes, and eosinophils (Al Qaraghuli et al. 2020). Additionally, the Fc domain of antibodies is considered to shield the antibodies from degradation by binding to Fc receptors. The mechanism of antibody-targeted polymeric micelles is shown in Fig. 11.2. For the conjugation of antibodies, polymeric micelles are an excellent platform nanocarrier because of their increased tumor extravasation and penetration (Ding et al. 2019; Cabral et al. 2018). These micelles have a large and diverse loading of bioactive compounds and their regulated release (Hussein and Youssry 2018).

By integrating antibody fragments, Ahn et al. developed polymeric micelles encapsulating oxaliplatin (1,2 diaminocyclohexane) platinum (II) (DACHPt) that has demonstrated excellent therapeutic effectiveness for treating numerous cancer models (anti-TF-Fav). DACHPt evaluated the anticancer effects of polymeric micelles containing anti-human TF targetable Fab fragments for pancreatic cancers.

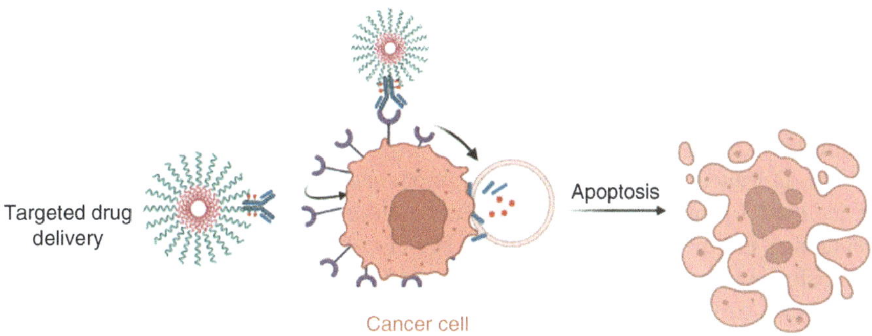

Fig. 11.2 Mechanism of antibody targeting polymeric micelles

Antibody-TF Fab'-DACHPt/m was successfully synthesized by incorporating maleimide-thiol chemistry. As compared to DACHPt/m, the capacity of anti-TF Fab' to recognize antigen aided in the binding and uptake of anti-TF Fab'-DACHPt/m in the cell more quickly. Anti-TF Fab'-DACHPt/m exhibited better antitumor effectiveness as seen by the slight weight loss of treated mice. The advantages of in vivo tumor targeting via an immune micellar system loaded with platinum drugs, which are necessary for any clinical anticancer therapy are established. If the parent micelles do not alter much in structure or size, it is possible to use this maleimide-thiol coupling approach to a wide range of different carrier peptides and antibodies, including therapeutically authorized tumor-directed peptides. The delivery of drugs into cancers with stroma-rich extravasation should provide universality and improve the therapeutic efficacy of carriers with stringent limits on extravasation (Ahn et al. 2015). Some examples of antibody conjugate polymeric micelles drug delivery systems that are developed are enlisted in Table 11.1.

11.3.2 Folate-Conjugated Polymeric Micelles

In the early 1990s, Kamen and colleagues discovered that receptor-mediated endocytic mechanisms can allow folate to enter tumor cells (Lee and Low 2000). A total of three isoforms of folate receptor (FRα, FRβ, FRγ) have been found in human tissues, including malignancies (Cai et al. 2017). Folate is used in anticancer medications as a targeting ligand to improve intracellular absorption into targeted tumor cells and decreases collateral damage to healthy cells (Zwicke et al. 2012). Folate has also been proven to penetrate deeper than conventional antibodies due to its small molecular weight. In addition, folate has a great affinity for folate receptors. It is possible to achieve this high-affinity characteristic with the right design of folate–drug conjugates, which may then be quickly absorbed via endocytosis. The folate-binding proteins that act as ligands on cancer cell surfaces are overexpressed in malignant cells (Farran et al. 2020). To improve the efficiency of anticancer treatment, these conjugates may be directed to tumor cells and taken up by the cells through a process called receptor-mediated endocytosis. Studies have shown that folic acid conjugates may either stay in recycling endosomes (pH 5–6) or make their way out into the cytoplasm and hence bypassing the multidrug efflux pumps present in cancer cells. With these features, intracellular absorption of the medicine may be increased, and harmful adverse effects of enzyme action can be avoided (Ebrahimnejad et al. 2022). The mechanism of folate conjugate polymeric micelles is enlisted in Fig. 11.3.

The anticancer medication 9-Nitro-camptothecin has low solubility, hence Han et al. tried to encapsulate it in polymeric micelles. Folate-conjugated micelles containing drugs were prepared using FA-PEG-DSPE9 (folate-polyethylene glycol-distearoyl phosphatidylethanolamine) and MPEG-DSPE (methoxy polyethylene glycol-distearoyl phosphatidylethanolamine) both of which were synthesized via chemical synthesis and exhibited high purity and smaller particle size range. MPEG-DSPE and FAPEG-DSPE at a molar ratio of 1:100 might evade macrophages and

Table 11.1 Some examples of antibodies conjugate polymeric micelles for drug delivery

S. no	Drug	Method	Outcome of study	Ref
1.	Oxaliplatin	Maleimide-thiol conjugation was used to implant antibody fragments into polymeric micelles which allowed for targeted delivery of oxaliplatin to pancreatic cancers. The tissue factor (TF)-targeting Fab′ was coupled by adjusting the maleimide surface density on the micelles	Compared to non-targeted micelles, fab conjugated oxaliplatin-loaded micelles enhanced cellular binding within 1 h and quick cellular internalization, leading to better in vitro cytotoxicity. The findings suggest that polymeric micelles with fab' implanted is a promising vehicle for effective medication delivery to solid tumors	Ahn et al. (2015)
2.	Curcumin, doxorubicin	In this study, researcher tested in vitro and in vivo therapeutic effectiveness of PEG-PE micelles loaded with curcumin (CUR) and doxorubicin (DOX) and conjugated with using an anti-GLUT1 antibody against HCT-116 human colorectal cancer cells (GLUT1)	Micelles loaded with doxorubicin (DOX) and CUR adorned with anti GLUT1 were highly effective in vitro at low concentrations. Tumor growth was not significantly inhibited by non-targeted CUR or CUR DOX micelles as compared to the control group at the levels assessed	Abouzeid et al. (2013)
3.	Streptavidin-conjugated monoclonal antibody	Researchers report the design, synthesis, and evaluation of a novel delivery system consisting of streptavidin-conjugated monoclonal antibody (mAb-SA) targeting CD22 that is capable of being internalized and (ii) a biotinylated diblock copolymer having positively charged siRNA and pH-responsive block to facilitate endosome release	Polymeric micelles that targeted CD22 were more successful at gene knockdown. DoHH2 cells treated with 15 nmol/L siRNA-targeting CD22 had 70% less gene expression. CD22-targeted polymer carrier may deliver siRNA to lymphoma cells	Palanca-Wessels et al. (2011)
4.	Doxorubicin	Incorporating a monoclonal antibody (mAb), researcher developed doxorubicin-based hydrazine polymer prodrug combination for self-assembled nanoparticles (PDNPs)	The results proved that pH-sensitive PDNPs are able to maintain a high degree of stability in PB 7.4 buffer solution. HepG2 cells were more responsive to mAb-PDNPs (IC50: 1.12 mg/L) than they were	Huang et al. (2021)

(continued)

Table 11.1 (continued)

S. no	Drug	Method	Outcome of study	Ref
		with pH-sensitive drug release, improved targeting, and tumor tissue aggregation. Researcher also investigated the PDNPs' and mAb-PDNPs' degradation resistance, in vitro drug behavior, and cell uptake	to PDNPs (IC50: 2.62 mg/L) because of the addition of the monoclonal antibody. CD147 mAb-modified peptide-DNA nanoparticles (mAb-PDNPs) are more stable and facilitate targeted medication delivery with greater efficiency	

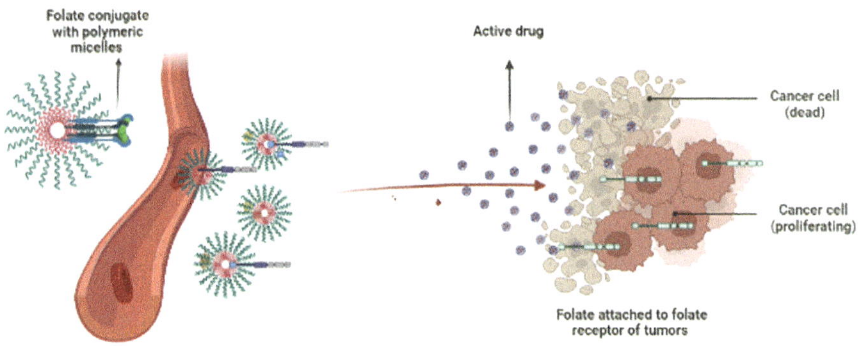

Fig. 11.3 Folate-conjugated polymeric micelles

express a highly specific targeting capability. When compared to blank micelles or without folate-conjugated micelles, the folate-conjugated micelles were able to target tumor cells more effectively with upregulation of folate receptors on tumor cell surface. For poorly soluble anticancer drugs that are difficult to encapsulate, polymeric micelle formulation and active targeting using folate as ligand presented in this study might be a viable carrier (Han et al. 2009).

Anticancer medication doxorubicin was encapsulated in PLGA-PEG-FOL carrier that was designed to target tumors by Zhao et al. Drug release in the endosomes is hastened by the addition of pH-sensitive poly b-amino ester (PAE) to PLGA-PEG-FOL to develop mixed micelles. According to these findings, the ionization of PAE can cause medication release at various pH levels. Micelles were made by combining PLGA-PEG-FOL and PAE-PEG-FOL in a multitude of weight ratios. By ionizing PAE, the drug release at endosomal pH can be induced, resulting in greater concentrations in the nucleus and better cytotoxicity due to increased buffering potential of the mixed micelles. The 20:80 mixed micelles had superior extracellular stability, drug release rate, and cytotoxicity when compared to single micelles (Zhao et al. 2010).

Ovarian cancer patients may benefit from photodynamic therapy (PDT), which is a cutting-edge treatment option that has gained popularity in recent years. Li et al. prepared a biodegradable polymer of PEG and PLA-folate was developed to create a photosensitizer nanocarrier that is both active targeting and water soluble. Excellent encapsulation efficiency and controlled, delayed release of therapeutics were seen in vitro and in vivo when hypocrellin B (HB) was encapsulated in FA-PEG-PLA micelles in comparison with free hypocrellin. More effective post-PDT cell death was achieved using HB/FA PEG-PLA micelles than with blank micelles as well as free medicines. In addition, SKOV3 cells were found to bind and take up folate-targeted micelles following intravenous injection into ascitic ovarian tumor bearing cancer rat models. Ascites tumor tissues show a 20-fold increase in drug concentrations proving the effectiveness of folate-targeted localized therapy (Li et al. 2018). Some examples of folate conjugate polymeric micelles drug delivery systems are enlisted in Table 11.2.

11.3.3 Chitosan Conjugated Polymeric Micelles

As a polysaccharide, chitosan is frequently used because of its positive zeta potential and ability to better adhere to tissues and cells (Fathi et al. 2018). Two types of D-glucopyranoses are found in chitosan, one of which is an unprocessed natural polysaccharide and the other which is semi-processed (Naskar et al. 2019). Mucoadhesive, biocompatibility, non-immunogenicity, and non-toxicity are only a few of its unique attributes. The polymer is primarily degraded by bacterial and lysozyme enzymes which are present in the colon; therefore, it does not pose a threat to human health (Sharifi-Rad et al. 2021; Thotakura et al. 2017).

Thotakura et al. prepared chitosan-biocompatible stearic acid polymeric micelles for the delivery of tamoxifen for the treatment of breast cancer. FT-IR and NMR were used to examine the conjugation, whereas micromeritics, surface charge, drug loading, and morphological characteristics were examined in the nanocarrier. By conducting in vitro MTT assays, as well as in vivo pharmacokinetic investigations, researchers were able to determine the drug's effectiveness and safety, as well as its biodistribution. Although the drug loading and release were enhanced, the pharmacokinetic profile was significantly modified, and the cytotoxicity against MCF-7 breast cancer cells was significantly boosted. As a result of the lower clearance and improved AUC, pharmacokinetic modification may be achieved using these nanocarriers (Thotakura et al. 2017). To improve the biopharmaceutical performance of medications that are not particularly water soluble, a novel polymer consisting of oleic acid grafting low-molecular weight carboxymethyl chitosan (OA-CMCS) was synthesized. Analytical methods such as 1H-NMR and FT-IR spectroscopy were used to provide a detailed characterization of this polymer after it was developed, synthesized, and thoroughly characterized via amidation process. There was a low critical micellar concentration of 1 g/mL for the OA-CMCS conjugate when it was suspended in water. Docetaxel (DTX), chemotherapeutic drug with low solubility in water, was chosen as a representative example. Following parameter optimization,

Table 11.2 Some examples of folate conjugate polymeric micelles for drug delivery

S. no.	Drug	Method	Outcome of study	Ref
1	Doxorubicin	To facilitate the administration of doxorubicin (DOX), researcher conjugate folic acid (FA) with amphiphilic copolymer MaPCL: Poly(−caprolactone) macromonomer) methacryloyloxyethyl phosphorylcholine)	Unloaded FA-P (MPC-co-MaPCL) micelles shows less cytotoxicity, while DOX-loaded FA-P (MPC-co-MaPCL) micelles are highly cytotoxic to HeLa cells. DOX-loaded FA-P(MPC-co-MaPCL) micelles absorb DOX more than unconjugated folic acid DOX-loaded P(MPC-co-MaPCL) micelles after 6 h of incubation in HeLa cells. These findings emphasize the system's potential as an anticancer drug delivery technique	Lu et al. (2019)
2	Dactolisib	Researchers employed platinum (II)-based linker chemistry to integrate dactolisib into the core of poly (ethylene glycol)-b-poly (acrylic acid) (PEG-b-PAA) polymeric micelles. Amphiphilic dactolisib-PEG-PAA monomers form, which then self-assemble into core-shell polymeric micelles in an aqueous environment	Dactolisib encapsulated polymeric micelles had excellent colloidal stability in water, and they secreted the linked drug in chloride and glutathione buffers. Micelles coated with folate showed targeted cellular cytotoxicity at 50–75 nM IC50 and were actively absorbed by positive folate-receptor KB cells	Shi et al. (2020)
3	Doxorubicin and paclitaxel	Researcher developed micellar system for the simultaneous delivery of DOX and PTX. The folate was conjugated to the doxorubicin using Schiff's base reaction. The prodrug was then encapsulated, and folate was grafted on polymeric micelles	Researchers found that polymeric micelles released DOX and PTX steadily over time, and lowering the pH of the media speed up the drug release rate. Endocytosis of polymeric micelles by SW1353 cells was detected using a	Lang et al. (2020)

(continued)

Table 11.2 (continued)

S. no.	Drug	Method	Outcome of study	Ref
			cellular uptake test, and the cellular absorption of folate-conjugated micelles was increased due to an active FR-mediated endocytosis pathway, as evidenced by their greater rcd fluorescence compared to that of non-folate micelles	
4	Silibinin	The purpose of this research was to develop folic acid conjugated pluronic F127 nanomicelle loaded with silibinin (SLB) for the treatment of liver cancer. SLB-F127-FA nanomicelles were developed by using the Steglich esterification method to conjugate folic acid with Pluronic F127 copolymer	In vitro cytotoxicity investigations demonstrated folate-targeted nano micelle increased drug cytotoxicity against HepG2 cancer cells. Overall, the proposed SLB-F127-FA polymeric nanomicelles demonstrated tremendous potential as an active-targeted drug delivery system for treatment of cancer cells	Ghalehkhondabi et al. (2021)

doxorubicin conjugate OA-CMCS micelles with an entrapment efficiency of 57.26 1.25 nm were formed. In an in vitro release investigation, DTX-loaded OA-CMCS micelles showed a gradual and persistent drug release behavior in a model of the body's fluid. Using these new nanocarriers, pharmacokinetic modification may be achieved without using standard lipid-based nanocarriers while also improving dosage delivery, tissue adhesion, and tissue adhesion (Kumar et al. 2020). Some examples of chitosan conjugate polymeric micelles drug delivery systems which are developed are enlisted in Table 11.3.

11.3.4 RGD Peptide Conjugated Polymeric Micelles

Many peptide sequences useful for targeting certain malignancies and cell types have been discovered via peptide phage display. More specifically peptides with RGD and NGR sequences have been studied for their potential as tumor-homing agents. These peptides are specific for the $\alpha v\beta 3$ integrin and aminopeptidase N

Table 11.3 Examples of chitosan conjugated polymeric micelles

S. no	Drug	Method	Outcome of study	Ref
1.	Paclitaxel	This study involved the synthesis of an amphiphilic compound of cholesterol-modified chitosan and mPEG (mPEG-CS-Hz-CH). Micelles made of mPEG-CS-Hz-CH that are encapsulated with paclitaxel (PTX) were developed using ultrasonic probes	Micelles loaded with paclitaxel (PTX) demonstrated enhanced cytotoxicity and strong selectivity for tumor cells in an in vitro cytotoxicity study. Micelles loaded with paclitaxel (PTX) were shown to increase PTX's therapeutic efficacy while decreasing its negative effects	Han et al. (2021)
2	Paclitaxel	For enhanced oral bioavailability of paclitaxel (PTX), l-carnitine-chitosan-stearic acid polymeric micelles were created	Lower micelle concentration improves water stability. Pharmacokinetic studies demonstrate that stearic acid-grafted chitosan increases PTX bioavailability. In vitro studies show that medication-containing micelles released the therapeutic slowly and at controlled release. The activity improved intestinal absorption, as evidenced by Caco-2 cell uptake in their own membranes	Yang et al. (2020)
3	Doxorubicin	With the goal of improving the oral bioavailability of doxorubicin (DOX), a quercetin-chitosan compound (QT-CS) was developed. This conjugation improves DOX's water solubility, opens tight junctions, and circumvents the P-glycoprotein (P-gp)	Micelles formed from the produced QT-CS encapsulated DOX efficiently due to their small size, high zeta potential, and high encapsulation rate. In GI simulation fluid, these micelles show sustained-release profile. The absorption of doxorubicin by cells was increased 2.2-fold when QT-CS micelles were used as a delivery vehicle. Doxorubicin's apparent permeability coefficient was 10.17 times greater than that of free doxorubicin	Mu et al. (2019)
4	Paclitaxel	Conjugation of αtocopherol succinate to the glycol chitosan resulted in the formation of an amphiphilic polymer called TS-GC,	In vitro and in vivo testing demonstrated the ability of the polymer as carriers for PTX to enhance their transport capabilities. The toxicity level and dosage of	Liang et al. (2018)

(continued)

Table 11.3 (continued)

S. no	Drug	Method	Outcome of study	Ref
		which was further characterized by NMR	the PTX-loaded TS-GC micelles were found to be superior to those of the Cremophor EL-based formulation	

receptors. Both are vastly expressed in tumor cells and tumor blood vessels (Shi et al. 2019). In recent years, targeting V3 integrin, a protein often found in tumors, has emerged as a promising and sought-after therapeutic approach. These heterodimeric receptors are a key component of a regulatory cascade that regulates angiogenesis and can function as target epitopes in drug vectorization systems by facilitating the adhesion of activated endothelial cells to extracellular matrix components. αvβ3receptors are overexpressed on the surface of endothelial cells and, in some circumstances on the tumor cells themselves in malignant tumors. Metastasis development in these cells is a multi-step process, and αvβ3, receptors play a crucial role in each. They take part in matrix remodeling, interstitial matrix invasion, micro-emboli development, and tumor cell extravasation. αvβ3 integrins are upregulated in both invasive cancer cells and angiogenic endothelium cells suggesting that targeting them may have considerable therapeutic benefits (Qiu et al. 2018). Some examples of RGD peptide conjugate polymeric micelles drug delivery systems which are developed are given in Table 11.4.

11.3.5 Transferrin and Epidermal Growth Factor Conjugated Polymeric Micelles

Since tumor cells have a greater number of transferrin receptors than healthy ones, this has led to their use as a ligand for the transport of anticancer drugs. Numerous initiatives are currently underway to investigate this putative ligand-receptor-mediated delivery pathway (Kedar et al. 2010). Blood plasma protein transferrin (Tf) binds to the transferrin-receptor on the surface of cell membrane and transports iron into the cell. Since TfRs are more expressed in a wide variety of human cancers, they are hypothesized to serve as a useful biological marker for the administration of anticancer medications and the detection of malignancies in individuals whose tumors show high levels of transferrin receptors. Transferrin conjugated drug delivery has been extensively researched in past several decades for its prospective use in the treatment of broad variety of cancers, including breast cancer, prostate cancer, myeloid leukemia, and others. In addition, the transport of medications across the blood–brain barrier for the treatment of brain cancers has been the focus of a lot of research and development work employing Tf or anti-TfR antibodies as the target moiety in creating Tf-TfR mediated drug delivery systems (such as glioma). Although Tf-drug conjugates are the most often reported kind of Tf-carrying drug

Table 11.4 Examples of RGD peptide conjugate polymeric micelles

S. no	Drug	Method	Outcomes	Ref
1	Paclitaxel	RGD-decorated PEG-paclitaxel disulfide-linked conjugates were created. Micelles (RGD micelles) self-assemble from amphiphilic PEG-PTX conjugates, which are degraded when glutathione (GSH) levels decrease, releasing PTX into the intracellular environment	In vitro experiments showed that RGD micelles target stomach cancer cells and cause apoptosis, limiting cell growth. RGD micelles prevent tumor development in vivo by traveling to the tumor site and releasing PTX within tumor cells. The micelles' high therapeutic effectiveness and negligible side effects may help tailor pharmaceutical delivery for stomach cancer	Lorenzi et al. (2022)
2	Monomethyl auristatin (MME)	To improve drug loading and targeted drug delivery of MMAE to HCT-116 colorectal tumor xenografts the use of cRGD-functionalized polylipopeptide micelles (cRGD-Lipep-Ms) was investigated	Most notably, cRGD-Lipep-Ms boosted MMAE loading content by 55 times MMAE-cRGD-Lipep-Ms showed excellent absorption and antiproliferative action. MMAE-cRGD-Lipep-Ms completely suppressed HCT-116 tumor xenograft development in mice, showing tenfold greater tolerability than free MMAE	Fang et al. (2017)
3	Cross-linked polymers	The researcher cross-linked polymeric (CCPM) micelles at their cores and conjugated them with cyclic arginine-glycine-aspartic acid (cRGD) peptide. Using various cell lines with varying expression levels of the integrin receptor, the advantage of conjugating CCPM with varying densities of cRGD was examined	RGD micelles were biologically stable since they were 50-nm spherical nanoparticles. Scientists investigated if GSH may stimulate micelle PTX release. In vitro experiments showed that RGD micelles target stomach cancer cells and cause apoptosis, limiting cell growth. RGD micelles prevent tumor development in vivo by traveling to the tumor site and releasing PTX within tumor cells	Marques and Kumar (2022)

(continued)

Table 11.4 (continued)

S. no	Drug	Method	Outcomes	Ref
4	Doxorubicin	Scientists have developed and evaluated a novel DOX-loaded polymeric micelle that self-assembles from RGD-terminated poly (ethylene glycol)-block-poly (trimethylene carbonate) (RGD-PEG-PTMC) amphiphilic biodegradable block copolymer to enhance the activity of DOX for treating osteosarcoma	The RGD-conjugated DOX-loaded polymeric micelle (RGD-DOX-PM) has a high drug loading efficiency. 63% of RGD-DOX-DOX PMs were released in 60 h. MTT experiments on MG-63 and MNNG/HOS osteosarcoma cells showed that conjugated RGD Dox has high cytotoxicity against tumor cells	Bergonzini et al. (2022)

delivery system, Tf-carrying nanocarriers including Tf-liposome and Tf-polymersome systems are also being developed and studied (Zhao et al. 2016).

Many kinds of cancer cells have abnormally high levels of the epidermal growth factor receptor (EGFR), a 170 kDa transmembrane protein featuring an intracellular tyrosine kinase domain. Cancer cell proliferation, apoptosis, angiogenesis, and metastasis all receive assistance from the epidermal growth factor receptor (EGFR) autocrine pathway (Chang et al. 2018). Considering the importance of EGFR in carcinogenesis, researchers have been looking for specific inhibitors of the EGFR signaling system. Targeting the epidermal growth factor receptor (EGFR) has shown promise in early clinical trials and is supported by the results of a huge body of preclinical research. Numerous treatment initiatives are being centered on the epidermal growth factor receptor (EGFR). Human epithelial tumors with overexpressed epidermal growth factor receptor (EGFR) tend to be more aggressive, have a worse prognosis for patients, and have a higher metastatic risk. EGFR-overexpressing breast cancer cells are demonstrated to undergo cell-cycle arrest in the G1 phase and subsequent activation of cell-type-specific apoptosis in response to drug-free EGF-conjugated micelles (Lee et al. 2007).

11.4 Conclusion

Over the last several years, polymeric micelles have emerged as a promising new delivery system for a wide range of biological macromolecules as well as conventional anticancer drugs like DNA- and siRNA-targeting small interfering RNAs (siRNAs) and antibodies like oligonucleotides. Complex micelles that include many modalities into a single carrier have been created by chemically modifying the structure of block copolymers that generate micelles. There is a definite move away from producing only single-therapeutic agent drug delivery system and toward micelles drug delivery system that incorporates several therapeutic payloads and can be changed for active targeting, delivery of imaging agents, and responsiveness to unique signals supplied either by the tumor microenvironment or outside, to regulate the release of encapsulated cargoes spatially and temporally. This chapter has discussed a broad range of polymeric micelle modifications, such as conjugation with active ligands for active targeting, ranging from passively targeted polymeric micelles that depend on the EPR effect to more sophisticated systems that incorporate targeting ligands. Polymeric micelles have promising future because of their wide range of uses and structural flexibility. Given their many practical applications and amenability to structural tweaks, polymeric micelles have a bright future ahead of them.

References

Abouzeid AH, Patel NR, Rachman IM, Senn S, Torchilin VP (2013) Anti-cancer activity of anti-GLUT1 antibody-targeted polymeric micelles co-loaded with curcumin and doxorubicin. J Drug Target 21(10):994–1000. https://doi.org/10.3109/1061186X.2013.840639

Ahmad A, Khan F, Mishra RK, Khan R (2019) Precision cancer nanotherapy: evolving role of multifunctional nanoparticles for cancer active targeting. J Med Chem 62(23):10475–10496. https://doi.org/10.1021/acs.jmedchem.9b00511

Ahn J, Miura Y, Yamada N, Chida T, Liu X, Kim A, Sato R, Tsumura R, Koga Y, Yasunaga M, Nishiyama N, Matsumura Y, Cabral H, Kataoka K (2015) Antibody fragment-conjugated polymeric micelles incorporating platinum drugs for targeted therapy of pancreatic cancer. Biomaterials 39:23–30. https://doi.org/10.1016/j.biomaterials.2014.10.069

Al Qaraghuli MM, Kubiak-Ossowska K, Ferro VA, Mulheran PA (2020) Antibody-protein binding and conformational changes: identifying allosteric signalling pathways to engineer a better effector response. Sci Rep 10(1):13696. https://doi.org/10.1038/s41598-020-70680-0

Anselmo AC, Mitragotri S (2014) An overview of clinical and commercial impact of drug delivery systems. J Control Release 190:15–28. https://doi.org/10.1016/j.jconrel.2014.03.053

Bergonzini C, Kroese K, Zweemer AJM, Danen EHJ (2022) Targeting integrins for cancer therapy—disappointments and opportunities. Front Cell Dev Biol 10:863850. https://doi.org/10.3389/fcell.2022.863850

Bharali DJ, Mousa SA (2010) Emerging nanomedicines for early cancer detection and improved treatment: current perspective and future promise. Pharmacol Ther 128(2):324–335. https://doi.org/10.1016/j.pharmthera.2010.07.007

Cabral H, Miyata K, Osada K, Kataoka K (2018) Block copolymer micelles in nanomedicine applications. Chem Rev 118(14):6844–6892. https://doi.org/10.1021/acs.chemrev.8b00199

Cai L, Michelakos T, Ferrone CR, Zhang L, Deshpande V, Shen Q, DeLeo A, Yamada T, Zhang G, Ferrone S, Wang X (2017) Expression status of folate receptor alpha is a predictor of survival in pancreatic ductal adenocarcinoma. Oncotarget 8(23):37646–37656. https://doi.org/10.18632/oncotarget.16841

Chang M-H, Pai C-L, Chen Y-C, Yu H-P, Hsu C-Y, Lai P-S (2018) Enhanced antitumor effects of epidermal growth factor receptor targetable Cetuximab-conjugated polymeric micelles for photodynamic therapy. Nanomaterials 8(2):121. https://doi.org/10.3390/nano8020121

Da Silva FLO, Marques MBDF, Kato KC, Carneiro G (2020) Nanonization techniques to overcome poor water-solubility with drugs. Expert Opin Drug Discovery 15(7):853–864. https://doi.org/10.1080/17460441.2020.1750591

De Lorenzi F, Rizzo LY, Daware R, Motta A, Baues M, Bartneck M, Vogt M, van Zandvoort M, Kaps L, Hu Q, Thewissen M, Casettari L, Rijcken CJF, Kiessling F, Sofias AM, Lammers T (2022) Profiling target engagement and cellular uptake of cRGD-decorated clinical-stage core-crosslinked polymeric micelles. Drug Deliv Transl Res. https://doi.org/10.1007/s13346-022-01204-8

Din FU, Aman W, Ullah I, Qureshi OS, Mustapha O, Shafique S, Zeb A (2017) Effective use of nanocarriers as drug delivery systems for the treatment of selected tumors. Int J Nanomedicine 12:7291–7309. https://doi.org/10.2147/IJN.S146315

Ding J, Chen J, Gao L, Jiang Z, Zhang Y, Li M, Xiao Q, Lee SS, Chen X (2019) Engineered nanomedicines with enhanced tumor penetration. Nano Today 29:100800. https://doi.org/10.1016/j.nantod.2019.100800

Ebrahimnejad P, Sodagar Taleghani A, Asare-Addo K, Nokhodchi A (2022) An updated review of folate-functionalized nanocarriers: a promising ligand in cancer. Drug Discov Today 27(2): 471–489. https://doi.org/10.1016/j.drudis.2021.11.011

Edis Z, Wang J, Waqas MK, Ijaz M, Ijaz M (2021) Nanocarriers-mediated drug delivery systems for anticancer agents: an overview and perspectives. Int J Nanomedicine 16:1313–1330. https://doi.org/10.2147/IJN.S289443

Ekladious I, Colson YL, Grinstaff MW (2019) Polymer–drug conjugate therapeutics: advances, insights and prospects. Nat Rev Drug Discov 18(4):273–294. https://doi.org/10.1038/s41573-018-0005-0

Fang Z, Sun Y, Xiao H, Li P, Liu M, Ding F, Kan W, Miao R (2017) Targeted osteosarcoma chemotherapy using RGD peptide-installed doxorubicin-loaded biodegradable polymeric micelle. Biomed Pharmacother 85:160–168. https://doi.org/10.1016/j.biopha.2016.11.132

Farran B, Montenegro RC, Kasa P, Pavitra E, Huh YS, Han Y-K, Kamal MA, Nagaraju GP, Rama Raju GS (2020) Folate-conjugated nanovehicles: strategies for cancer therapy. Mater Sci Eng C 107:110341. https://doi.org/10.1016/j.msec.2019.110341

Fathi M, Majidi S, Zangabad PS, Barar J, Erfan-Niya H, Omidi Y (2018) Chitosan-based multifunctional nanomedicines and theranostics for targeted therapy of cancer. Med Res Rev 38(6): 2110–2136. https://doi.org/10.1002/med.21506

Fernandes C, Suares D, Yergeri MC (2018) Tumor microenvironment targeted nanotherapy. Front Pharmacol 9:1230. https://doi.org/10.3389/fphar.2018.01230

Fluksman A, Benny O (2019) A robust method for critical micelle concentration determination using coumarin-6 as a fluorescent probe. Anal Methods 11(30):3810–3818. https://doi.org/10.1039/C9AY00577C

Ghalehkhondabi V, Soleymani M, Fazlali A (2021) Folate-targeted nanomicelles containing silibinin as an active drug delivery system for liver cancer therapy. J Drug Deliv Sci Technol 61:102157. https://doi.org/10.1016/j.jddst.2020.102157

Gu W, Meng F, Haag R, Zhong Z (2021) Actively targeted nanomedicines for precision cancer therapy: concept, construction, challenges and clinical translation. J Control Release 329:676–695. https://doi.org/10.1016/j.jconrel.2020.10.003

Han X, Liu J, Liu M, Xie C, Zhan C, Gu B, Liu Y, Feng L, Lu W (2009) 9-NC-loaded folate-conjugated polymer micelles as tumor targeted drug delivery system: preparation and evaluation in vitro. Int J Pharm 372(1–2):125–131. https://doi.org/10.1016/j.ijpharm.2008.12.035

Han Y, Pan J, Liang N, Gong X, Sun S (2021) A pH-sensitive polymeric micellar system based on chitosan derivative for efficient delivery of paclitaxel. Int J Mol Sci 22(13):6659. https://doi.org/10.3390/ijms22136659

Hari SK, Gauba A, Shrivastava N, Tripathi RM, Jain SK, Pandey AK (2022) Polymeric micelles and cancer therapy: an ingenious multimodal tumor-targeted drug delivery system. Drug Deliv Transl Res 13:135. https://doi.org/10.1007/s13346-022-01197-4

Heuer-Jungemann A, Feliu N, Bakaimi I, Hamaly M, Alkilany A, Chakraborty I, Masood A, Casula MF, Kostopoulou A, Oh E, Susumu K, Stewart MH, Medintz IL, Stratakis E, Parak WJ, Kanaras AG (2019) The role of ligands in the chemical synthesis and applications of inorganic nanoparticles. Chem Rev 119(8):4819–4880. https://doi.org/10.1021/acs.chemrev.8b00733

Huang Y-K, Tian H-R, Zhang M-Z, He J-L, Liu J, Ni P-H (2021) Monoclonal antibody-conjugated Polyphosphoester-hyd-DOX prodrug nanoparticles for targeted chemotherapy of liver cancer cells. Chin J Polym Sci 39(11):1392–1402. https://doi.org/10.1007/s10118-021-2582-3

Hussein Y, Youssry M (2018) Polymeric micelles of biodegradable Diblock copolymers: enhanced encapsulation of hydrophobic drugs. Materials 11(5):688. https://doi.org/10.3390/ma11050688

Hwang D, Ramsey JD, Kabanov AV (2020) Polymeric micelles for the delivery of poorly soluble drugs: from nanoformulation to clinical approval. Adv Drug Deliv Rev 156:80–118. https://doi.org/10.1016/j.addr.2020.09.009
Kadkhoda J, Akrami-Hasan-Kohal M, Tohidkia MR, Khaledi S, Davaran S, Aghanejad A (2021) Advances in antibody nanoconjugates for diagnosis and therapy: a review of recent studies and trends. Int J Biol Macromol 185:664–678. https://doi.org/10.1016/j.ijbiomac.2021.06.191
Kataoka K, Harada A, Wakebayashi D, Nagasaki Y (1999) Polyion complex micelles with reactive aldehyde groups on their surface from plasmid DNA and end-functionalized charged block copolymers. Macromolecules 32(20):6892–6894. https://doi.org/10.1021/ma990973n
Kedar U, Phutane P, Shidhaye S, Kadam V (2010) Advances in polymeric micelles for drug delivery and tumor targeting. Nanomedicine 6(6):714–729. https://doi.org/10.1016/j.nano.2010.05.005
Kumar R, Sirvi A, Kaur S, Samal SK, Roy S, Sangamwar AT (2020) Polymeric micelles based on amphiphilic oleic acid modified carboxymethyl chitosan for oral drug delivery of bcs class iv compound: intestinal permeability and pharmacokinetic evaluation. Eur J Pharm Sci 153:105466. https://doi.org/10.1016/j.ejps.2020.105466
Lang X, Wang T, Sun M, Chen X, Liu Y (2020) Advances and applications of chitosan-based nanomaterials as oral delivery carriers: a review. Int J Biol Macromol 154:433–445. https://doi.org/10.1016/j.ijbiomac.2020.03.148
Large DE, Soucy JR, Hebert J, Auguste DT (2019) Advances in receptor-mediated, tumor-targeted drug delivery. Adv Ther 2(1):1800091. https://doi.org/10.1002/adtp.201800091
Lee RJ, Low PS (2000) Folate as a targeting device for proteins utilizing folate receptor-mediated endocytosis. In: Francis GE, Delgado C (eds) Drug targeting, vol 25. Humana Press, Totova, pp 69–76. https://doi.org/10.1385/1-59259-075-6:69
Lee H, Hu M, Reilly RM, Allen C (2007) Apoptotic epidermal growth factor (EGF)-conjugated block copolymer micelles as a nanotechnology platform for targeted combination therapy. Mol Pharm 4(5):769–781. https://doi.org/10.1021/mp0700559
Lei W, Yang C, Wu Y, Ru G, He X, Tong X, Wang S (2022) Nanocarriers surface engineered with cell membranes for cancer targeted chemotherapy. J Nanobiotechnol 20(1):45. https://doi.org/10.1186/s12951-022-01251-w
Liang N, Sun S, Gong X, Li Q, Yan P, Cui F (2018) Polymeric micelles based on modified glycol chitosan for paclitaxel delivery: preparation, characterization and evaluation. Int J Mol Sci 19(6):1550. https://doi.org/10.3390/ijms19061550
Li J, Yao S, Wang K, Lu Z, Su X, Li L, Yuan C, Feng J, Yan S, Kong B, Song K (2018) Hypocrellin B-loaded, folate-conjugated polymeric micelle for intraperitoneal targeting of ovarian cancer in vitro and in vivo. Cancer Sci 109(6):1958–1969. https://doi.org/10.1111/cas.13605
Lu Q, Yi M, Zhang M, Shi Z, Zhang S (2019) Folate-conjugated cell membrane mimetic polymer micelles for tumor-cell-targeted delivery of doxorubicin. Langmuir 35(2):504–512. https://doi.org/10.1021/acs.langmuir.8b03693
Majumder N, Das NG, Das SK (2020) Polymeric micelles for anticancer drug delivery. Ther Deliv 11(10):613–635. https://doi.org/10.4155/tde-2020-0008
Marques SM, Kumar L (2022) Mechanism of polymeric micelles for drug targeting to brain tumors. In: Nanocarriers for drug-targeting brain tumors. Elsevier, Amsterdam, pp 367–399. https://doi.org/10.1016/B978-0-323-90773-6.00022-1
Mu Y, Fu Y, Li J, Yu X, Li Y, Wang Y, Wu X, Zhang K, Kong M, Feng C, Chen X (2019) Multifunctional quercetin conjugated chitosan nano-micelles with P-gp inhibition and permeation enhancement of anticancer drug. Carbohydr Polym 203:10–18. https://doi.org/10.1016/j.carbpol.2018.09.020
Naskar S, Kuotsu K, Sharma S (2019) Chitosan-based nanoparticles as drug delivery systems: a review on two decades of research. J Drug Target 27(4):379–393. https://doi.org/10.1080/1061186X.2018.1512112
Palanca-Wessels MC, Convertine AJ, Cutler-Strom R, Booth GC, Lee F, Berguig GY, Stayton PS, Press OW (2011) Anti-CD22 antibody targeting of pH-responsive micelles enhances small interfering RNA delivery and gene silencing in lymphoma cells. Mol Ther 19(8):1529–1537. https://doi.org/10.1038/mt.2011.104

Parra A, Jarak I, Santos A, Veiga F, Figueiras A (2021) Polymeric micelles: a promising pathway for dermal drug delivery. Materials 14(23):7278. https://doi.org/10.3390/ma14237278

Qiu M, Wang X, Sun H, Zhang J, Deng C, Zhong Z (2018) Cyclic RGD-peptide-functionalized polylipopeptide micelles for enhanced loading and targeted delivery of monomethyl Auristatin E. Mol Pharm 15(11):4854–4861. https://doi.org/10.1021/acs.molpharmaceut.8b00498

Salahpour Anarjan F (2019) Active targeting drug delivery nanocarriers: ligands. Nano Struct Nano Objects 19:100370. https://doi.org/10.1016/j.nanoso.2019.100370

Senapati S, Mahanta AK, Kumar S, Maiti P (2018) Controlled drug delivery vehicles for cancer treatment and their performance. Signal Transduct Target Ther 3(1):7. https://doi.org/10.1038/s41392-017-0004-3

Sharifi-Rad J, Quispe C, Butnariu M, Rotariu LS, Sytar O, Sestito S, Rapposelli S, Akram M, Iqbal M, Krishna A, Kumar NVA, Braga SS, Cardoso SM, Jafernik K, Ekiert H, Cruz-Martins-N, Szopa A, Villagran M, Mardones L, Calina D (2021) Chitosan nanoparticles as a promising tool in nanomedicine with particular emphasis on oncological treatment. Cancer Cell Int 21(1): 318. https://doi.org/10.1186/s12935-021-02025-4

Shi J, Liu S, Yu Y, He C, Tan L, Shen YM (2019) RGD peptide-decorated micelles assembled from polymer–paclitaxel conjugates towards gastric cancer therapy. Colloids Surf B Biointerfaces 180:58–67. https://doi.org/10.1016/j.colsurfb.2019.04.042

Shi H, van Steenbergen MJ, Lou B, Liu Y, Hennink WE, Kok RJ (2020) Folate decorated polymeric micelles for targeted delivery of the kinase inhibitor dactolisib to cancer cells. Int J Pharm 582:119305. https://doi.org/10.1016/j.ijpharm.2020.119305

Sutradhar KB, Amin ML (2014) Nanotechnology in cancer drug delivery and selective targeting. ISRN Nanotechnol 2014:1–12. https://doi.org/10.1155/2014/939378

Thotakura N, Dadarwal M, Kumar P, Sharma G, Guru SK, Bhushan S, Raza K, Katare OP (2017) Chitosan-stearic acid based polymeric micelles for the effective delivery of tamoxifen: cytotoxic and pharmacokinetic evaluation. AAPS PharmSciTech 18(3):759–768. https://doi.org/10.1208/s12249-016-0563-6

van Elsas MJ, van Hall T, van der Burg SH (2020) Future challenges in cancer resistance to immunotherapy. Cancer 12(4):935. https://doi.org/10.3390/cancers12040935

Veselov VV, Nosyrev AE, Jicsinszky L, Alyautdin RN, Cravotto G (2022) Targeted delivery methods for anticancer drugs. Cancer 14(3):622. https://doi.org/10.3390/cancers14030622

Wen H, Jung H, Li X (2015) Drug delivery approaches in addressing clinical pharmacology-related issues: opportunities and challenges. AAPS J 17(6):1327–1340. https://doi.org/10.1208/s12248-015-9814-9

Yang T, Feng J, Zhang Q, Wu W, Mo H, Huang L, Zhang W (2020) L-carnitine conjugated chitosan-stearic acid polymeric micelles for improving the oral bioavailability of paclitaxel. Drug Deliv 27(1):575–584. https://doi.org/10.1080/10717544.2020.1748762

Yu G, Ning Q, Mo Z, Tang S (2019) Intelligent polymeric micelles for multidrug co-delivery and cancer therapy. Artif Cells Nanomed Biotechnol 47(1):1476–1487. https://doi.org/10.1080/21691401.2019.1601104

Zhao H, Duong HHP, Yung LYL (2010) Folate-conjugated polymer micelles with pH-triggered drug release properties: folate-conjugated polymer micelles with pH-triggered. Macromol Rapid Commun 31(13):1163–1169. https://doi.org/10.1002/marc.200900876

Zhao H, Li Q, Hong Z (2016) Paclitaxel-loaded mixed micelles enhance ovarian cancer therapy through extracellular pH-triggered PEG detachment and endosomal escape. Mol Pharm 13(7): 2411–2422. https://doi.org/10.1021/acs.molpharmaceut.6b00164

Zwicke GL, Ali Mansoori G, Jeffery CJ (2012) Utilizing the folate receptor for active targeting of cancer nanotherapeutics. Nanotechnol Rev 3(1):18496. https://doi.org/10.3402/nano.v3i0.18496

Polymeric Micelles in the Delivery of Proteins

12

Sumel Ashique, Ashish Garg, Afzal Hussain, Shvetank Bhatt, Ankur Agrawal, and Neeraj Mishra

Abstract

Polymeric micelles (PM) developed in solution by self-assembling of amphiphilic polymers and represents an advanced tool to overcome several issues related to drug delivery like low aqueous solubility drug that may lead to low drug permeability across biological barriers. PM have several benefits, such as their tiny size, high solubility, ease of sterilizing, and controlled release of pharmaceuticals. Since the integrated medicine may be released rapidly in vivo, the physical stability of this carrier is a crucial concern. To effectively develop micelles that can carry drugs to their sites of action, the significant effort still must be done in understanding how PM interact with plasmatic and cellular components. This chapter describes the advantages and challenges of PM for

S. Ashique
Department of Pharmaceutics, Bharat Institute of Technology (BIT), School of Pharmacy, Meerut, UP, India

A. Garg
Department of P.G. Studies and Research in Chemistry and Pharmacy, Rani Durgavati University, Jabalpur, MP, India

A. Hussain
Department of Pharmaceutics, College of Pharmacy, King Saud University, Riyadh, Saudi Arabia

S. Bhatt
Department of Pharmacology, Amity Institute of Pharmacy, Amity University of Madhya Pradesh (AUMP), Gwalior, Madhya Pradesh, India

A. Agrawal
School of Pharmacy ITM University, Gwalior, Madhya Pradesh, India

N. Mishra (✉)
Department of Pharmaceutics, Amity Institute of Pharmacy, Amity University Madhya Pradesh (AUMP), Gwalior, Madhya Pradesh, India

S. K. Singh et al. (eds.), *Polymeric Micelles: Principles, Perspectives and Practices*, https://doi.org/10.1007/978-981-99-0361-0_12

the delivery of proteins, oral uptake of PM, types of polymers utilized in the administration of micellar drugs, and PM for multiple functionality and protein delivery.

Keywords

Polymeric micelles · SiRNA · Proteins · Ligands · Nanoparticles · Oral delivery

12.1 Introduction

Although protein medicines offer a lot of potential as targeted therapeutics, their use is hampered by issues including instability, a brief half-life, and unfavourable immune reactions. As a result, ways of effectively increasing the availability and activation of proteins in specific tissues may be found in protein delivery approaches based on stimuli-responsive nanocarriers. Thus, substantial research is being done on PM that may encapsulate proteins (Tao et al. 2020). The core-shell structure of PM for developing nanoscale drug delivery device was created via the self-assembling of amphiphilic block copolymers after dispersion in the aqueous phase. Amphiphilic di-block-copolymers such as polystyrene and poly(ethylene glycol) (PEG) and triblock-copolymers (poloxamers) are common and frequently used polymeric materials for the development of micelles, but graft (like G-chitosan) and ionic [like PEG-poly(-caprolactone)-g-polyethyleneimine] copolymers are also used (Yadav et al. 2019). As excipients for conventional pharmaceutical formulations, biocompatible polymers have been widely used in pharmaceutical research (Bharate et al. 2016; Cabral et al. 2018). More recently, they have been used in nanomedicines to improve the therapeutic results of strong pharmaceuticals. The development of micelle-based delivery methods as possible human nanomedicines has been suggested for several new block copolymers (Lee et al. 2008; Kim et al. 2004). The delivery of medicinal compounds has been greatly improved because of several important developments in PM. Due to the functions the polymer confers on the formulation, standard PM formulations are anticipated to improve the therapeutic efficacy of the medication encapsulated in a protein. Because PM can release their cargo from the core in a regulated way throughout systemic circulation, their PK profiles for therapeutic drugs differ from those of native substances. PM structural characteristics, such as their hydrophilic shell, aid in preventing opsonization by the complement system and unanticipated drug loss from serum components, both of which usually lead to the fast clearance of medicines from systemic circulation (Owens and Peppas 2006). Due to their many advantageous properties, such as their capability to access the affected area with compromised vasculature, longevity, safety, long term drug release, improved stability (in vivo and in vitro), and capacity to successfully emulsify various poor soluble drugs, they have become very popular. Additionally, by modifying the surface of these micelles with different ligands and cell-penetrating moieties, new activities may be added to enable the targeted delivery and intracellular accumulation of protein-based therapeutics (Jhaveri and Torchilin 2014). PM are the most effective alternative drug carriers when compared to other micellar systems. The advantages of mixed PM include the incorporation of noticeably higher medicine doses, longer blood circulation durations, and thermodynamic stability. The PM core has been designed to have the best durability and drug loading capacity. The length of hydrophobic blocks and the kind of substituents found in the core have the biggest effect on PM capacity to carry medicines. The insoluble medications can be contained in the micellar core via either chemical conjugation or physical trapping.

12.2 Polymeric Micelles (PM) as Therapeutic Carriers for Protein Delivery

The use of proteins and peptides as PM targeting ligands is also quite common. The creation of transferrin-targeted nanocarriers is made possible by the transferrin receptor (TfR), which is overexpressed in many malignancies (Singh 1999). Transferrin (Tf), an endogenous ligand for PM, or antibodies against TfR may be used to modify them (Torchilin 2006). Sawant et al. developed micelle of transferrin modified PEG-PE (polyethylene glycol-phosphatidylethanolamine) to deliver a CDK inhibitor "R457." The drug targeting efficacy and cell toxicity were evaluated in vitro and in vivo using A2780 ovarian cancer cell lines (ovarian carcinomas). Antitumour activity was improved as compared to free drug (Sawant et al. 2013). Water is distributed in PM in an anisotropic manner, with the amount of water decreasing as it moves from the surface to the hydrophobic centre. Thus, the polarity of the medication determines where it will be located within micelles: hydrophobic APIs (active pharmaceutical ingredients) will locate within core (lipophilic) or on the surface depending upon drug polarity and hydrophilicity. Generally, hydrophilic candidates are lodged within centre or surface of the micellar structure, whereas lipophilic drug candidate settled within lipophilic core of micelle (Rangel-Yagui et al. 2005). Most of the time, unimers and polyion complex micelles are used to chemically conjugate hydrophilic molecules, or they use electrostatic interactions to load hydrophilic molecules. For instance, RNA is often loaded into amphiphilic block copolymers by the insertion of polycations and then employed for RNA condensation. Amphiphile micellar solutions are an efficient method of delivering medications to their targets. Water-insoluble medicines are easily soluble in the hydrophobic environment of the micelle core and loaded for distribution to the necessary sites. Targeted protein drug delivery systems are created to ensure minimal drug loss and degradation, avoid negative side effects, boost the bioavailability of protein pharmaceuticals, and raise the concentration of medications in the desired zone of interest. Insoluble polymers (synesthetic and natural), soluble polymers, liposomes, micelles, microparticles, cell ghosts, cells, and amphiphilic polymers are just a few of the numerous drug carriers that are often used (Yokoyama 1998).

12.3 Advantages and Challenges of Polymeric Micelles (PM) for the Delivery of Proteins

Due to the potential of PM to boost the solubility and stability of aquaphobic medications and their in vivo therapeutic efficacy, which is equivalent to or better than that of the free drug, PM have received scientific attention in recent years. PM may be made to be big enough to prevent early clearance owing to fast glomerular filtration, which prolongs circulation time, by regulating their size. The particle size is kept tiny enough to safely pass through the smallest veins at the same time. The properties of the PM can improve the cellular localization of the drug loaded micelles and provide a different pathway for endosomal internalization. Because the drug delivery is selective and targeted to tissues, these properties aid PM in having a better mean residence time (MRT), which may lead to a lowered dose, higher bioavailability, as well as a potential reduction in the risk of nonspecific organ toxicity. To improve medication delivery to particular sites and penetrate tumours more effectively, PM have been found to have a higher therapeutic index. Thus, PM may be useful for enhancing biodistribution while posing little threat of accumulation and persistent toxicity in the body. While research on PM has advanced significantly as excellent nanocarriers for pharmaceuticals, particularly hydrophilic medications, their development has been hampered by issues with poor stability and limited drug loading. Nevertheless, due to the chemical adaptability provided by amphiphilic block copolymers, PM may be designed to circumvent these difficulties. Modifying the micellar core is one method for raising the drug loading capacity and micellar stability, and it has been studied in several studies as one method for improving the loading efficiency of PM. Wan et al. loaded paclitaxel and cisplatin into micelles made of amphiphilic copolymers for the purpose of targeting ovarian and breast cancers, resulting in a noticeably higher loading efficiency (Wan et al. 2019). In addition to the numerous uses for PM, there are several difficulties that need to be overcome before they may be considered as viable drug carriers. These include enhancing drug loading effectiveness even further, stabilizing blood aerosol injection, and facilitating transport across cell membranes.

They physically entrap sparingly soluble medications, improve their bioavailability, and transport them to the intended site of action at concentrations that are greater than its intrinsic water solubility. The addition of micelles also improves the drug's stability. Additionally, compared to free pharmaceuticals, there are less negative side effects due to reduced interaction of the medication with inactivating species, such as enzymes found in bodily fluids (Torchilin 2001). Micellar delivery systems' tiny size (10–30 nm) and narrow size distribution are by far their most distinctive characteristics, setting them apart from other particulate drug carriers (Florence and Hussain 2001). Nonionic surfactant-based micelles are often employed as drug carrier for controlled drug delivery (Bardelmeijer et al. 2002). Since most organizations established their individual micelle system made from distinct hydrophilic-lipophilic combinations because of the high degree of activity, this field now has a significant lot of diversity. Figure 12.1 depicts the schematic representation of PM.

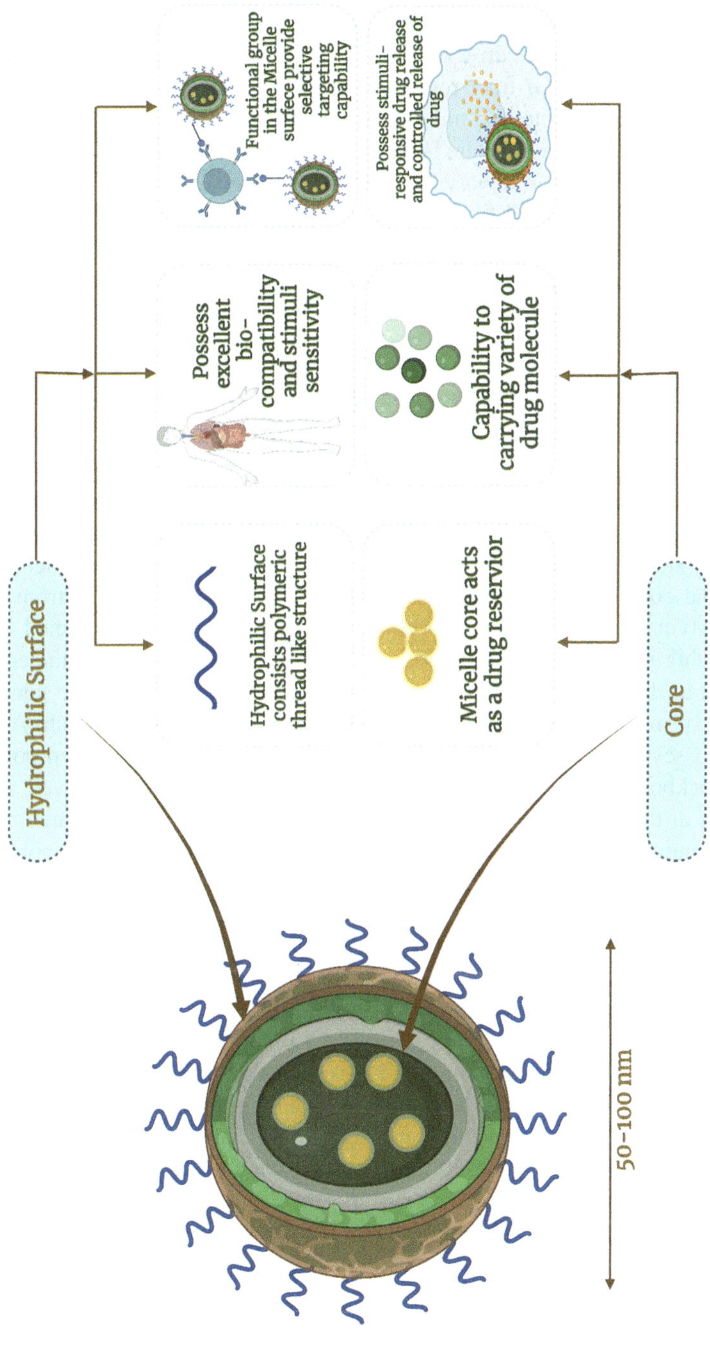

Fig. 12.1 Schematic representation of advantage of micelles

12.4 Oral Uptake of Polymeric Micelles (PM)

PM have been tested for oral administration for a variety of therapeutic goals, including increasing apparent drug solubility in GI fluids and facilitating absorption, penetrating pathological GI tract regions for locoregional treatment, carrying the drug directly to the bloodstream, minimizing pre-systemic losses, and targeting the drug to precise tissues or cells in the body after oral absorption. PM have been tested for oral administration for a variety of therapeutic goals, including increasing apparent drug solubility in GI fluids and facilitating absorption, penetrating pathological GI tract regions for locoregional treatment, carrying the drug directly to the bloodstream, minimizing pre-systemic losses, and targeting the drug to specific tissue or cells in the body after oral absorption (Simões et al. 2015).

12.5 Types of Polymers Utilized in the Administration of Micellar Drugs

Graft copolymers, amphiphilic block copolymers, and triblock copolymers are frequently used to create PM. According to its compatibility with the integrated medication, stability, drug release profile, and toxicity, the block polymer of the hydrophobic core is selected. One can modify the medication release from these block copolymers. Micelles can also be further altered by having the shell cross-linked, having the PM surfaces functionalized, and linking ligands (aptamers and antibodies) to the surface for active targeting. Multigraft copolymers are generally comprised of minimum three homopolymers joined by a common branch. A graft polymer possessed multiple polymer chain acting as both the side grafted polymer and the backbone of the polymer. They gain from the transplant as well as the advantages of the backbone (Kulthe et al. 2012). A hyperbranched polymer with a significant number of functional end group belongs to another class of polymers. Because they have so many end groups, they may readily have their characteristics changed, making them an ideal choice for drug release under certain responsive stimuli such as pH, electrolytic strength, and temperature. Several varieties of manufactured polyester amide and poly(urea-urethane) hyperbranched amphiphilic polymers can be developed and altered to be utilized as a carrier for hydrophobic medicines, such as anticancer therapy (Gao and Yan 2004). Since each polymer has a special benefit, choosing the right polymer is crucial for extending the circulation period and ensuring a regulated release of the medicine. Since they have all been given FDA approval for biomedical uses in humans, polymers including PLGA [poly(lactic-co-glycolic acid], PCL [poly(-caprolactone)], and PLA [poly(lactic acid)] are frequently utilized to form the lipophilic core (Cagel et al. 2017). Figure 12.2 describes the drug loaded PM with several targeting functions.

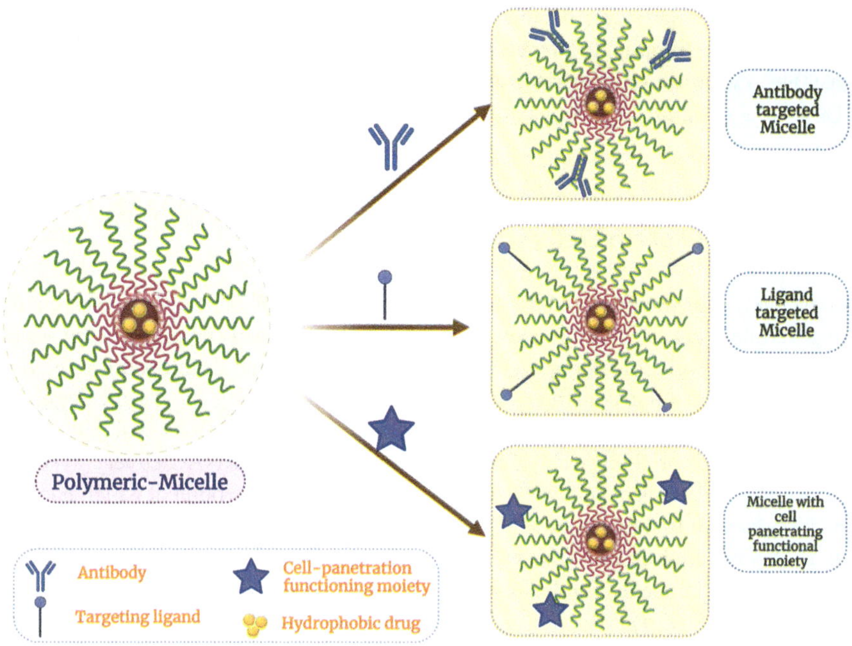

Fig. 12.2 Drug loaded polymeric micelles with several targeting functions

12.6 Stimuli Sensitive Micelles for Delivery of Protein

When exposed to high temperatures, thermo-responsive PM alter structurally; this characteristic may be used to direct the deposition of protein drugs precisely where it is desired. Now, scientists utilized the concept of thermoresponsive polymer for the drug release. Therefore, a drug can be allowed to release loaded payload due to temperature triggered transition in the thermosensitive polymer based micelle above or below critical solution temperature (CST) (Rijcken et al. 2007). Above the LCST (low CST), these thermosensitive polymers solubilize, releasing the medication from the carrier. Therefore, in this form of micellar system, LCST is the most important factor. Another external stimulus that has been employed to cause the release of drugs from stimuli-sensitive multifunctional PM is a magnetic field. A thermo-sensitive star-block copolymer was used in one study to create magnetothermally responsive drug loaded micelles by fusing the concepts of temperature-triggered drug release with the application of a magnetic field (Ji et al. 2014). An external magnetic stimulation was used to locate the micelles, and after that, the temperature rose, causing the thermoresponsive micelles to release the medication. To obtain desired release of payloads, PM may be designed to react to a variety of stimuli (intrinsic or extrinsic) of various sources (chemical, physical, and biological sources) (Cheng et al. 2013). The therapeutic payloads of PM that are "environmentally sensitive" or "smart" can be released by altering their structural composition in response to the stimuli. As a result of the reaction, micelles may degrade or become unstable, isomerize, polymerize, or aggregate supramolecular (Fleige et al. 2012) (Table 12.1). Figure 12.3 depicts the multifunctional PM for oral delivery of drugs.

12.7 Polymeric Micelles (PM) for Multiple Functionality and Protein Delivery

At present, multifunctional PM have been the focus of substantial research for localized delivery drug and nucleic acid such as RNA. Over the past several years, various amphiphilic block copolymers were synthesized and developed for delivery of siRNA in micellar construct. These fundamental building blocks are regularly enhanced to get the most out of them by adding certain ligands or ethically sound building blocks or linkages. A fundamental goal of all siRNA delivery systems, including micelles, is to stop siRNA from degrading from the point at which it enters the bloodstream to the point at which it travels via the endocytic route for intracellular trafficking and escapes the endosome. In their paper, several authors utilized multifunctional micelles for delivery either drug or nucleic acid, or co-administration of drug and NA by allowing self-assembling of RNA with block copolymer (s) (Christie et al. 2012) (Table 12.2). Figure 12.4 shows the schematic representation of the phenomena regulating in vivo drug delivery, in terms of both rate and location, from PM.

12.8 Co-delivery of Drugs and siRNA Using Multifunctional Micelles

The siRNA therapy based on PM has shown remarkable promise and is presently the focus of intensive study. However, it is extremely possible that tumours will have genetic changes, which might lower the usefulness of siRNA as a solitary agent in the therapy of malignancies (Liu et al. 2013). Traditional anticancer medications also have issues with off-target effect and MDR which is mainly responsible for substantial impairment in cancer treatment. Due to the nonspecific character of these inhibitors, there has been relatively little clinical success in the creation of compounds that block the action of drug transporter proteins like P-glycoprotein (Pg-p), which is produced by the MDR1-gene, to sensitize tumour cells to anticancer drugs (Shukla et al. 2008). In these circumstances, it would be more advantageous to treat cancer using RNAi to directly reduce Pg-p expression rather than only its function by suppressing the expression of MDR genes after conventional chemotherapy (Wu et al. 2003). Studies show that pre-treating cancer cells with siRNAs prior to treating them with conventional anticancer medications can greatly increase the cells' sensitivity to the drug and increase the effectiveness of therapy (Spankuch et al. 2007; Macdiarmid et al. 2009). To have the greatest effect in vivo, however, siRNA and medicine must be given to the same tumour cell at the same time

Table 12.1 Mechanistic perspective of some polymeric micelles (PM) for drug release in vitro under specific responsive stimuli

Drug loaded micelles	Mechanism of drug release	References
DOX loaded chondroitin sulfate-histamine based micelle	pH (acidic) triggered release of DIX from micelle due to changed conjugate of histamine residue by protonation	Yu et al. (2013)
Camptothecin and Nile red based poly(ketal adipate)-co-PEG (PKA-PEG) micelle	The polymeric co-block of ketal linkage is cleaved from the backbone of polymer under acidic condition and resulting in the drug release	Lee et al. (2013)
siRNA loaded mPEG–PCL–CH2R4H2C (cell-penetrating peptide) micelle. C, H, and R represent cysteine, histidine, and arginine	siRNA responsible to express VEGR is loaded in the micelle. Cysteine is a sulfide amino acid. Therefore, cleavage of S-S linkage in the cytoplasm results in the drug release (GSH, redox)	Tanaka et al. (2013)
siRNA loaded siRNA-SS-PLGA)/linear PEI (polyethyleneimine)	GSH (redox) Reduction of S-S bond (disulfide bond breaking) causes GFP siRNA release in the cytoplasm	Lee et al. (2011)
Gemcitabine loaded PEG-b-poly(2-methyl-2-carboxylpropylene carbonate)-g-gemcitabine-g-dodecanol (PEG-b-PCC-g-GC-g-DC) micelle	Cathepsin-B (an enzyme) based hydrolysis of amide bond to disintegrate micelles and subsequent release of the drug	Chitkara et al. (2013)
Cisplatin and paclitaxel loaded PEG-b-poly(L-glutamic acid)-b-poly(L-phenylalanine) (PEG-b-PGlu-b-PPhe) micelle	Polypeptide bond is sensitive to cathepsin-B enzyme. Micelles containing polypeptide bond in building block of polymer were cleaved by the same enzyme at low pH (protected in lysozyme)	Desale et al. (2013)
DOX loaded pluronic F105 (PEO-PPO-PEO) based micelle	Ultrasound creates a cavitation and the generated transient cavitation by applying impulse of 70 kHz disrupts micelles and subsequent release of DOX	Husseini et al. (2013)
siRNA loaded nanobubbles (NB) through hetero-assembling of mPEG-b-P(L-lysine). Micelles of siRNA and gas loaded in liposomes to tailor siRNA nanobubbles (NB)	Ultrasound creates a cavitation and the generated transient cavitation by applying impulse of low frequency (kHz) NB and subsequent release of siRNA (sonoporation effect)	Yin et al. (2013)
DOX probed folic acid/dextran-retinoic acid magnetic nanoparticles.	Localized internalization of micelle responsive to an external magnetic field (0.42 T). Magnetic effect produced by magnetic NPs	Varshosaz et al. (2013)
DOX loaded PEG-b-PCL and SPION	SPION generates hyperthermia due to heating and subsequent release of DOX from micelle	Glover et al. (2013)

(continued)

Table 12.1 (continued)

Drug loaded micelles	Mechanism of drug release	References
DOX loaded P-(NIPAAm-co-NHMAAm)-b-PCL micelles	The backbone of poly-(NIPAAm-co-HMAAm) shell undergoes from hydrophilic to hydrophobic transition and release of DOX above the lowest critical solution temperature (LCST) of 38 °C, after collapsing micelle structure	Wang et al. (2014)
Two graft biocompatible copolymers such as chitosan-g-poly (N-isopropylacrylamide) (CS-g-PNIPAM) and carboxymethyl cellulose-g-poly(N-isopropylacrylamide) (CMC-g-PNIPAM)	Thermoresponsive polyelectrolyte complex micelle loaded 5-fluorouracil (5-FU) for controlled release of 5-FU above LCST (37 °C) Core was formed of positively charged chitosan and negatively charged CMC by electrolyte interaction Hydrogen bonding between 5-FU and micelle increased the drug loading Micelles were deformed under the changed electrolytic ionic strength, pH, and temperature	Li et al. (2013a)
Rifampicin and paclitaxel loaded PEO-b-P(LGA-co-COU) based micelle	Photoresponsive drug release mechanism. Near infrared radiation caused shift in hydrophilic-hydrophobic balance due to the light absorption by the coumarin moiety which caused instability of micellar structure and subsequent release of the drugs	Kumar et al. (2012)
Curcumin loaded dialkyl cyano stilbene polymethacrylate-b-PEO (PDACS-b-PEO) micelle	Under UV light, stilbene exhibits trans-cis photo-isomerization transition which results in low polymeric hydrophobicity and cleaves micelles to release loaded curcumin	Menon et al. (2011)

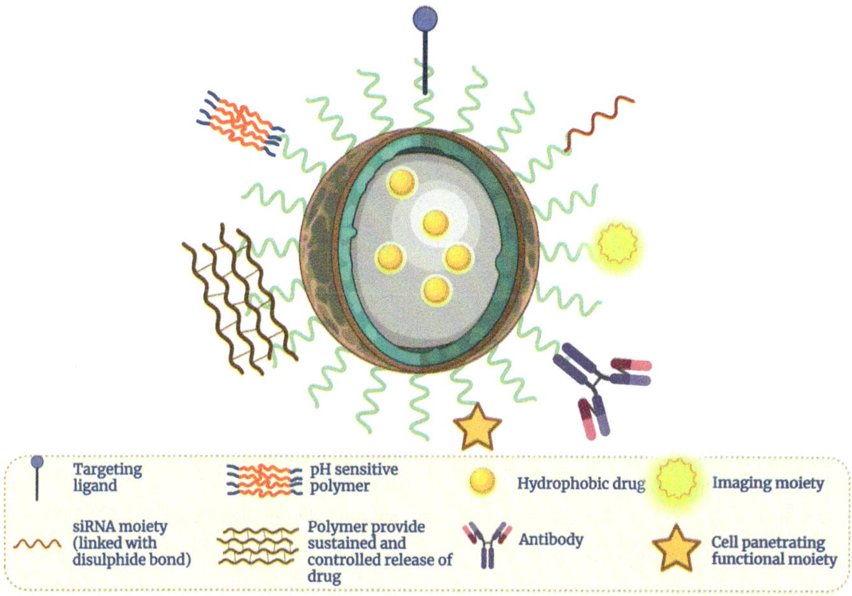

Fig. 12.3 Multifunctional polymeric micelle

following systemic distribution. For the best potential collaboration, they ought to also be distributed optimally among cells (Sun et al. 2011). In this chapter, we examine many examples of such PM, which include siRNA and medications inside of a single nanocarrier. Multifunctional PEO-b-PCL block copolymer micelles having functional modifications on both blocks. These micelles may carry out a number of functions, including pH-triggered drug release in endosomes, siRNA and DOX co-delivery, facilitated cell membrane translocation, passive and active targeting, and siRNA and DOX administration. These micelles may carry out a range of functions, such as pH-triggered drug release in endosomes, siRNA and DOX co-delivery, simpler cell membrane translocation, passive and active targeting, and siRNA and DOX administration. Using a pH-sensitive hydrazone linkage, the PCL core of the micelles may conjugate short polyamines [spermine (SP)] to chemically combine MDR1 siRNA and DOX, as well as fluorescent imaging probes to monitor micelles in vitro and in vivo. Two ligands were added to the virus-like shell of these micelles: a cell-penetrating TAT-peptide to aid in intracellular absorption and an active targeting ligand, RGD4C specific for integrin (αvβ3) receptors (Table 12.3).

Table 12.2 Various block polymers, drug, ligands, and their major finding in drug delivery using micelle

Main composition	Drug and ligand for targeting	Major findings	References
Poly(N-isopropylacrylamide-co-acrylamide)-block-poly (ε-caprolactone) random block copolymer, ferromagnetic nanoparticles, integrin β4	DOX and integrin β4	• Multifunctional micelle responsive to heat and magnet • Developed SPION for sustained release of the drug in response to magnetic heat to treat cancer. Produced high MRI contrast due to SPIO loaded micelle	Kim et al. (2013)
Cross-linked folate-poly (ethylene glycol)-b-poly[N-(N ′,N′-diisopropylaminoethyl) glutamine] (folated-PEG-P [GA-DIP]) amphiphilic block copolymer, DIP: N,N-diisopropyl tertiary amine for conjugation with –COOH for polymer	DOX	• Nanomicelle for targeted drug delivery to control cancer and MRI imaging contrast. PEG-P(GA-DIP) could be self-assembled into nanomicelle by ultrasonic emulsification for DOX delivery and SPION in aqueous • The resulted nanomicelle was subjected to surface functionalization by folic acid	Li et al. (2013b)
Supramagnetic iron platinum NPs, biotin functionalized phospholipid, PSMA antibodies	PTX (paclitaxel)	• Supramagnetic Fe-Pt nanoparticles and PTX were encapsulated in polyethylene glycolated biotin functionalized phospholipid based micelle (SMP) for targeting prostate and MRI contrast imaging • Release was unique in serum (50% in 30.2 h) and saline (100% in 30 min). SPM and PTX were equally cytotoxic at the end of 72 h of incubation	Taylor and Sillerud (2012)
Poly(ethylene glycol)-block-poly(ε-caprolactone) (PEG-PCL), Cetuximab, superparamagnetic iron oxide	DOX	• Aimed to target EGFR overexpressing tumour cells • Cetuximab linked immunomicelles were tailored to load DOX loaded SPION to facilitate interaction between micelle and A431 cell lines • MRI and CLSM (confocal laser scanning microscopy) confirmed cell uptake of micelle loaded with SPION and DOX • Immunomicelles showed effective inhibition of cell proliferation as compared to the non-targeted counterparts	Liao et al. (2011)

(continued)

Table 12.2 (continued)

Main composition	Drug and ligand for targeting	Major findings	References
Manganese (Mn as oxide) based paramagnetic nanoparticles as an alternative gadolinium contrast agent, phosphatidylethanolamine (PE), PEG-2000, DC-cholesterol, dioleoyl-PE, oleic acid (OA)	DNA and DOX delivery	• Gadolinium is a toxic contrast agent and nephrotoxic in T1 MRI • Synthesized multifunctional lipidic micelle NPs for DNA and drug delivery. OA coated micelle was used to encapsulate Mn-oxide NPs • The particles were effectively taken up by the Lewis lungs carcinoma, A549, and human embryonic kidney cell lines	Howell et al. (2013)
OEGMA, NIPAM, and NBA are oligo(ethylene glycol) monomethyl ether methacrylate, N-isopropylacrylamide, and o-nitrobenzyl acrylate, respectively	DOX	• POEGMA-b-P(NIPAM-co-NBA-co-Gd) diblock copolymer covalently labelled with Gd3+ complex (Gd) in the light (UV)-responsive block • In situ monitoring of biodistribution of chemotherapeutic agent for image guided therapy • Diblock copolymer based micelle exhibited light responsive hydrophilic-lipophilic transition inside micellar core for the release of encapsulated drug. 64% of the drug was released within 12 h upon UV irradiation	Li et al. (2012)
MePEG-b-PCL, COOH-PEG-b-PCL, and NH2-PEG-b-PCL, N-hydroxysuccinimide-PEG-b-PCL (NHS-PEG-b-PCL), copolymer radiolabelled-DTPA-PEG-b-PCL, copolymer bearing NLS and trastuzumab fab (NLS-TmAb-fab-PEG-b-PCL)	Trastuzumab fab fragments	• Block copolymer micelle (BCM) to target solid breast cancer cell (nucleus targeting). The study elucidated 31 nm sized micelle to target overexpressing breast cancer cell nucleus (HER2) • Active (5 times improved cell uptake) and passive targeting in mice • Active targeting of BCM improved cellular and subcellular uptake	Hoang et al. (2013)
Poly(amidoamine) [core: 1,4-diaminobutane; {dendri-poly(amidoamine)-(OH)64};	DOX	• DOX was encapsulated in unimolecular micelle composed of PAMA and PLA	Guo et al. (2013)

(continued)

Table 12.2 (continued)

Main composition	Drug and ligand for targeting	Major findings	References
(G = 4)] dendrimer, poly (ethylene glycol) (PEG) derivatives, HOOC–PEG–maleimide (HOOC–PEG–Mal), and HOOC–PEG–OCH_3		through physical method • The carrier exhibited small uniform size and pH responsive drug release • TRC105-conjugated micelle has more uptake as compared to non-conjugated counterparts	
Superparamagnetic iron oxide NPs, polymeric micelle, folate functionalized micelle, N-hydroxysuccinimide, dicyclohexylcarbodiimide, folate-PEG-PCL	Sorafenib	• Sorafenib and SPION were encapsulated into polymeric micelle • Folate functionalized micelle for active targeting to hepatic carcinoma • Improved inhibitory effect	Zhang et al. (2013)

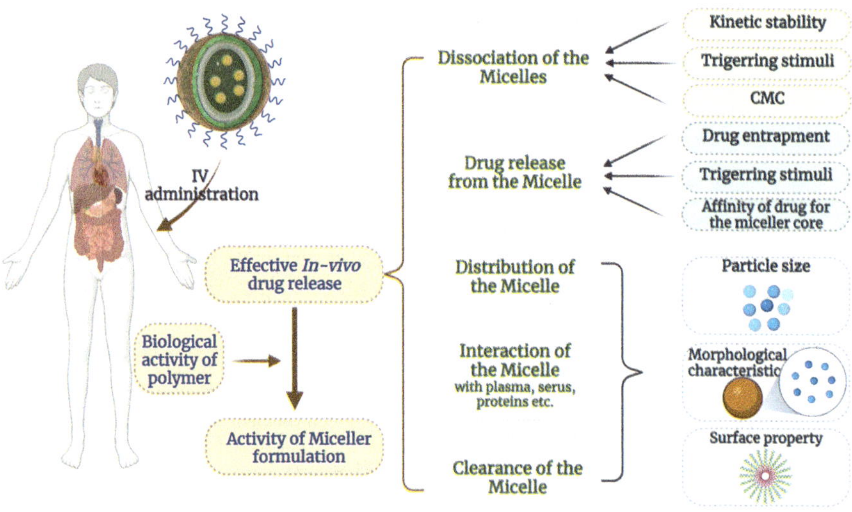

Fig. 12.4 Schematic depiction of the process regulating in vivo drug delivery, in terms of both rate and location, from polymeric micelles

Table 12.3 A summary of a few multifunctional micelles for delivery of drug and siRNA responsive to certain stimuli

Polymers	Delivery	Major findings	Stimulus	References
Poly(ethylene oxide)-block-poly (ε-caprolactone) (PEO-b-PCL) block copolymers	Co-delivery of doxorubicin and MDR-1 siRNA	• Developed traceable multifunctional polymeric nanocarriers to target (passive and active) cancer with imposed physicochemical properties and pH-triggered release. • Two virus mimetic shells were attached. One for active targeting (integrin αvβ3-specific ligand, RGD4C) and other for cell penetration (TAT protein) • This improved DOX efficacy against resistant type of MDA-MB-435	pH responsive	Xiong et al. (2010)
Poly(ethylene glycol) and poly(L-lysine) grafted with polyethyleneimine through reduced disulfide linkage	siRNA delivery in copolymer	• Synthesized ternary copolymer using mPEG-b-PLL-g-(ss-IPEI) • PLL block and –ss– Linkage rendered the carrier degradability whereas IPEI brought about buffering capacity and reduced cationic toxicity • Copolymer featured for high siRNA transfection efficiency, low toxicity, and ease in degradation to control tumour growth	pH and grafting using imine	Li et al. (2014)
The PICMs comprised of poly (amidoamine) dendrimer (PAMAM)–NA core and a removable PEG-block-poly(propyl methacrylate-co-methacrylic acid) [PEG-b-P(PrMA-co-MAA)] shell	Delivery of short chain interfering RNA and antisense oligonucleotides NA = nucleic acid	• Novel pH responsive polyion complex micelle (PIMS) • PIMS with size of 50–70 nm, neutral surface and narrow size distribution • PIMS can be lyophilized,	pH responsive polyion nanocomplex	Elsabahy et al. (2009)

(continued)

Table 12.3 (continued)

Polymers	Delivery	Major findings	Stimulus	References
		reconstituted without change in colloidal properties, and complexation efficiency • Greater transfection and cellular uptake into cancerous cells • Nonspecific toxicity was lower than the PAMAM		
Six-armed-PEG conjugate of siRNA (6PEG-siRNA), Hph1, KALA peptide	Multifunctional siRNA delivery system	• Sequential synthesis of 6PEG-siRNA, 6PEG-siRNA-Hph1, and 6PEG-siRNA-Hph1-cl-KALA through conjugation electrostatic • Showed superior physical stability and resistant to enzyme degradation	Reducible and multifunctional electrolyte micelles	Choi et al. (2010)
Poly(2-(dimethylamino)ethyl methacrylate)-block-poly (2-(diisopropylamino)ethyl methacrylate) (PDMA-b-PDPA) diblock copolymers	SiRNA and amphotericin-B	• To overcome endosomal barrier for intracellular delivery of siRNA • Prepared dual responsive PDMA-b-PDPA micelleplexes • At pH 7.4, amphotericin-B was loaded with PDPA core and siRNA was complexed with cationic PDMA to construct micelleplexes • Significant increase in luciferase knockdown efficiency	pH responsive	Yu et al. (2011)
N-acetylgalactosamine (NAG)	siRNA to target liver	• NAG functionalized mixed micellar nanoparticles (Gal-MNP) to deliver siRNA into liver (hepatocytes) and silence the target gene expression after systemic administration • Gal-MNP was assembled into	Aqueous solution	Wang et al. (2013)

		aqueous system from Gal-MNP-b-PCL and cationic PLC-b-poly (2-aminoethyl ethylene phosphate) (PCL-b-PPEEA) • Downregulation of apolipoprotein-B expression was achieved • Study was conducted in mice model and mouse model		
Thermo responsive poly (N-isopropylacrylamide-co-acrylamide)-block-poly (ε-caprolactone) random block copolymer micelles and magnetic superparamagnetic NPs	DOX and imaging agent	• SPION (Superparamagnetic nanoparticles) and thermoresponsive polymeric micelle to deliver DOX and imaging agent • Unique multiple functionality • Modulated for controlled drug release • Surface functionalization with the integrin β4 antibody resulted in accumulation of micelle on the cell surface of tumour • Moreover, SPION loaded micelle improved MRI contrast	Responsive to magnetic heating	Kim et al. (2013)
Folic acid functionalized folated-PEG-P[GA-DIP] amphiphilic block copolymer, DIP: N,N-diisopropyl tertiary amine	DOX	• Acid triggered core of folated-PEG-P[GA-DIP] nanomicelle for targeted drug delivery to control liver cancer using SPION loaded with DOX • Nanomicelle improved loading efficiency and cellular uptake of DOX and SPION when DIP was linked with PGA [poly(L-glutamic acid)] at neutral pH • In acidic medium (endosomes), core is disassembled for DOX release	Acid responsive nanomicelle for DOX delivery and MRI contrast	Li et al. (2012)

12.9 Conclusion

PM appear to be a perfect carrier for poorly water-soluble drugs due to their benefits, such as tiny size, high solubility, ease of sterilizing, and controlled release of drugs. Since the integrated medicine may be released rapidly in vivo, the physical stability of this carrier is a crucial concern. To effectively develop micelles that can carry drugs to their sites of action, significant effort still must be done in understanding how PM interact with plasmatic and cellular components. PM have made significant strides in recent years in the delivery of a wide range of payloads, including traditional anticancer medications and biological macromolecules including antibodies, siRNA, DNA, and oligonucleotides. Chemically modified structure of various block copolymers generated micelles which enabled the formation of complex micelles that integrate numerous modalities inside a single carrier. Purposely, multifunctional micelles were fabricated for targeted delivery to the infected sites (by surface functionalization with ligands), alteration caused for imaging contrast in disease diagnosis, and responsive drug delivery using magnetic field, pH, electrolyte strength and temperature provided externally or internally (tumour microenvironment). These imposed physical characteristics innate in polymer control drug release from micelle due to deformation or transition in construct cargoes or drug effectiveness at the site of action (tumour lesion, lysozymes, and enzymatic degradation). The fact that they may be tweaked and altered to meet demands gives them a clear edge over conventional drug delivery methods. Although appearing straightforward at first glance, it is obvious that PM constitute a far more complicated system than what is initially believed. Clinical studies for many PM formulations have begun after achieving some promising preclinical results, but only a small number of them have obtained regulatory permission for usage in humans. Numerous obstacles have made it difficult for them to follow the regulatory road. Clinically significant PM formulations serve as a supply path for insoluble small molecules, and due to reason that the existing PM formulation systems must develop further to act as efficient carriers. A detailed analysis of drug encapsulation and the successive drug release profile in systemic circulation will reveal how novel PM systems for human usage could be developed in the future.

References

Bardelmeijer HA, Ouwehand M, Malingré MM, Schellens JH, Beijnen JH, van Tellingen O (2002) Entrapment by Cremophor EL decreases the absorption of paclitaxel from the gut. Cancer Chemother Pharmacol 49(2):119–125

Bharate SS, Bharate SB, Bajaj A (2016) Interactions and incompatibilities of pharmaceutical excipients with active pharmaceutical ingredients: a comprehensive review. J Excip Food Chem 1:1131

Cabral H, Miyata K, Osada K, Kataoka K (2018) Block copolymer micelles in nanomedicine applications. Chem Rev 118:6844–6892

Cagel M, Tesan FC, Bernabeu E et al (2017) Polymeric mixed micelles as nanomedicines: achievements and perspectives. Eur J Pharma Biopharm 113:211–228

Cheng R, Meng F, Deng C, Klok HA, Zhong Z (2013) Dual and multistimuli responsive polymeric nanoparticles for programmed site-specific drug delivery. Biomaterials 34:3647–3657. https://doi.org/10.1016/j.biomaterials.2013.01.084

Chitkara D, Mittal A, Behrman SW, Kumar N, Mahato RI (2013) Self-assembling, amphiphilic polymer-gemcitabine conjugate shows enhanced antitumor efficacy against human pancreatic adenocarcinoma. Bioconjug Chem 24:1161–1173. https://doi.org/10.1021/bc400032x

Choi SW, Lee SH, Mok H, Park TG (2010) Multifunctional siRNA delivery system: polyelectrolyte complex micelles of six-arm PEG conjugate of siRNA and cell penetrating peptide with crosslinked fusogenic peptide. Biotechnol Prog 26:57–63. https://doi.org/10.1002/btpr.310

Christie RJ, Matsumoto Y, Miyata K, Nomoto T, Fukushima S, Osada K et al (2012) Targeted polymeric micelles for siRNA treatment of experimental cancer by intravenous injection. ACS Nano 6:5174–5189. https://doi.org/10.1021/nn300942b

Desale SS, Cohen SM, Zhao Y, Kabanov AV, Bronich TK (2013) Biodegradable hybrid polymer micelles for combination drug therapy in ovarian cancer. J Control Release 171:339–348. https://doi.org/10.1016/j.jconrel.2013.04.026

Elsabahy M, Wazen N, Bayó-Puxan N, Deleavey G, Servant M, Damha MJ et al (2009) Delivery of nucleic acids through the controlled disassembly of multifunctional nanocomplexes. Adv Funct Mater 19:3862–3867. https://doi.org/10.1002/adfm.200901139

Fleige E, Quadir MA, Haag R (2012) Stimuli-responsive polymeric nanocarriers for the controlled transport of active compounds: concepts and applications. Adv Drug Deliv Rev 64:866–884. https://doi.org/10.1016/j.addr.2012.01.020

Florence AT, Hussain N (2001) Transcytosis of nanoparticle and dendrimer delivery systems: evolving vistas. Adv Drug Deliv Rev 50:S69–S89

Gao C, Yan D (2004) Hyperbranched polymers: from synthesis to applications. Prog Polym Sci 29(3):183–275

Glover AL, Bennett JB, Pritchett JS, Nikles SM, Nikles DE, Nikles JA et al (2013) Magnetic heating of iron oxide nanoparticles and magnetic micelles for cancer therapy. IEEE Trans Magn 49:231–235. https://doi.org/10.1109/TMAG.2012.2222359

Guo J, Hong H, Chen G, Shi S, Zheng Q, Zhang Y et al (2013) Image guided and tumor-targeted drug delivery with radiolabeled unimolecular micelles. Biomaterials 34:8323–8332. https://doi.org/10.1016/j.biomaterials.2013.07.085

Hoang B, Ekdawi SN, Reilly RM, Allen C (2013) Active targeting of block copolymer micelles with trastuzumab fab fragments and nuclear localization signal leads to increased tumor uptake and nuclear localization in HER2-overexpressing xenografts. Mol Pharm 10:4229–4241. https://doi.org/10.1021/mp400315p

Howell M, Mallela J, Wang C, Ravi S, Dixit S, Garapati U et al (2013) Manganese-loaded lipid-micellar theranostics for simultaneous drug and gene delivery to lungs. J Control Release 167:210–218. https://doi.org/10.1016/j.jconrel.2013.01.029

Husseini GA, Velluto D, Kherbeck L, Pitt WG, Hubbell JA, Christensen DA (2013) Investigating the acoustic release of doxorubicin from targeted micelles. Colloids Surf B Biointerfaces 101:153–155. https://doi.org/10.1016/j.colsurfb.2012.05.025

Jhaveri AM, Torchilin VP (2014) Multifunctional polymeric micelles for delivery of drugs and siRNA. Front Pharmacol 5:77

Ji G, Yang J, Chen J (2014) Preparation of novel curcumin-loaded multifunctional nanodroplets for combining ultrasonic development and targeted chemotherapy. Int J Pharm 466(1–2):314–320

Kim TY, Kim DW, Chung JY, Shin SG, Kim SC, Heo DS, Kim NK, Bang YJ (2004) Phase I and pharmacokinetic study of Genexol-PM, a cremophor-free, polymeric micelle-formulated paclitaxel, in patients with advanced malignancies. Clin Cancer Res 10:3708–3716

Kim DH, Vitol EA, Liu J, Balasubramanian S, Gosztola DJ, Cohen EE et al (2013) Stimuli-responsive magnetic nanomicelles as multifunctional heat and cargo delivery vehicles. Langmuir 29:7425–7432. https://doi.org/10.1021/la3044158

Kulthe SS, Choudhari YM, Inamdar NN, Mourya V (2012) Polymeric micelles: authoritative aspects for drug delivery. Des Monomer Polym 15(5):465–521

Kumar S, Allard J-F, Morris D, Dory YL, Lepage M, Zhao Y (2012) Near-infrared light sensitive polypeptide block copolymer micelles for drug delivery. J Mater Chem 22:7252–7257. https://doi.org/10.1039/c2jm16380b
Lee KS, Chung HC, Im SA, Park YH, Kim CS, Kim SB, Rha SY, Lee MY, Ro J (2008) Multicenter phase II trial of Genexol-PM, a Cremophor-free, polymeric micelle formulation of paclitaxel, in patients with metastatic breast cancer. Breast Cancer Res Treat 108:241–250
Lee SH, Mok H, Lee Y, Park TG (2011) Self-assembled siRNA-PLGA conjugate micelles for gene silencing. J Control Release 152:152–158. https://doi.org/10.1016/j.jconrel.2010.12.007
Lee I, Park M, Kim Y, Hwang O, Khang G, Lee D (2013) Ketal containing amphiphilic block copolymer micelles as pH-sensitive drug carriers. Int J Pharm 448:259–266. https://doi.org/10.1016/j.ijpharm.2013.03.017
Li Y, Qian Y, Liu T, Zhang G, Liu S (2012) Light-triggered concomitant enhancement of magnetic resonance imaging contrast performance and drug release rate of functionalized amphiphilic diblock copolymer micelles. Biomacromolecules 13:3877–3886. https://doi.org/10.1021/bm301425j
Li G, Meng Y, Guo L, Zhang T, Liu J (2013a) Formation of thermo-sensitive polyelectrolyte complex micelles from two biocompatible graft copolymers for drug delivery. J Biomed Mater Res A 102(7):2163–2172. https://doi.org/10.1002/jbm.a.34894
Li X, Li H, Yi W, Chen J, Liang B (2013b) Acid-triggered core crosslinked nanomicelles for targeted drug delivery and magnetic resonance imaging in liver cancer cells. Int J Nanomedicine 8:3019–3031. https://doi.org/10.2147/IJN.S45767
Li J, Cheng D, Yin T, Chen W, Lin Y, Chen J et al (2014) Copolymer of poly(ethylene glycol) and poly(l-lysine) grafting polyethylenimine through a reducible disulfide linkage for siRNA delivery. Nanoscale 6:1732–1740. https://doi.org/10.1039/c3nr05024f
Liao C, Sun Q, Liang B, Shen J, Shuai X (2011) Targeting EGFR overexpressing tumor cells using Cetuximab-immunomicelles loaded with doxorubicin and superparamagnetic iron oxide. Eur J Radiol 80:699–705. https://doi.org/10.1016/j.ejrad.2010.08.005
Liu X-Q, Sun C-Y, Yang X-Z, Wang J (2013) Polymeric-micelle-based nanomedicine for siRNA delivery. Part Part Syst Charact 30:211–228. https://doi.org/10.1002/ppsc.201200061
Macdiarmid JA, Amaro-Mugridge NB, Madrid-Weiss J, Sedliarou I, Wetzel S, Kochar K et al (2009) Sequential treatment of drug-resistant tumors with targeted minicells containing siRNA or a cytotoxic drug. Nat Biotechnol 27:643–651. https://doi.org/10.1038/nbt.1547
Menon S, Thekkayil R, Varghese S, Das S (2011) Photoresponsive soft materials: synthesis and photophysical studies of a stilbene-based diblock copolymer. J Polym Sci A Polym Chem 49:5063–5073. https://doi.org/10.1002/pola.24973
Owens DE 3rd, Peppas NA (2006) Opsonization, biodistribution, and pharmacokinetics of polymeric nanoparticles. Int J Pharm 307:93–102
Rangel-Yagui C, Pessoa A, Tavares L (2005) Micellar solubilization of drugs. J Pharm Pharm Sci 8:147–165
Rijcken CJ, Snel CJ, Schiffelers RM, Van Nostrum CF, Hennink WE (2007) Hydrolysable core-crosslinked thermosensitive polymeric micelles: synthesis, characterisation and in vivo studies. Biomaterials 28(36):5581–5593
Sawant RR, Jhaveri AM, Koshkaryev A, Zhu L, Qureshi F, Torchilin VP (2013) Targeted transferrin-modified polymeric micelles: enhanced efficacy in vitro and in vivo in ovarian carcinoma. Mol Pharm 11:375–381. https://doi.org/10.1021/mp300633f
Shukla S, Wu CP, Ambudkar SV (2008) Development of inhibitors of ATP-binding cassette drug transporters: present status and challenges. Expert Opin Drug Metab Toxicol 4:205–223. https://doi.org/10.1517/17425255.4.2.205
Simões SM, Figueiras AR, Veiga F, Concheiro A, Alvarez-Lorenzo C (2015) Polymeric micelles for oral drug administration enabling locoregional and systemic treatments. Expert Opin Drug Deliv 12(2):297–318
Singh M (1999) Transferrin as a targeting ligand for liposomes and anticancer drugs. Curr Pharm Des 5:443–451
Spankuch B, Kurunci-Csacsko E, Kaufmann M, Strebhardt K (2007) Rational combinations of siRNAs targeting Plk1 with breast cancer drugs. Oncogene 26:5793–5807. https://doi.org/10.1038/sj.onc.1210355

Sun TM, Du JZ, Yao YD, Mao CQ, Dou S, Huang SY et al (2011) Simultaneous delivery of siRNA and paclitaxel via a "two-in-one" micelleplex promotes synergistic tumor suppression. ACS Nano 5:1483–1494. https://doi.org/10.1021/nn103349h

Tanaka K, Kanazawa T, Horiuchi S, Ando T, Sugawara K, Takashima Y et al (2013) Cytoplasm-responsive nanocarriers conjugated with a functional cell penetrating peptide for systemic siRNA delivery. Int J Pharm 455:40–47. https://doi.org/10.1016/j.ijpharm.2013.07.069

Tao A, Huang GL, Igarashi K, Hong T, Liao S, Stellacci F, Matsumoto Y, Yamasoba T, Kataoka K, Cabral H (2020) Polymeric micelles loading proteins through concurrent ion complexation and pH-cleavable covalent bonding for in vivo delivery. Macromol Biosci 20(1):1900161

Taylor RM, Sillerud LO (2012) Paclitaxel-loaded iron platinum stealth immunomicelles are potent MRI imaging agents that prevent prostate cancer growth in a PSMA-dependent manner. Int J Nanomedicine 7:4341–4352. https://doi.org/10.2147/IJN.S34381

Torchilin VP (2001) Structure and design of polymeric surfactant-based drug delivery systems. J Control Release 73(2–3):137–172

Torchilin VP (2006) Multifunctional nanocarriers. Adv Drug Deliv Rev 58:1532–1555. https://doi.org/10.1016/j.addr.2006.09.009

Varshosaz J, Sadeghi-Aliabadi H, Ghasemi S, Behdadfar B (2013) Use of magnetic folate-dextran-retinoic acid micelles for dual targeting of doxorubicin in breast cancer. Biomed Res Int 2013: 680712. https://doi.org/10.1155/2013/680712

Wan X, Beaudoin JJ, Vinod N, Min Y, Makita N, Bludau H, Jordan R, Wang A, Sokolsky M, Kabanov AV (2019) Co-delivery of paclitaxel and cisplatin in poly (2-oxazoline) polymeric micelles: implications for drug loading, release, pharmacokinetics and outcome of ovarian and breast cancer treatments. Biomaterials 192:1–14

Wang H-X, Xiong M-H, Wang Y-C, Zhu J, Wang J (2013) N-acetylgalactosamine functionalized mixed micellar nanoparticles for targeted delivery of siRNA to liver. J Control Release 166: 106–114. https://doi.org/10.1016/j.jconrel.2012.12.017

Wang X, Li S, Wan Z, Quan Z, Tan Q (2014) Investigation of thermo-sensitive amphiphilic micelles as drug carriers for chemotherapy in cholangiocarcinoma in vitro and in vivo. Int J Pharm 463:81–88. https://doi.org/10.1016/j.ijpharm.2013.12.046

Wu H, Hait WN, Yang JM (2003) Small interfering RNA-induced suppression of MDR1 (P-glycoprotein) restores sensitivity to multidrug-resistant cancer cells. Cancer Res 63:1515–1519

Xiong XB, Uludag H, Lavasanifar A (2010) Virus-mimetic polymeric micelles for targeted siRNA delivery. Biomaterials 31:5886–5893. https://doi.org/10.1016/j.biomaterials.2010.03.075

Yadav HKS, Almokdad AA, Shaluf SIM, Debe MS (2019) Chapter 17—polymerbased nanomaterials for drug-delivery carriers. In: Mohapatra SS, Ranjan S, Dasgupta N, Mishra RK, Thomas S (eds) Nanocarriers for drug delivery. Elsevier, Amsterdam, pp 531–556

Yin T, Wang P, Li J, Zheng R, Zheng B, Cheng D et al (2013) Ultrasound sensitive siRNA-loaded nanobubbles formed by hetero-assembly of polymeric micelles and liposomes and their therapeutic effect in gliomas. Biomaterials 34:4532–4543. https://doi.org/10.1016/j.biomaterials.2013.02.067

Yokoyama M (1998) Novel passive targetable drug delivery with polymeric micelles. Academic, San Diego, p 193

Yu H, Zou Y, Wang Y, Huang X, Huang G, Sumer BD et al (2011) Overcoming endosomal barrier by amphotericin B-loaded dual pH-responsive PDMA-b-PDPA micelleplexes for siRNA delivery. ACS Nano 5:9246–9255. https://doi.org/10.1021/nn203503h

Yu C, Gao C, Lu S, Chen C, Yang J, Di X et al (2013) Facile preparation of pH-sensitive micelles self-assembled from amphiphilic chondroitin sulfate histamine conjugate for triggered intracellular drug release. Colloids Surf B Biointerfaces 115C:331–339. https://doi.org/10.1016/j.colsurfb.2013.12.023

Zhang L, Gong F, Zhang F, Ma J, Zhang P, Shen J (2013) Targeted therapy for human hepatic carcinoma cells using folate-functionalized polymeric micelles loaded with superparamagnetic iron oxide and sorafenib in vitro. Int J Nanomedicine 8:1517–1524. https://doi.org/10.2147/IJN.S43263

Regulatory Aspects for Polymeric Micelles 13

Anshita Gupta Soni, Renjil Joshi, Deependra Soni, Ujala Gupta, Mayur Aalhate, and Pankaj Kumar Singh

Abstract

In recent decades, nanoformulations have been at the forefront of pharmaceutical research, posing new problems for the scientific community, businesses, and regulators. The rapid development of scientific and technological instruments is strongly desired to fulfill unmet medical requirements and enhance human health care and quality of life. The fields of biomaterials and nanotechnology have made enormous strides, which has led to their usage as promising methods to address significant limitations, largely related to the non-therapeutic response of traditional medicinal techniques. But the broad range of uses for nanoformulation necessitates a thorough understanding and categorization of these intricate goods. To prevent unexpected consequences on patients, such as possible immunological reactivity, their features must be thoroughly understood.

Additionally, a lot of micelle formulations are either accepted for use in clinical trials or are currently undergoing them. However, several obstacles have prevented this sector's use in the field of health, the most significant of which is the absence of unified standards for regulating and uniform procedures for testing safety and quality. This emphasizes the necessity of investigating new regulations and rules for nanocarriers jointly. The regulation of nanomicelles, their significance, and the current regulatory frameworks will all be covered in this chapter, along with the initiatives being developed by the industry and

A. G. Soni · R. Joshi
Shri Rawatpura Sarkar Institute of Pharmacy, Kumhari, Durg, Chhattisgarh, India

D. Soni
Faculty of Pharmacy, MATS University, Aarng, Raipur, Chhattisgarh, India

U. Gupta · M. Aalhate · P. K. Singh (✉)
Department of Pharmaceutics, National Institute of Pharmaceutical Education and Research (NIPER) Hyderabad, Hyderabad, Telangana, India

S. K. Singh et al. (eds.), *Polymeric Micelles: Principles, Perspectives and Practices*,
https://doi.org/10.1007/978-981-99-0361-0_13

regulatory bodies to support their commercialization. The development of nanomedicine more quickly and safely will be facilitated by regulatory science issues, which will be a major concern in the next few years.

Keywords

Nanoformulation · Polymeric micelles · Regulatory approaches

13.1 Introduction

Polymeric micelles, which block copolymer self-assemblies, have become carriers or vectors of choice for the delivery of medicines and genes. Various clinical and preclinical studies have shown their excellent drug-encapsulating efficiency. Polymeric micelles may have various advantages over long-circulating liposomes, including regulated drug release, tissue penetrating capabilities, and decreased toxicity such as hand-foot syndrome and hypersensitive reaction (Choi and Han 2018). The nature and design of block copolymers can be used to alter important characteristics of polymeric micelles as drug carriers, such as particle size, stability, loading capacity, and drug release kinetics (Nishiyama and Kataoka 2006a). Recent advancements in drug design have enabled block copolymer nanoengineering with increased target specificity and higher physiochemical sensitivity (Nishiyama and Kataoka 2006b). Therefore, polymeric micelles are carrier systems based on nanotechnology that may exhibit the action of powerful bioactive chemicals in a site-directed manner, assuring their efficacy and safety in the clinical setting (Sahoo and Labhasetwar 2003).

13.2 The Importance of Regulation

After a certain dose, all chemicals become poisonous. The appropriate dosage distinguishes between a poison and a treatment. Paracelsus has stated in his theory no drug is safe but the right amount makes it a medicine. Regulatory affairs have made it possible to ensure the quality, safety, and efficacy of a drug or therapeutically active chemical entity in the same phase of time through proper regulations and documentation.

A properly systematically validated and trusted regulation guideline is required to confirm the QSE of the drug. Along with the appropriateness and suitability of the publically accessible drug information. To speed up the creation and delivery of safe and efficient healthcare goods to people all over the world, regulatory organizations provide strategic, tactical, and operational direction and assistance for working within regulations. Every nation has its regulatory bodies that are in charge of all domestic regulations about drug substances (Struble et al. 2013; Clark 1930; Abueva 1959; Otto 2017).

Regulatory authorities serve as a spectacular who confirms the efficacy, safety, and quality of pharmaceuticals that are made available to the general public. They analyze pitfalls and loopholes of drug regulation and suggest ways to strengthen it. They also play a crucial role in ensuring and boosting the application of regulations in unregulated regions of the world for the protection of those who live there. International regulatory organizations are crucial in many facets of pharmaceutical regulation, including registration, manufacture, distribution, pricing control, marketing, and intellectual property protection (Sparrow 2011; Jasanoff 1998; Geest et al. 1996; Meyboom et al. 1999; Avorn 2008).

When events or problems arise, the regulatory authorities have a specific need to communicate. However, communication in these "crisis" situations is likely to be far more effective if a connection has previously been built via routine, everyday conversations. As a result, proactive communication on safety and regulations when there are no accidents to record can be just as crucial as communication in response to events.

13.2.1 Properties and Recent Updates of PMs

Polyethylene glycol (PFG) is the most widely used hydrophilic component, while other new polymers such as poly(vinylpyrrolidone) and poly (trimethylene carbonate). Polyesters, copolymers of glycolic and lactic acids, or poly(propylene oxide) are examples of hydrophobic materials that are frequently utilized. Furthermore, by experimenting with different combinations and ratios of hydrophobic/hydrophilic components, the copolymer blocks can be tailored to meet the end use (Rasal et al. 2010). The size of PMs contributes to the drug's increased bioavailability, target-specific action, controlled release, and prolonged circulation time, decreasing adverse drug reactions. Due to their association with aberrant vasculature, nano-PMs of a specific size (30–100 nm) can passively aggregate and remain within tumors, providing the desired outcomes. Additionally, the neutral and hydrophilic nature of surfaces can aid to prolong the circulation time of micelles by having weak interactions with proteins, which contributes to the lengthy circulation of micelles (Nofar et al. 2019). Positively charged surfaces are necessary for the oral administration of micelles because they promote penetration and mucoadhesive qualities and improve drug transport across biological barriers. However, because of non-specific protein interaction, which can cause aggregation in vivo, positively charged surfaces are also linked to poor stability in biological contexts (Gunatillake et al. 2006). To create stable micelles, it is crucial to maintain a proper balance of charge on their surface. The micellar structure can be used to capture several medications, enhancing the therapeutic efficacy of the nanosystem (Kwon and Furgeson 2007). Drug loading may be assisted by micelle polymer chemical conjugation or physical trapping. Chemically conjugated medications are typically released by surface erosion or total breakdown of the PMs, whereas pharmaceuticals loaded using a physical entrapment approach are typically released by simple diffusion (Auras et al. 2011).

Techniques like nuclear magnetic resonance (NMR) and Fourier transform infrared spectroscopy (FTIR) can be used to determine the primary structural and physicochemical properties of the polymers, which will help to understand the nature of amphiphilic components to help predict the properties of the micellar system. Contrary to water or a buffer, biological fluids such as plasma have a distinct CMC. So it would be possible to forecast its stability by determining CMC in various solvent systems and storage circumstances (Caputo et al. 2020). Surface tension calculation techniques, measurements of fluorescence intensity in solutions, and dynamic light scattering can all be used to determine the CMC. A study that used aliphatic dicarboxylic acids, such as glutaric acid, adipic acid, pimelic acid, and suberic acid as hydrophobic blocks and the conductivity meter of amphiphilic poly (ethylene glycols) (PEGs) as a hydrophilic block to analyze CMC. The PMs' reported diameters ranged from 127.5 to 354 nm, and their CMC values were between 112 and 155 mg/L. Azelaic acid, sebacic acid, dimethyl isophthalate acid, and dimethyl terephthalate were used as hydrophobic blocks, and PEGs as hydrophilic blocks in an amphiphilic block-co-polymer investigation. The PM size ranged for aliphatic polymers from 51.6 to 174 nm, with CMC values between 95 and 130 mg/L, and for aromatic polymers from 135.5 to 371 nm, with CMC values between 420 and 1500 mg/L. Several experimental methods, including cryo-transmission electron microscopy (cryo-TEM) and transmission electron microscopy, can be used to analyze the physical characteristics of PM (TEM). A typical TEM image of empty micelles and drug-loaded micelles ranged between 25–30 nm in diameter (Zamani and Khoee 2012). These methods can also be used to assess morphological alterations brought on by interactions with biological systems in polymeric micelles. The physicochemical characteristics of PMs as well as a deeper knowledge of the structural organization of the outside and inner architecture of PMs loaded with pharmaceuticals may be studied using more sophisticated techniques, such as X-ray scattering. Additionally, to forecast biological reactions, X-ray scattering tests can offer details on the behavior of PMs in the tissue microenvironment. Fluorescence resonance energy transfer (FRET), another intriguing method, helps us comprehend the fluctuating nature of the medicinal chemical contained within the micelles (Ziaee et al. 2017). The stability of the PM structure and drug release profile under physiological settings can be predicted with the help of this phenomenon (Fig. 13.1).

13.3 Perspective from a Regulatory Standpoint on the Creation of Nanomedicines or Polymeric Micelles

Recent advances in drug delivery had shown various delivery systems from nano to micro range for ease of drug development. Despite numerous attempts, there has yet to be a clear definition of global regulatory trends. Regulatory agencies still need to work together more closely, but significant progress has been made over the past 5 years (Sainz et al. 2015). As an alternative, the methods used to develop conventional pharmaceutical products have frequently been adopted to assess the

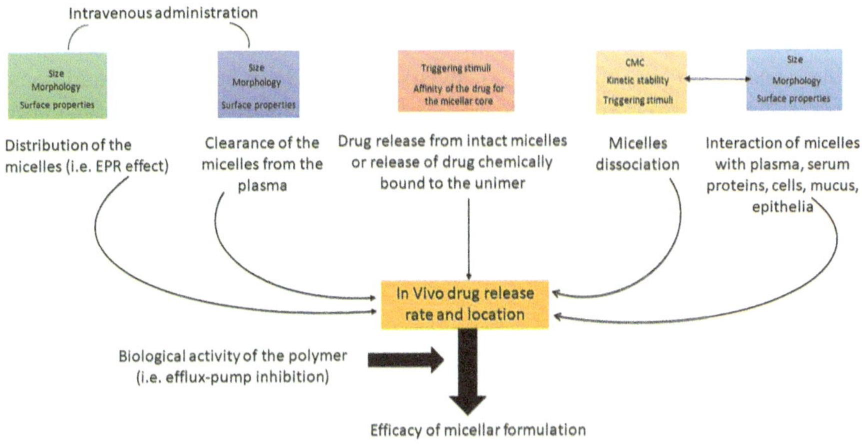

Fig. 13.1 Polymeric micelles' mechanisms are shown schematically

compatibility, safety, and toxicity of nanomedicines. According to the general public, the bioactive moieties of nanoformulation determine the specifications that must be examined in the context of regulatory analysis. The innovative product must adhere to the rules established for therapeutic medicinal agents and novel chemical entities when it comes to biological entities like proteins, peptides, or antibodies (NCEs) (Lembo et al. 2018).

In summary, one of the barriers to the regulation of nanomedicines is connected to their unique properties. Given the substantial body of research on polymer, liposomes polymer therapeutics, and other polymeric approaches (such as polymeric micelles), as well as the controlling concerns surrounding their conception and growth, the clinical application of this sophisticated and complex nanoformulation is heavily reliant on a thorough evaluation, categorization, and knowledge of key characteristics (Cabral et al. 2018). In actuality, tiny adjustments to the manufacturing process as well as minor alterations to the raw materials can quickly modify their qualities. And while these modifications would just slightly alter the structure, they might drastically impact biological characteristics and biodistribution patterns. There is an urgent need to develop more précised and validated quality control parameters and techniques for screening new drugs, prodrug ligands, and targeting moieties before they are converted into nanotherapeutics for intended or external use. Various pharmacokinetic parameters should be made specific to ensure the drug release, cellular uptake, and other basic characteristics of a drug or delivery system.

Nanoformulation is called to hold the capacity to react with immunological cells and to adsorb plasma proteins, liable on their size range and physicochemical characteristics. Biocompatibility and immunotoxicity must be taken into account throughout the preclinical evaluation. The dose regimen, therapeutic index, method of administration, and targeted disease environment should all be considered while evaluating toxicity during the development process. As substances are confined and delivered for their purposes, nanoformulation has demonstrated safety, biocompatibility, and even the capacity to overcome the harmfulness of traditional pharmaceuticals. However, further worries have been raised by several categories of suggested nanoformulation like quantum dots, dendrimers, and carbon nanotubes. Due to their possible toxicity and immunological negative action, the clinical application of this novel substance may be seriously hampered for several years (Tang et al. 2012; De Jong and Borm 2008; Yang et al. 2020).

Adapting the manufacturing procedures has been a barrier to the research and clinical translation of this nanoformulation. Due primarily to the wide variety of features of novel materials, recent pharmaceutical invention at fabrication facilities faces significant manufacturing problems that pose a threat to their ability to scale up. Finding and managing the important points during each production process is essential. The use of "quality-by-design" principles and tools like process analytical technologies (PAT) will guarantee online/at-online quality evaluation methods. Knowing and foreseeing the most crucial production points makes it easier to implement automated methods to address issues as they arise in line.

The ICH Q8, Q9, and Q10 regulations for innovative pharmaceutical improvement were created as a result of these creative notions. In addition to having a significant impact on already available medical products, they aim to foster the advancement of nanomedicines and manufacturing techniques in the future. Watchdogs and industry from Japan, Europe, and the United States of America (US) have generally worked to build thorough controlling systems through the ICH. Even so, some opposing viewpoints continue to underlie several actions taken by the European Medicines Agency and the US Food and Drug Administration (FDA) (EMA). It is necessary to use a variety of standardized assays and methodologies to examine factors that are known to directly affect the safety and effectiveness of nanomedicines in vivo. The application of delicate tests to identify low quantities of nanocarriers, to distinguish those from made aggregates, or to separate intact from metabolic forms presents the most challenges. To get over these restrictions, different imaging approaches including fluorescence or cellular imaging techniques have been suggested and investigated.

The type of data that must be provided previously and during the substance lifecycles, necessitating in vivo animal and clinical research, is another barrier to the regulation of nanomedicines. To identify a lesser quantity of nanoformulation, distinguish them from formed aggregates, or separate intact from metabolized forms, the Marketing Authorization Application (MAA) is used in Europe. To get over these restrictions, different imaging techniques including fluorescence or cellular imaging methods have been suggested and investigated. In Europe, the legal framework for Marketing Authorization Applications (MAA) offers applicants the chance to receive "scientific counseling" from regulators as early as the R&D stage. This could swiftly assist in the coordinated growth of breakthrough nanoformulation, minimizing the impression of significant unnecessary roadblocks along the way. This can help advance pharmaceutical development to become more standardized and less affected by significant roadblocks when they are developed.

A working group has recently been established by EMA to examine issues directly related to the QSE of nanoformulation. Additionally, this cluster has created "orientation documents" that relate to crucial factors that applicants should take into account when developing nanomedicines. Even with the lack of particular procedures for nanoformulation, the regulatory bodies from the EU (EMA), the USA (the FDA), and Japan (PDMA/MHLW) have been cooperating since 2009–2010 to build shared viewpoints in the field of developing nanomedicines. The interest in "proof of concept" and clinical development of these intricate organizations has surged among the world's largest pharmaceutical companies as well. Together, these factors will support procedures that are more clearly defined for determining the QSE of this nanoformulation. Pharmacoeconomic studies will be necessary to show the social and economic added value of this innovative substance when compared to conventional therapies. In the development of these novel and advanced nanomedicines, significant indications such as the growth in QALYs (quality-adjusted life expectancy years) or expenses related to future successive hospitalizations must be taken into account.

13.4 A Versatile Drug Delivery Carrier: Polymeric Micelles

Amphiphilic copolymers are incredibly adaptable parts of medication delivery systems. It is possible to create a variety of self-assembled structures by adjusting the hydrophilic and hydrophobic building blocks (Verma and Hassan 2013). These structures can host pharmaceuticals, improve their apparent solubility and stability, penetrate biological barriers, and deliver the drugs to the desired location. Additionally, some block copolymers have demonstrated the ability to function as active molecules that can improve the therapeutic effects of the medications (Fig. 13.2) (Schmidt and Barner-Kowollik 2017). The range of applications in the field of drug delivery is expanded even further when block copolymers and cyclodextrins are combined. The rheological characteristics of the formulations and the drug release kinetics can be adjusted, thanks to the production of poly(pseudo)rotaxane (Xing and Zhao 2016) (Table 13.1).

13.4.1 Regulatory Aspects

Despite a large number of studies on nanomedicines being available, very less nano-based preparations have been converted from the research setting to clinical trials, and very less have been sanctioned for clinical usage against diverse diseases. Agencies established by individual nations or regions, such as the US Food and Drug Administration (US FDA), the European Medicines Agency (EMA), the UK Medicines and Healthcare Products Regulatory Agency (MHRA), and the Japanese Pharmaceuticals and Medical Devices Agency, heavily weigh and evaluate whether to approve or reject such a formulation (PMDA). The regulations and evaluation of nano techniques may differ from one agency to another (see Fig. 13.3).

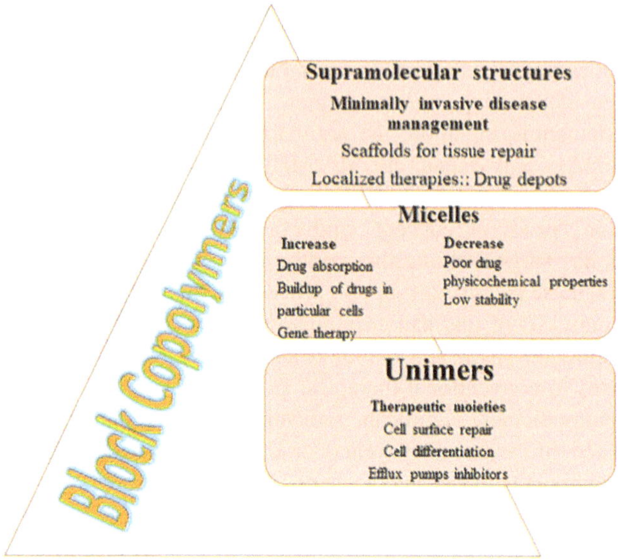

Fig. 13.2 The development of block copolymer structure. Supramolecular structure and polymeric micelles have several benefits

Table 13.1 Examples of polymeric micelles formulation (Choi and Han 2018)

Polymer	Active ingredients	Therapeutic uses
PEG-b-PLA	PTX, DOX	Breast cancer
PEG-b-PPBA	PTX	Breast cancer
PEG-PDLLA	DTX PTX	Breast cancer Breast cancer
XR-17block polymer	DTX	Breast cancer
PEG-b-poly (glutamate)	NK012 and carboplatin	Metastasis triple negative breast cancer
Poloxamer 407	Oleanolic acid	Anti-aging
Pluronic P123	Resveratrol	Psoriasis
Pluronic P123	Terconazole	Anti-fungal
Block copolymeric micelles	Camptothecin	Increase in systemic exposure
Block copolymeric micelles	Doxorubicin	Increase in systemic exposure
Block copolymeric micelles	Paclitaxel	Increase in systemic exposure
Block copolymeric micelles	Pilocarpine	Glaucoma
Self-micellizing solid dispersion	Tranilast	Increase of oral F
Pluronic F127	Benzoyl peroxide	Acne vulgaris
Poly(ethylene glycol)	Docetaxel HCL	Antineoplastic
Polyvinylpyrrolidone	Fenofibrate	Lipid regulation

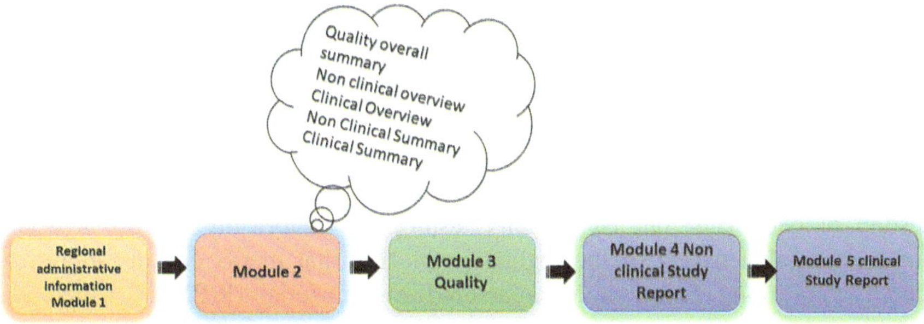

Fig. 13.3 Technical document common. Five modules make up the eCTD: *1*. Administrative details and prescription details. This regional module is unique to each country or region and is country-specific; *2*. Synopses of typical technical documents. This is a standard module across all areas; *3*. Excellence; *4*. Reports on nonclinical studies; *5*. Reports on clinical studies

13.4.2 Regulatory Guidelines, eCTD Submission in Various Countries

Due to a dearth of research meeting the fundamental requirements for clinical trials, namely affluence of manufacture, reproducibility, and long-term stability of the nano techniques, significantly created novel nanotechnology fails the efficacy criteria. As a result, several of the confirmed preparations for oncology therapy are still at an early stage (Yager et al. 2006). The majority of nanoparticles used now are liposome-based and intended to treat cancer. Because of this, there are particular recommendations available for nanoformulations or liposomal formulations. These regulations, however, prevent the introduction of new and distinct varieties of nanoparticles, which may have genuine potential (James and Gambhir 2012). A wide range of heterogeneous medication compositions is offered by monotherapies. In contrast to conventional drug formulations, each new nanoparticle technique produces various sizes, surface characteristics, morphologies, and other attributes that alter in vivo drug behavior (Kandachar and Halme 2017). Therefore, detailed records on QSE, as well as the risk-benefit assessment, which becomes more complicated and requires a significant time investment, are crucial for regulatory bodies' evaluation and decision-making when it comes to evaluating nanotechnology. Regulatory bodies might make use of the database of nanoparticles to expedite the evaluation of NPs. Data-based nanomedicine can be used globally and can display details on nanoparticles linked to physical characterization, in vitro studies, in vivo outcomes, and clinical-stage observation, similar to gene databases (Fursov and Miles 2013).

13.5 Nanotechnology Regulatory Challenges

Experts consider the FDA's current position on polymeric micelles to be unclear. Simply because changing facts show that larger equivalents of nanoparticles (NPs) have altered features. For instance, research has found that polymeric micelles' pharmacokinetic profiles differ from those of their traditional counterparts. The experts advise treating the polymeric micelles of already-approved medications as novel molecular entities (NMEs), which would need the submission of a brand-new regulatory application (Bus 2008).

The fact that the OECD (Organization for Economic Co-operation and Development) recommendations cover acute, sub-acute, and chronic studies for toxicity testing of chemicals, but NP toxicity testing has historically been a neglected topic, is another aspect of the regulation of NPs that merits consideration. The OECD's Working Party on Manufactured Nanomaterial (WPMN), which was established in 2006 to evaluate the safety implications of nanomaterials, is one example of the efforts that are now being made in this direction. To improve the safety and efficiency of the usage of nanomaterials, rules for their advancement are still urgently needed (Saifi et al. 2018).

Although ex vivo lab analyses and in vivo animal studies aid in the laboratory evaluation of nanomaterials, they do not have the freedom to dispel concerns about their safety when administered to patients during phase 1 of a clinical trial. Regulations and regulatory guidelines demand that the potential risk be appropriate given the potential rewards (Emanuel et al. 2000). One of the most important and difficult areas of the regulation of polymeric micelles is risk management, risk assessment, and risk communication (Resnik and Tinkle 2007).

A meticulous multistep manufacturing method that includes crucial parameter-affecting aspects with guaranteed control to create a consistent desired final product results in the production of biological and nonbiological polymeric micelles (Tinkle et al. 2014; Holloway et al. 2012). The production method and control of crucial factors determine the predominant properties of polymeric micelles that fall within the acceptable range for the needed therapeutic performance. Therefore, just like with biologicals, there are still therapeutic and regulatory questions regarding how comparable nanosimilars are and what their range of reversible use is. Furthermore, there is still disagreement regarding what term best describes nanomaterials (Sandhiya et al. 2009; Schellekens et al. 2011).

13.6 Regulation of Nanotechnology-Based Pharmaceuticals

The FDA has not yet provided a regulatory definition of the term "nanotechnology," which indicates the ambiguities and complications that surround it. The FDA attempts to assess each application of drug items incorporating nanomaterials on its own merits and not as specified by the descriptors because it is well aware that nanoparticles are present in the products it regulates. A tremendous effort is put to meet the need for a unique regulatory approach to nanotechnology (Hamburg 2012).

However, over the past few decades, there has been a remarkable advancement in the field of nanomaterials. Well-written basic guidelines have been developed for industry-related nano-sized liposomal medicinal products (Department of Health and Human Services United States et al. 2018). The study of nanomaterials is the study of substances at the nanoscale with more or less comparable chemical

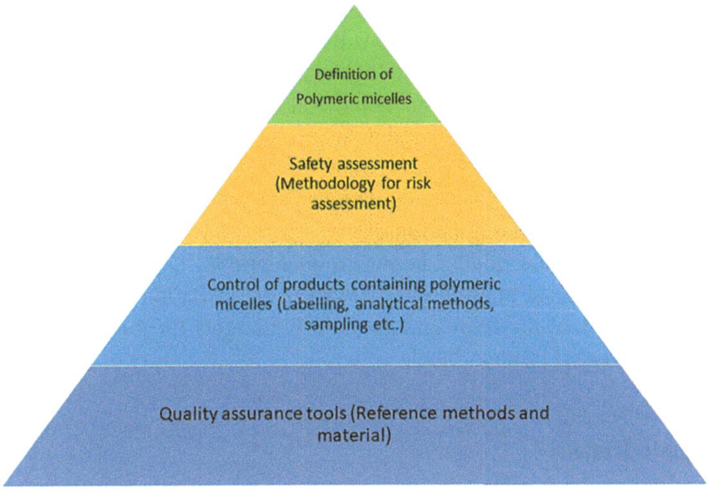

Fig. 13.4 Major obstacles in the regulation of polymeric micelles

properties but with some unexpected outcomes. This idea is comprehensive in that it acknowledges that not all chemicals are harmful, even though certain of them may be dangerous and require a legal warning on consumer items.

Nanomaterials are no different from the idea that not all chemicals are dangerous, even though some nanomaterials with known risks are ingested globally. In such circumstances, the law ought to step in right away. But since it turned out to be an interdisciplinary study, only the chemicals are relevant in this situation. The question of whether the existing legal rules governing chemicals and chemical management will be enough and appropriate in the case of nanoparticles or whether a new law should be developed is still up for debate among the international community.

Researchers and regulators concur that it is best to understand the legal and regulatory concerns relating to nanomaterials from their lifecycle, i.e. from the laboratory through consumer items, to disposal in the environment. If we can examine the lifecycle of nanomaterials, it will become clear that there are many aspects of regulation that could be discussed for nanotechnology, including laws relating to occupational health, factories, chemicals, toxins, consumers, waste products, environment, food and agriculture, fisheries, etc. (Fig. 13.4).

13.6.1 Specified Polymeric Micelles for PMs and Injectables

To mine the essential data and assign nanostructure-based annotations, a well-designed nanomaterial database with current information is required. This information aids in the creation of nanoparticles with specific or nearly specific characteristics for use in biological systems. With this as their goal, the European Union, the United States, and a few other nations launched the Nanoinformatics

Roadmap 2030 in 2018. The objective of this endeavor was to integrate all the data on nanomaterials from a few other nations in 2018. By storing and exchanging data, this endeavor set out to integrate all the information about nanomaterials. The goal of this program is to create a consensus about nanomaterials and any potential health risks among academics, regulators, and industry. As a result, by using this technique, all the most recent data on nanoparticles will be accessible on a single platform, assisting in the creation of rigorous regulations for nanomaterials and facilitating the clinical translation of those regulations.

When determining the quality target product profile, factors including micelle formulation, sterility, and pharmaceutical quality are taken into account. According to the specifications of the relevant agencies, the targets for the qualities were set based on the marketed product label.

13.7 Conclusion

PMs are superior drug nanocarriers when compared to other micellar systems. Smaller than 100 nm in size, polymeric micelles are easier to penetrate and aggregate at the sites of disease because they are more stable than surfactant micelles. Longer blood circulation periods, thermodynamic stability, and the incorporation of considerably more drugs are all characteristics of them. The PM core is engineered with the aim to maximize drug loading capacity and blood circulation half-life. Studies have revealed that micelleplexes have a variety of advantages, including the ability to penetrate micelles in various cell membranes. Many studies have shown the benefits of micelleplexes, including their capacity to penetrate neoplastic cell membranes, biodegradability, ability to enhance the stability of encapsulated genetic material, and high efficacy in the transport and delivery of drugs and genetic material to target cells.

References

Abueva JV (1959) Formulation of the Philippine community development program. University of Michigan, Ann Arbor, MI

Auras RA, Lim LT, Selke SE, Tsuji H (eds) (2011) Poly (lactic acid): synthesis, structures, properties, processing, and applications. Wiley, Hoboken, NJ

Avorn J (2008) Powerful medicines: the benefits, risks, and costs of prescription drugs. Vintage, New York

Bus BR (2008) Nanoparticle-based therapeutics in humans: a survey. Nanotech L & Bus 5:135–155

Cabral H, Miyata K, Osada K, Kataoka K (2018) Block copolymer micelles in nanomedicine applications. Chem Rev 118(14):6844–6892

Caputo P, Shaikhah D, Porto M, Loise V, De Santo MP, Oliviero RC (2020) NMR diffusiometry spectroscopy, a novel technique for monitoring the micro-modifications in bitumen ageing. Appl Sci 10(16):5409

Choi YH, Han HK (2018) Nanomedicines: current status and future perspectives in aspect of drug delivery and pharmacokinetics. J Pharm Investig 48(1):43–60

Clark CA (1930) The Korean church and the Nevius methods. Doctoral dissertation, The University of Chicago

De Jong WH, Borm PJ (2008) Drug delivery and nanoparticles: applications and hazards. Int J Nanomedicine 3(2):133

Department of Health and Human Services United States, Food and Drug Administration, Center for Drug Evaluation and Research (2018) Liposome drug products chemistry, manufacturing, and controls human pharmacokinetics and bioavailability; and labeling documentation guidance for industry guidance for industry. Center for Drug Evaluation and Research, Silver Spring, MD

Emanuel EJ, Wendler D, Grady C (2000) What makes clinical research ethical? JAMA 283(20): 2701–2711

Fursov KS, Miles I (2013) Framing emerging nanotechnologies: steps towards a forward-looking analysis of skills. Higher School of Economics Research Paper No. WP BRP. 15

Geest SV, Whyte SR, Hardon A (1996) The anthropology of pharmaceuticals: a biographical approach. Annu Rev Anthropol 25(1):153–178

Gunatillake P, Mayadunne R, Adhikari R (2006) Recent developments in biodegradable synthetic polymers. Biotechnol Annu Rev 12:301–347

Hamburg MA (2012) Science and regulation. FDA's approach to regulation of products of nanotechnology. Science 336(6079):299–300

Holloway C, Mueller-Berghaus J, Lima BS et al (2012) Scientific considerations for complex drugs in light of established and emerging regulatory guidance. Ann N Y Acad Sci 1276(1):26–36

James ML, Gambhir SS (2012) A molecular imaging primer: modalities, imaging agents, and applications. Physiol Rev 92(2):897–965

Jasanoff S (1998) The fifth branch: science advisers as policymakers. Harvard University Press, Cambridge, MA

Kandachar P, Halme M (eds) (2017) Sustainability challenges and solutions at the base of the pyramid: business, technology and the poor. Routledge, Abingdon-on-Thames, Oxfordshire

Kwon GS, Furgeson DY (2007) Biodegradable polymers for drug delivery systems. In: Biomedical polymers. Woodhead Publishing, Sawston, Cambridge, pp 83–110

Lembo D, Donalisio M, Civra A, Argenziano M, Cavalli R (2018) Nanomedicine formulations for the delivery of antiviral drugs: a promising solution for the treatment of viral infections. Expert Opin Drug Deliv 15(1):93–114

Meyboom RH, Egberts AC, Gribnau FW, Hekster YA (1999) Pharmacovigilance in perspective. Drug Saf 21(6):429–447

Nishiyama N, Kataoka K (2006a) Nanostructured devices based on block copolymer assemblies for drug delivery: designing structures for enhanced drug function. In: Polymer therapeutics II. Springer, Berlin, pp 67–101

Nishiyama N, Kataoka K (2006b) Current state, achievements, and future prospects of polymeric micelles as nanocarriers for drug and gene delivery. Pharmacol Ther 112(3):630–648

Nofar M, Sacligil D, Carreau PJ, Kamal MR, Heuzey MC (2019) Poly (lactic acid) blends: processing, properties and applications. Int J Biol Macromol 125:307–360

Otto GA (2017) A theoretical framework for strategic leadership. Doctoral dissertation, Union Institute and University

Rasal RM, Janorkar AV, Hirt DE (2010) Poly (lactic acid) modifications. Prog Polym Sci 35(3): 338–356

Resnik DB, Tinkle SS (2007) Ethical issues in clinical trials involving nanomedicine. Contemp Clin Trials 28(4):433–441

Sahoo SK, Labhasetwar V (2003) Nanotech approaches to drug delivery and imaging. Drug Discov Today 8(24):1112–1120

Saifi MA, Khan W, Godugu C (2018) Cytotoxicity of nanomaterials: using nanotoxicology to address the safety concerns of nanoparticles. Pharm Nanotechnol 6(1):3–16

Sainz V, Conniot J, Matos AI, Peres C, Zupan??i? E, Moura L, Silva LC, Florindo HF, Gaspar RS (2015) Regulatory aspects on nanomedicines. Biochem Biophys Res Commun 468(3):504–510

Sandhiya S, Dkhar SA, Surendiran A (2009) Emerging trends of nanomedicine—an overview. Fundam Clin Pharmacol 23(3):263–269
Schellekens H, Klinger E, Mühlebach S, Brin J-F, Storm G, Crommelin DJ (2011) The therapeutic equivalence of complex drugs. Regul Toxicol Pharmacol 59(1):176–183
Schmidt BV, Barner-Kowollik C (2017) Dynamic macromolecular material design—the versatility of cyclodextrin-based host–guest chemistry. Angew Chem Int Ed 56(29):8350–8369
Sparrow MK (2011) The regulatory craft: controlling risks, solving problems, and managing compliance. Brookings Institution Press, Washington, DC
Struble EB, Cipollo JF, Kimchi-Sarfaty C, Sauna ZE, Ragheb JA, Marszal E (2013) Perspectives from good manufacturing practices and regulatory guidance. In: Narhi NL (ed) Biophysics for therapeutic protein development, vol 4. Springer, New York, p 261
Tang F, Li L, Chen D (2012) Mesoporous silica nanoparticles: synthesis, biocompatibility and drug delivery. Adv Mater 24(12):1504–1534
Tinkle S, McNeil SE, Mühlebach S et al (2014) Nanomedicines: addressing the scientific and regulatory gap. Ann N Y Acad Sci 1313(1):35–56
Verma G, Hassan PA (2013) Self-assembled materials: design strategies and drug delivery perspectives. Phys Chem Chem Phys 15(40):17016–17028
Xing P, Zhao Y (2016) Multifunctional nanoparticles self-assembled from small organic building blocks for biomedicine. Adv Mater 28(34):7304–7339
Yager P, Edwards T, Fu E, Helton K, Nelson K, Tam MR, Weigl BH (2006) Microfluidic diagnostic technologies for global public health. Nature 442(7101):412–418
Yang W, Veroniaina H, Qi X, Chen P, Li F, Ke PC (2020) Soft and condensed nanoparticles and nanoformulations for cancer drug delivery and repurpose. Adv Ther 3(1):1900102
Zamani S, Khoee S (2012) Preparation of core–shell chitosan/PCL-PEG triblock copolymer nanoparticles with ABA and BAB morphologies: effect of intraparticle interactions on physicochemical properties. Polymer 53(25):5723–5736
Ziaee A, Albadarin AB, Padrela L, Faucher A, O'Reilly E, Walker G (2017) Spray drying ternary amorphous solid dispersions of ibuprofen—an investigation into critical formulation and processing parameters. Eur J Pharm Biopharm 120:43–51

14 Toxicological and Regulatory Challenges in Design and Development of Polymeric Micelles

Ankur Kaul, Bhavna Kumar, and Dimple Sethi Chopra

Abstract

Polymeric micelles have great potential as nanoscaffolds for drug delivery by virtue of their smaller size and higher drug pay-loading potential highly suitable for cancer therapeutics. They can be designed to be "smart nanocarriers" for varied applications in theranostics. The regulatory considerations are vital to the development of such targeted drug delivery systems as the nano-based formulations are to be designed to achieve the maximum effectiveness of the drug without compromising on its safety profile in preclinical and clinical settings.

Keywords

Polymeric micelles · Toxicity · Preclinical studies · Clinical studies · Regulatory guidelines

A. Kaul
Division of Cyclotron and Radiopharmaceutical Sciences, Institute of Nuclear Medicine and Allied Sciences (INMAS), Delhi, India
e-mail: ankurkaul.inmas@gov.in

Bhavna Kumar
Faculty of Pharmacy, DIT University, Dehradun, India
e-mail: bhavna@dituniversity.edu.in

D. S. Chopra (✉)
Department of Pharmaceutical Sciences and Drug Research, Punjabi University, Patiala, India

S. K. Singh et al. (eds.), *Polymeric Micelles: Principles, Perspectives and Practices*,
https://doi.org/10.1007/978-981-99-0361-0_14

14.1 Introduction to Nanotechnology in Drug Delivery

Richard Feyman envisioned the immense scope of nanotechnology as early as in nineteenth century and in fact, in recent times, the nanotechnological advancements have been offering promising leads in design of newer drug delivery systems (Feynman 1961; Farokhzad and Langer 2009). These "nanoscale range" drug delivery systems modify the pharmacokinetics of the drug and steer it through physiological membranes and deliver the drug to the target site in a well-defined manner (Patra et al. 2018). This has led to development of targeted drug delivery platforms for personalized medicine. Nanomedicine has numerous applications in gene therapy, onco-chemotherapeutics, and immunotherapeutics (Manzari et al. 2021). Various design strategies in development of nanoparticles are employed to increase efficacy, tolerability, or therapeutic index of drug in question, thereby improving overall human well-being. In fact, the nanomaterials have a crucial role to play in attainment of 17 global sustainable development goal (United Nations 2018), especially SDG3# which emphasizes on well-being and healthy living for all. As per FDA guidelines, drug delivery system is termed as "nano"when one of its dimensions in nanometer ranges between 1 and 100 nm. Such nanoscale sized particles could be an integral part of a biosensor, microfluidic system, targeted drug delivery systems, and tissue engineering (FDA n.d.). As the size of material is reduced from micro to nano range, the chemical, structural, magnetic, mechanical, electrical, and even the biological properties may differ distinctively. Hence, the objective in development of such nanoformulations is target-specific delivery of drugs in vivo, thus improving its bioavailability and reducing its toxicity (Choi and Han 2018). Existing nano-based drug delivery platforms can be categorized into various classes on the basis of their basic composition: polymeric, metallic, lipid-based nanoparticles (Table 14.1).

Recently, drug loaded nanoplatforms are being tagged physically or chemically with many molecular imaging ligands for theranostic application. Such theranostic nanoformulations have been extensively investigated for novel drug development (Yetisgin et al. 2020).

14.2 Polymeric Nanoparticles as Drug Delivery System

Polymeric nanoparticles can deliver drug payload in a controlled pattern at a specific site. They can be designed to be stimulus-responsive with desirable long circulation times and improved intracellular administration. The method of synthesis and choice of polymer determine the critical quality attributes of nanoparticles such as mean particle size, polydispersity, and drug encapsulation efficiency. Natural biopolymers such as chitosan, collagen, albumin, and gelatin by virtue of their inherent properties like biocompatibility, biodegradability, low immunogenicity, and antibacterial activity possess high application potential, whereas synthetic polymers like polylactic-co-glycolic acid (PLGA), polylactic acid (PLA), polyglycolic acid (PGA) need to be

Table 14.1 Classes of nanoparticle-based drug delivery platforms

Class of NP	Types	Unique characteristics	References
Polymeric	Polymersome	Ease of surface functionalization high payload for both hydrophilic and hydrophobic drugs, prone to aggregation	Hasannia et al. (2022)
	Dendrimer		Wang et al. (2022a, b)
	Polymeric micelle		Kotta et al. (2022)
	Nanosphere		Tian et al. (2022)
Inorganic	Silica NP	Distinct magnetic, optical or electrical property, variable size/geometry, toxicity issues	Vallet-Regí et al. (2022)
	Quantum Dot		Jain et al. (2022)
	Gold NP		Bhattacharya et al. (2022)
	Iron oxide NP		Caldera et al. (2022)
	Carbon NP		Holmannova et al. (2022)
Lipid based	Liposome	Better bioavailable, simple to formulate, low encapsulation ability	Wang et al. (2022a, b)
	Emulsion		Karpuz and Silindir-Gunay (2022)
	Lipid NP		Xu et al. (2022)

chemically modified to obtain optimum specificity, improved bioavailability, reduced toxicity, and desirable pharmacokinetics. Stimuli responsive polymers are known as smart as they spontaneously respond to changes in pH, temperature, and external stimulus. Vinyl esters, hydrazones respond to acidic pH of inflamed, neoplastic, and lysosomal membrane with release of active ingredient (Castro et al. 2022).

Polymeric nanoparticles can be formulated from preformed synthetic or natural biopolymers by emulsification solvent evaporation/diffusion, ionic gelation, salting-out method. Polymeric nanoparticles can also be fabricated from monomer of choice using polymerization techniques such as interfacial polymerization resulting in preparation of microemulsion, nanoemulsion, or surfactant-free emulsion. The choice of method of synthesis of polymeric nanoparticles depends upon polarity of the active constituent. However, the production of polymeric nanoparticles using preformed polymers offers better control and higher yield compared to polymerization of monomers. Stimulus-sensitive nanocarriers respond explicitly to pathological triggers and markedly result in reduction of side effects. The sources of stimulus can be external (light, magnetic or electric field, ultrasound, among others), internal (pH, temperature, enzyme activity, host recognitions, antigen-antibody interactions), or a combination of both. Caldorera-Moore et al. 2019 prepared poly (methacrylic acid-grafted-ethylene glycol) pH sensitive nanoparticles which protected IFN-α in the acidic pH of the stomach but resulted in, its controlled release in the upper small intestine with almost neutral pH.

14.3 Clinically Approved Polymeric Nano Formulation

Polymeric nanosystems score over other delivery system owing to their biodegradability, good aqueous solubility, biocompatibility, and ease of surface modification (Blanco et al. 2015). In recent years, several nanomedicines have reached the clinics after FDA clearance and several nanotherapeutic agents are currently undergoing clinical trials. It is reported that about 20% of the nano-based drug delivery systems are for oncological application (Anselmo and Mitragotri 2019).

14.4 Development of Polymeric Micelles

14.4.1 Design Considerations

The design of polymeric micelles should be efficient enough to result in optimum endosomal escape efficiency and minimum toxicity of bioactive agents (Butt et al. 2022). The direct access of polymeric micelles to target sites in the cytoplasm accelerates therapeutic effect and minimizes side effects (Wolfram et al. 2015). Cellular uptake of polymeric micelles takes place mainly by clathrin- and caveolae-mediated endocytosis (Oh and Park 2014). The nanocarrier gets entrapped in the lysosome where it gets eventually degraded and as a result very less amount of drug reaches the target organelle. The surface functionalization of polymeric micelles should enable it to escape from the membrane vesicles of the endocytic pathway and hence protect its drug payload. The caveolae-mediated endocytosis bypasses lysosomes and protects nanocarriers from lysosomal degradation (Ahmad et al. 2019). Although viral vectors have been developed to facilitate endosomal escape, their use is limited because of toxicity and immunogenicity issues. Thus, there is an urgent demand for development of non-viral vectors that can escape endosomes and achieve same magnitude of therapeutic effects that are obtained in vitro (Varkouhi et al. 2011). Recently, polymeric micelles have been prepared using amphiphilic block copolymers for CNS delivery of drugs and biologics. Their unique physicochemical properties like self-assembly, flexibility, nano-size, charge modulation, stimuli responsiveness facilitate them to overcome limitation of conventional drug delivery. Incorporation of stimuli-sensitive functional groups in copolymer chains, association of bioactive molecules with the core by suitable linkers, and attachment of smart ligands and imaging molecules with the block copolymers creating corona are some of the novel techniques used for functionalization of polymeric micelle. The ease of surface functionalization, facilitating penetration and thus retention in brain tissues makes them highly suitable for diagnosis of CNS diseases (Kaur et al. 2022).

14.4.2 Toxicity Issues

Polymeric micelles initially act as massive particulates that avoid renal excretion while being delivered to their destination but end up breaking down into solitary polymeric chains that can be eliminated from the blood at the kidney site. As a result, polymeric micelles with dual large and small dimensions offer excellent targeting efficiency while being exempted from toxicities since carriers cannot be excreted by the kidney. The rate at which the polymeric micelles disassemble is crucial for striking a compromise between low toxicity and effective targeting. Since single polymer chains that are produced when polymeric micelles are disintegrated and can be eliminated from the organisms through renal route, polymeric micelles are therefore thought to demonstrate little risk of chronic toxicities. This is one compelling factor in the intensive research being done on polymeric micelles for drug targeting (Thotakura et al. 2021). In one instance, Kawaguhi studied the toxicity studies on the polymeric carrier without incorporating the drug to the formulation and discovered no anomalies, even at very high concentrations. However, it was found that the spleen and liver have activated mononuclear phagocyte systems (MPS). But the polymer's toxicity decreased when used to transport anticancer medications because the drugs were toxic to MPS (Kawaguchi et al. 2009). There are numerous studies supporting the developed carriers' high safety ratings in publications. The safety profile of the copolymer must be thoroughly investigated by the customized polymers. Researchers typically compare the copolymer's potential biocompatibility and safety to the use of less toxic and biocompatible monomers. Many of the time, this works precisely, but still, the copolymer should be subjected through detailed safety and biocompatibility testing (Zhao et al. 2013).

14.4.3 Ethical Issues in Clinical Trials

Ethical considerations in clinical trials of nanomedicine are based on the quantum of risk involved in the administration. Metal nanoparticles imbibed onto diagnostic device pose lesser health risk as compared to intravenous administration of same metallic nano-based drug delivery system for therapy. (Tinkle et al. 2011; Resnik et al. 2007).

Hence, for each nanoparticle-based delivery system, robust research methodology has to be devised starting from good review of literature, in vitro assays, unbiased preclinical validation to proper recording of the observations throughout the study until the clinical trial study is completed. The polymeric micelle as nanosystems is being researched extensively for application in cancer chemotherapy (Table. 14.2), but they have similar toxicity profiles as the corresponding chemotherapeutic drug, hence eventually, they fail in clinical trials (Yamamoto et al. 2007).

Table 14.2 Polymeric Micelles based anticancer chemotherapeutic drug delivery systems under trials (Source: https://beta.clinicaltrials.gov/study/PM)

S. no.	Name of clinical trial	Interventions	Drug
1	Docetaxel-polymeric micelles (PM) and oxaliplatin for esophageal carcinoma	Esophagus squamous cell carcinoma (SCC) Metastatic cancer	Drug: Docetaxel-PM Drug: Oxaliplatin
2	A study of docetaxel-polymeric micelles for injection in patients with advanced solid tumors	Advanced solid tumors	Drug: Docetaxel-polymeric micelles for injection
3	A clinical trial of paclitaxel-loaded polymeric micelle in patients with taxane-pretreated recurrent breast cancer	Recurrent breast cancer	Drug: Paclitaxel-loaded polymeric micelle
4	Study to evaluate the efficacy and safety of docetaxel-polymeric micelle (PM) in recurrent or metastatic HNSCC	Head and neck squamous cell carcinoma	Drug: Docetaxel-PM
5	Paclitaxel-loaded polymeric micelle and carboplatin as first-line therapy in treating patients with advanced ovarian cancer	Ovarian cancer	Drug: Carboplatin Drug: Paclitaxel-loaded polymeric micelle
6	Study of Genexol-PM in patients with advanced urothelial cancer previously treated with gemcitabine and platinum	Bladder cancer Ureter cancer	Drug: Genexol-PM
7	A trial of paclitaxel (Genexol®) and cisplatin versus paclitaxel-loaded polymeric micelle (Genexol-PM®) and cisplatin in advanced non-small cell lung cancer	Non-small cell lung cancer	Drug: Paclitaxel (Genexol®) Drug: Paclitaxel-loaded polymeric micelle (Genexol-PM®)

A typical challenge in formulation development is ever increasing number of poorly aqueous soluble drugs being discovered. The PM-based delivery systems for such molecule can aid in not only better solubilization and reduction in inherent toxicity of the active pharmaceutical ingredient, but also in an overall improvement in its therapeutic efficacy (Kwon 2003). However, the critical bottlenecks of clinical development of any polymeric micellar systems still remain their thermodynamic stability and formulation specific characterization studies. Hope the pharmaceutical scientists and clinicians together fill the research gaps and utilize the potential of polymeric micelles in nanomedicine.

The risk assessment study following the exposure to nanomaterials can also be conducted at the several international agencies like Environment Protection Agency (EPA), National Science Foundation (NSF), National Institute of Occupational Safety and Health (NIOSH), and National Cancer Institute (NCI).

The ethical concerns in the clinical trials of polymeric micelles revolve around assessing the probable risk, managing the risk (establishing good risk-to-benefit ratio) and communicating the potential risk to study participants. It is mandatory to obtain informed consent from the recruited participants. The study investigator and his team should document all the clinical observations. After appropriate statistical evaluation, the investigated nanomedicine must exhibit significant difference in comparison to existing drug therapies.

14.4.4 Regulatory Aspects

Like other nanoscale formulations, polymeric micelles are needed to follow strict regulations, to ensure patient safety which is well documented in ICH pharmaceutical development guidelines (Sainz et al. 2015). The long-term stability and specific characterization are valid bottlenecks to successful clinical translation of the polymeric micelles as drug delivery systems. (Ghezzi et al. 2021). Hence, various regulatory agencies like FDA, EMA, Japan MHLM, Therapeutic Goods Authority, CDSCO across the globe have their similar regulatory requirements for nanomedicine development. Generally, for the polymeric micelle, API used determines the types of validation studies required. The FDA assessment starts with product specific regulatory policy so that issues in critical process parameters are addressed at the early development stage itself, followed by premarketing review where the additional need for safety data is ruled out before marketing approval and it continues to monitor the nanomedicine safety profile and its proven effectiveness track record before they are ready for marketing approval. Nanotechnological Task Force specific to nanotechnology (NTIG) recruited by FDA addresses key issues pertaining to nano-based products, to facilitate better overall coordination between all the stakeholders (NTF Reports 2020).

The EMA aims to scrutinize polymeric micelles based on the benefit to risk considerations including environmental risk assessment while apart of similar regulations on nanomedicines. TGA also conducts nanobased training programs to impart knowledge on comparative risk assessment of nano versus conventional ones (Sainz et al. 2015). One of the major joint initiatives by MHLW/EMA includes set of regulatory guidance document in form of reflection paper specific to polymeric micelles (European Medicines Agency 2013).

14.5 Conclusion

The multidisciplinary approach in nanotechnology-based drug delivery has led to significant development in both basic and clinical research. But the pitfalls of increased toxicity concerns have increased pressure on the regulatory authorities, researchers, and stakeholders in industry to work in higher cooperation and in knowledge sharing mode so as to achieve better clinical outcome. Thus, the key to successful clinical translation of nanomedicine begins with extensive basic research which can pave toward exceptional good real-world evidence under the stringent regulatory expertise.

References

Ahmad A, Khan JM, Haque S (2019) Strategies in the design of endosomolytic agents for facilitating endosomal escape in nanoparticles. Biochimie 160:61–75
Anselmo AC, Mitragotri S (2019) Nanoparticles in the clinic: an update. Bioeng Transl Med 4:1–16
Bhattacharya T, Das D, Borges E, Soares GA, Chakrabarti P, Ai Z, Chopra H, Hasan MA, Cavalu S (2022) Novel green approaches for the preparation of gold nanoparticles and their promising potential in oncology. Processes 10(2):426
Blanco E, Shen H, Ferrari M (2015) Principles of nanoparticle design for overcoming biological barriers to drug delivery. Nat Biotechnol 33:941–951
Butt AM, Abdullah N, Rani NN, Ahmad N, Amin MC (2022) Endosomal escape of bioactives deployed via nanocarriers: insights into the design of polymeric micelles. Pharm Res 39:1–8
Caldera F, Nisticò R, Magnacca G, Matencio A, Khazaei Monfared Y, Trotta F (2022) Magnetic composites of dextrin-based carbonate nanosponges and iron oxide nanoparticles with potential application in targeted drug delivery. Nanomaterials 12(5):754
Caldorera-Moore M, Vela Ramirez JE, Peppas NA (2019) Transport and delivery of interferon-α through epithelial tight junctions via pH-responsive poly (Methacrylic acid-grafted-ethylene glycol) nanoparticles. J Drug Target 27:582–589
Castro KC, Costa JM, Campos MG (2022) Drug-loaded polymeric nanoparticles: a review. Int J Polym Mater Polym Biomater 71(1):1–3
Choi YH, Han H-K (2018) Nanomedicines: current status and future perspectives in aspect of drug delivery and pharmacokinetics. J Pharm Investig 48:60
European Medicines Agency (2013) Joint MHLW/EMA reflection paper on the development of block copolymer micelle medicinal products. EMA, London
Farokhzad OC, Langer R (2009) Impact of nanotechnology on drug delivery. ACS Nano 3(1): 16–20
FDA (n.d.) Drug products, including biological products, that contain nanomaterials guidance for industry. https://www.fda.gov/drugs/guidance-compliance-regulatory-information/guidances-drugs
Feynman RP (1961) There's plenty of room at the bottom: an invitation to enter a new field of physics. Miniaturization. Reinhold, New York
Ghezzi M, Pescina S, Padula C, Santi P, Del Favero E, Cantù L, Nicoli S (2021) Polymeric micelles in drug delivery: an insight of the techniques for their characterization and assessment in biorelevant conditions. J Control Release 332:312–336
Hasannia M, Aliabadi A, Abnous K, Taghdisi SM, Ramezani M, Alibolandi M (2022) Synthesis of block copolymers used in polymersome fabrication: application in drug delivery. J Control Release 341:95–117
Holmannova D, Borsky P, Svadlakova T, Borska L, Fiala Z (2022) Carbon nanoparticles and their biomedical applications. Appl Sci 12(15):7865
Jain NS, Somanna P, Patil AB (2022) Application of quantum dots in drug delivery. Nanosci Nanotechnol Asia 12(1):16–31
Karpuz M, Silindir-Gunay M (2022) Lipid-based drug delivery systems and their role in infection and inflammation imaging. In: Nanoengineering of biomaterials. Wiley, Hoboken, pp 469–503
Kaur J, Gulati M, Kapoor B, Jha NK, Gupta PK, Gupta G, Chellappan DK, Devkota HP, Prasher P, Ansari MS, Alkhayl FF (2022) Advances in designing of polymeric micelles for biomedical application in brain related diseases. Chem Biol Interact 361:109960
Kawaguchi T, Honda T, Nishihara M, Yamamoto T, Yokoyama M (2009) Histological study on side effects and tumor targeting of a block copolymer micelle on rats. J Control Release 136(3): 240–246, 97
Kotta S, Aldawsari HM, Badr-Eldin SM, Nair AB, Yt K (2022) Progress in polymeric micelles for drug delivery applications. Pharmaceutics 14(8):1636
Kwon GS (2003) Polymeric micelles for delivery of poorly water-soluble compounds. Crit Rev Ther Drug Carrier Syst 20:357–403

Manzari MT, Shamay Y, Kiguchi H, Rosen N, Scaltriti M, Heller DA (2021) Targeted drug delivery strategies for precision medicines. Nat Rev Mater 6(4):351–370

NTF Reports (2020) Nanotechnology—a report of the U.S. Food and Drug Administration nanotechnology task force, 25 July 2007. https://www.fda.gov/science-research/nanotechnology-programsfda/nanotechnology-task-force-report-2007. Revised NTF reports 2020

Oh N, Park JH (2014) Endocytosis and exocytosis of nanoparticles in mammalian cells. Int J Nanomedicine 9(Suppl 1):51–63

Patra JK, Das G, Fraceto LF, Campos EV, Rodriguez-Torres MD, Acosta-Torres LS, Diaz-Torres LA, Grillo R, Swamy MK, Sharma S, Habtemariam S (2018) Nano based drug delivery systems: recent developments and future prospects. J Nanobiotechnol 16(1):1–33

Resnik DB et al (2007) Ethical issues in clinical trials involving nanomedicine. Comtemp Clin Trials 28(4):433–441

Sainz V, Conniot J, Matos AI, Peres C, Zupancic E, Moura L, Silva LC, Florindo HF, Gaspar RS (2015) Guideline IH. Pharmaceutical development. Q8 (2R). Biochem Biophys Res Commun 468:504–510. As revised in August. 2009 Aug 23

Thotakura N, Parashar P, Raza K (2021) Assessing the pharmacokinetics and toxicology of polymeric micelle conjugated therapeutics. Expert Opin Drug Metab Toxicol 17(3):323–332. https://doi.org/10.1080/17425255.2021.1862085

Tian W, Gao X, Zhang J, Yu J, Zhang J (2022 Feb) Cellulose nanosphere: preparation and applications of the novel nanocellulose. Carbohydr Polym 1(277):118863

Tinkle S et al (2011) Nanomedicines: addressing the scientific and regulatory gaps. Ann N Y Acad Sci 131(1):35–36

United Nations (2018) The 2030 agenda and the sustainable development goals: an opportunity for Latin America and the Caribbean (LC/G. 2681-P/Rev. 3). United Nations, Santiago

Vallet-Regí M, Schüth F, Lozano D, Colilla M, Manzano M (2022) Engineering mesoporous silica nanoparticles for drug delivery: where are we after two decades? Chem Soc Rev 51:5365

Varkouhi AK, Scholte M, Storm G, Haisma HJ (2011) Endosomal escape pathways for delivery of biologicals. J Control Release 151:220–228

Wang J, Gong J, Wei Z (2022a) Strategies for liposome drug delivery systems to improve tumor treatment efficacy. AAPS PharmSciTech 23(1):1–4

Wang J, Li B, Qiu L, Qiao X, Yang H (2022b) Dendrimer-based drug delivery systems: history, challenges, and latest developments. J Biol Eng 16(1):1–12

Wolfram J, Zhu M, Yang Y, Shen J, Gentile E, Paolino D, Fresta M, Nie G, Chen C, Shen H, Ferrari M (2015) Safety of nanoparticles in medicine. Curr Drug Targets 16(14):1671–1681

Xu L, Wang X, Liu Y, Yang G, Falconer RJ, Zhao CX (2022) Lipid nanoparticles for drug delivery. Adv NanoBiomed Res 2(2):2100109

Yamamoto T, Yokoyama M, Maitani Y et al (2007) What are determining factors for stable drug incorporation into polymeric micelle carriers? Consideration on physical and chemical characters of the micelle inner core. J Control Release 123:11–18

Yetisgin AA, Cetinel S, Zuvin M, Kosar A, Kutlu O (2020) Therapeutic nanoparticles, and their targeted delivery applications. Molecules 25(9):2193

Zhao B, Wang XQ, Wang XY et al (2013) Nanotoxicity comparison of four amphiphilic polymeric micelles with similar hydrophilic or hydrophobic structure. Part Fibre Toxicol 10(1):47

Stability of Polymeric Micelles and Their Regulatory Status

15

Indhumathi Thirugnanasambandham, Kalaiselvi Aasaithambi, Imrankhan Nizam, and Gowthamarajan Kuppusamy

Abstract

Polymeric micelles (PMs) is an innovation approach toward advancement of nanodrug delivery systems. These PMs overcome the issues related to hydrophobic drugs such as solubility, bioavailability, and permeability in crossing biological barriers. PMs are prepared in margins of cost-effective and have special characters such as smaller in size, easy to sterile compared to other nanocarriers. The two biggest challenges to the clinical development of polymeric micelles are stability and regulatory requirements. Although PMs have brought significant benefits to the pharma market, their limited regulatory guidelines and challenges related to thermodynamics and kinetic stability have resulted in a lack of reproducibility. A comprehensive implementation of valid guidelines and regulations can bring greater attention to PMs as a promising option for targeted drug delivery systems. Their insights and potential could therefore boost the future of the pharma market.

Keywords

Polymeric micelles · EMA regulations · Safety profiling · Stability

I. Thirugnanasambandham · G. Kuppusamy (✉)
Department of Pharmaceutics, JSS College of Pharmacy, JSS Academy of Higher Education and Research, Ooty, Nilgiris, Tamil Nadu, India
e-mail: gowthamsang@jssuni.edu.in

K. Aasaithambi · I. Nizam
DBT-BUILDER, JSS Academy of Higher Education and Research, Mysuru, Karnataka, India

S. K. Singh et al. (eds.), *Polymeric Micelles: Principles, Perspectives and Practices*, https://doi.org/10.1007/978-981-99-0361-0_15

Abbreviations

ATP	Adenotriphosphate
CMC	Critical micellar concentration
DLS	Dynamic light scattering
DMF	Dimethylformamide
DSPE	Di-stearoyl phosphatidylethanolamine
EMA	European Medicines Agency
Hpbc	Hydrophobic chains
HPLC	High-pressure liquid chromatography
ICH	International Council of Harmonization
MDR	Multidrug resistance
PCL	Polycaprolactone
PD	Pharmacodynamics
PDI	Polydispersity index
PDLLA	Poly-D,L-lactic acid
PEEP	Poly ethyl ethylene phosphate
PEG	Polyethylene glycol
PG	Propylene glycol
PK	Pharmacokinetics
PMs	Polymeric micelles
SEC	Size exclusion chromatography
Tg	Glass transition temperature
THF	Tetrahydrofuran
TPGS	Tocopheryl polyethylene glycol succinate

15.1 Introduction

Combinatorial chemistry and rapid processing and screening of raw materials have substantially expanded the pool of identifying probable remedial medicine, which has greatly advanced drug development (Jachak and Saklani 2007). However, over one-third of the chemicals are systemically hazardous and have limited water solubility (Lipinski 2000; Kim and Park 2010). Solubility has a direct impact on absorption and physiological distribution because therapeutic medications are circulated through the blood (Van De Waterbeemd et al. 2001). As a tool for improving the solubility and lowering the toxicity of hydrophobic medicines, self-assembling copolymers into micelles is being investigated. Among these, a special class of amphiphilic polymer aggregations formulates nanoscale carriers (between 10 and 200 nm) that can encompass hydrophobic medications in the crux while enabling friendly alterations at the interface between two phases (Masayuki et al. 1990; Kabanov et al. 1992; Yokoyama et al. 1992). A paradigm for using polymeric micelles (PMs) to encapsulate medicinal medicines dates back a head three decades

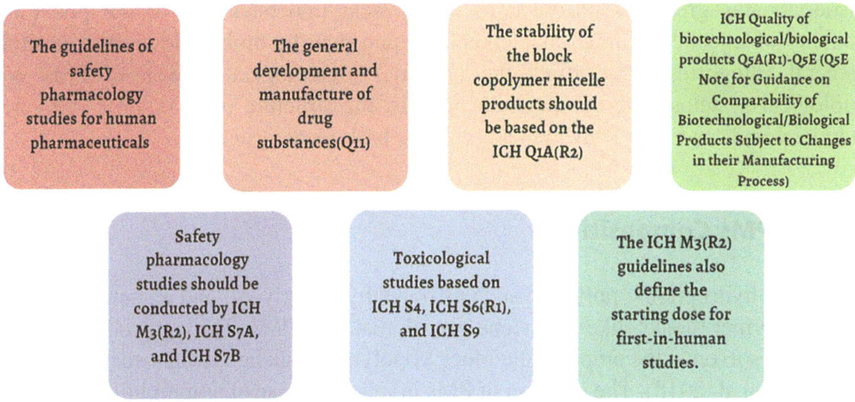

Fig. 15.1 General ICH specifications (ICH 1996, 1998, 2000, 2003, 2005, 2008, 2011, 2013a, b)

(Fig. 15.1), Ringsdorf et al. (Masayuki et al. 1990). Even though the synthesis of micelles from block copolymers was well recognized in that era (Li et al. 2005; Halperin and Alexander 1989), the bulk of micelles was prepared using substances that were not compatible with the human body. A novel class of amphiphilic carriers was needed for therapeutic delivery. These were purposely made of relatively bio-inert materials and were created to resemble natural vectors, like viruses. The hydrophobic polymer served as the micelle's core. The hydrophilic polymer served as its corona as the micelles assembled themselves with the copolymers in the aqueous environment. These amphiphilic PMs contained hydrophobic small molecule medicines that were soluble in the hydrophobic core. Oncology professionals anticipated that the use of PMs would revolutionize the treatment of cancer patients. Although most experts in the field imply evidently that PMs are still in their outset and inadequate certainty on regulation for clinical use is seriously impeding their translation research of micelles, the expectation has not lived up to the early hype (Torchilin 2006; Aliabadi and Lavasanifar 2006). Even though several challenges are faced in the area of pharmacokinetics, pharmacodynamics, and polymeric micelle toxicity, there are several potential advantages to using this technology. However, regulatory advice in this sector is imperatively needed and it is crucial to provide legal certainty to manufacturers, legislators, healthcare practitioners, and the general public. There is still much anticipation surrounding the subject of nanomaterials in the field of medicine and their impact on pharma markets.

15.2 Safety of PMs

The safety of nanoscale goods in general and PMs, in particular, cannot be easily determined. There has been a significant amount of research done to determine the possible toxicity of PMs, but considerably less has been done to evaluate the safety and toxicity of micelles. As a result, research on various other kinds of nanoscale materials has greatly contributed to our present understanding of the potential toxicity of micelles. Numerous elements, which may be broken down into several general categories, may affect the potential toxicological effects of PMs on biological systems: The composition of PMs, the structure of PMs, how PMs are administered; and how PMs interact with the biological systems.

15.2.1 PMs Composition

A core of hydrophobic polymer, such as polyethylene glycol, and a shell of hydrophilic polymer blocks, like propylene glycol, make up PMs. Through self-assembly in certain solvents, the amphiphilic block copolymers can create nanoscale micelles (Kondiah et al. 2018). The creation of PMs involves the use of amphiphilic polymers with both hydrophilic and lipophilic building components. Both diblock and triblock amphiphilic polymers are possible. Predominantly, the diblock polymers utilization was AB type, consisting of two polymers, one of which is a hydrophilic block and the other a lipophilic block. The two (ABA) or three (ABC) polymers that make triblock polymers are similar. Drug delivery processing frequently employs block polymers of the ABA type (Zhulina and Borisov 2012). In micellar production, the most commonly used polymer blocks are the amphiphilic diblock copolymers, such as polystyrene and poly (ethylene glycol), and triblock copolymers, like poloxamers. Graft and ionic (poly (ethylene glycol) Poly(-caprolactone)-g-polyethyleneimine) copolymers are also employed for micelle formulation (Yadav et al. 2019; Mandal et al. 2017; Kulthe et al. 2012; Jiang et al. 2006; Li et al. 2015). Polyethylene glycol (PEG) is typically used for the hydrophilic portion, but other polymers can also be used such as poly (vinyl pyrrolidone), poly (acryloyl morpholine), or poly (trimethylene carbonate); the hydrophobic portion shall be composed of poly (propylene oxide), polyesters like poly(-caprolactone), or polymers and copolymers of glycolic and lactic acids (Mandal et al. 2017).

Zhao et al. investigated the toxicity studies of four distinct categories of PMs made of poly (ethylene glycol) and polyglycerol poly(-caprolactone), poly (ethyl ethylene phosphate)-co-poly(-caprolactone), poly (ethylene glycol) and poly (ethylene glycol)-poly(-caprolactone), and poly (ethylene glycol)-di-stearoyl-sn-glycerophosphate (PEG-DSPE). They compared the nanotoxicity of different micelles on various cell types and an animal model. The micelles tested were PEG-PG-PCL, PEEP-PCL, PEG-PCL, and PEG-DSPE. According to the evidence, all micelle systems altered the factors responsible for inflammation, which results in the elevation of ROS levels. Cell volume was increased by PEEP-PCL and PEG-PG-PCL micelles. The size of the PEGPG-PCL micellar system and the polyphosphoester structure in PEEP-PCL are probably connected to these phenomena. Additionally, PEG-DSPE micelles prevented Eahy.926 cells from proliferating by triggering apoptosis. After treatment with these micelles, no indication of alterations to the cell membrane was discovered. These nanocarriers were discovered to influence blood components differently, most likely as a result of the direct injection into veins. However, the pathology and inflammatory markers in the major organs of the mouse model did not significantly alter the blood components. The toxicities observed in the in vivo model may not be necessarily seen in the in vivo model as the animals owing to their biological system can ward off some toxic substances. Conversely, some of the toxicities that were not observed in the in vivo studies might be observed in the in vivo studies owing to some unknown mechanisms. We find it challenging to determine if the observed alterations are significant or insignificant because there are currently no accepted criteria for nanotoxicity. Overall, it is shown that the polymeric micellar system studied here exhibits variable toxicity associated in regard to its structural characteristics, and its safety in relation to the biological system differs in different cellular models. This research will undoubtedly advance our understanding of amphiphilic PMs nanotoxicity (Zhao et al. 2013).

15.2.2 PMs Structure

The toxicity and safety of PMs are influenced by particle size, shape, and charge (ζ-potential). Highly permeable tumors may easily acquire micellar systems that range from 30 to 100 nm, but poorly permeable tumors can only be penetrated with micellar systems with the size of 30 nm, demonstrating the crucial importance of size (Cabral et al. 2011). This might theoretically also be the case when considering the absorption into epithelial cells or the passage across the mucosal layer, as has been shown for other nano-systems (Murgia et al. 2016; Rossi et al. 2019; Salatin et al. 2015). However, to the best of our knowledge, there are not any statistics about micelles in the literature. In this regard, it is also crucial to emphasize that micelles can disassemble due to their nature when they correspond to the epithelial cell network or mucus. Hence in these conditions, the size of the play is very important but rather depends on the properties of the unimers and how specifically they interact with cell membranes or mucus. Micelles' surface properties, which control their stability after intravenous administration and their interactions with mucosal and epithelial layers, also have a significant bearing on how they behave. The construction of the protein corona is reduced in the presence of a water affinity environment and a neutral surface lengthens the period of blood circulation following intravenous injection, which was investigated in solid polymeric nanoparticles. However, because of comprehensive binding to the proteins along with enhanced aggregation in vivo, micelles with positive zeta potential have relatively limited stability in biological fluids (Vllasaliu et al. 2014; Logie et al. 2014; Zhu et al. 2018; Honary and Zahir 2013). The mucus-penetrating characteristics are additionally enhanced by a neutral and hydrophilic surface, which is particularly important for oral delivery. Positively charged nanocarriers are well acknowledged for their muco-adhesiveness, improved interactions with epithelial cells, and also for enhanced transportation of the drugs through biological barriers (Li et al. 2015; Taipaleenmäki et al. 2017, 2020; Bandi et al. 2020; Hoeller et al. 2009; Jubeh et al. 2004). Lastly, active targeting of molecules can be done by functionalizing the surface of the micelles with well-defined molecules, which is highly helpful for chemotherapy (Mi et al. 2020; Wang et al. 2020). At certain times, a rod, worm, or disk-like arrangement is visible in the micelles, even though they are typically pictured as spherical (Owen et al. 2012; Zhong and Pochan 2010; Truong et al. 2015). The structure of the polymers employed and the surrounding environment's characteristics, including temperature, PH, and composition, are the key factors influencing the variances of micellar form (Kuntsche et al. 2011). Since the shape of these nanocarriers has a significant impact on circulation time, biodistribution, and cellular absorption, it has been recognized that studying their morphology is essential to anticipate their performance in vivo (Truong et al. 2015; Discher 2014). For instance, due to the elongated morphology, their flexible nature, and amplitude of fragmentation, the filomicelles demonstrate decreased clearance and increased blood circulation compared to the spherical micelles (Christian et al. 2009; Oltra et al. 2013, 2014). Recent research shows that shorter filomicelles (length: 180 nm) show the highest penetrative effect in tumors and higher cellular uptake compared to the longer filomicelles (length: 2.5 μm) and spherical micelles (Ke et al. 2020).

15.2.3 Interaction of PMs with the Biological Systems

PMs interact with biological systems and exhibit their toxicity and safety profile. Numerous systematic types of research on the interactions of micelles with biological systems have been conducted during the past few years. Studies on the interactions of micelles with biological systems, however, are scarce. Given the crucial function of the micelles in subcellular drug delivery to a particular target, hence, several gaps need to be filled (Nelemans and Gurevich 2020; Maysinger et al. 2007). Endocytosis is the primary mechanism by which micelles are internalized. It is based on the contact of the micelle with the cellular membrane and involves uptake by cells and transport through endosomes before reaching the cytoplasm of the cells (Nelemans and Gurevich 2020; Chen et al. 2019). In reality, the micellar system is rarely swallowed as a whole; instead, they typically undergo cell membrane disassembly if not destroyed in lysosomes (Chen et al. 2019). Therefore, micellar systems transport drugs into the cells, it is not necessary that it is taken as a whole but can also be ejected outside or taken up as unimers, which leads the drug to accumulate in multiple places as in cell membrane or distinct cell compartments (Lee et al. 2013; Cui et al. 2013). Because their ingestion by endocytosis results in bypassing of efflux pumps, and decreasing resistance of single or multidrug, the interest in micelles has intensified (MDR) (Rapoport et al. 2002). Pluronic®, D-Tocopheryl Polyethylene Glycol 1000 Succinate (TPGS), and Soluplus® are examples of polymers that serve as direct efflux pump inhibitors. Micelles are formed with polymers by conjugating them with compounds (e.g. quercetin) that have a deterrent effect on these pumps (Miller et al. 1999; Dabholkar et al. 2006; Jin et al. 2015; Mu et al. 2019). Earlier, the polymer inhibitory impact is directly related to unimers rather than micelles, which by increasing membrane fluidity might result in an ATP depletion and a subsequent decline in ATPase activity. In general, the maximum inhibition of efflux pumps is enumerated at unimers concentrations below the CMC (Pepić et al. 2013).

15.2.4 Administration Routes of PMs

Micellar systems are intriguing transporters for various routes of administration because of their small size, simple preparation, and great solubility. Micellar systems increase the bioavailability of drugs and release the medicine in a controlled and targeted fashion, which minimizes the side effects (Ambade et al. 2005; Mikhail and Allen 2009). The intravenous injection/infusion is the route of administration for micelles that have received the most attention, and it is mostly used for chemotherapy, although oral administration has also shown some very intriguing findings in terms of enhanced drug bioavailability (Zhang et al. 2014; Cho et al. 2015; Sabra et al. 2017; Gaucher et al. 2010; Khan et al. 2017; Grimaudo et al. 2019). Micelles are produced through reversible limits and can be destroyed by several destabilizing conditions as thermodynamically self-assembled structures. Depending on the method of delivery, micelles may encounter various issues such as pressure during i.v injection, extreme dilution, serum interactions, and potential protein corona formation, and therefore require careful consideration. Drug delivery through the skin and mucosa promotes the interactions with sebum and mucus and helps in the diffusion in the epithelial layer. The micellar behavior will then be resolved owing to the morphology as well as their general attributes. To decide the micellar system's fate and the ensuing bioavailability of the drug, all of these properties are vital.

15.2.5 Regulatory Aspects of PMs

The ICH Pharmaceutical Development Q8(R2) guideline contains rules on the pharmaceutical development of a novel formulation. PMs should go through additional particular standards because of their nanoscale size to ensure the safety of products and minimize hazards connected to unexpected effects on patients (Sainz et al. 2015; Choi and Han 2018). A particular EMA Reflection Paper released in 2013 on the creation of block copolymer micelle pharmaceutical products addresses the in-depth appraisal, with extensive interpretation and knowledge of important features, of nanocarriers because of their distinctiveness (CHMP 2013).

The document offers fundamental details for first-in-human investigations as well as information on the appraisal of quality, specifications of the product, stability, manufacturing, non-clinical pharmacokinetics (PK), and pharmacodynamics (PD) studies. Although the study primarily focuses on intravenous products, it is made apparent that the presented ideas can also be applied to other routes of administration. The document recommends evaluating Chemistry, manufacturing, controls, micellar morphology, charge on the surface, plasma stability or another admissible medium, and drug release in bio-relevant settings in addition to the stability and impurity data of polymers and active compounds. The characterization of micelles serves as the primary foundation for creating product requirements, conducting stability studies, and designing pharmacokinetic and pharmacodynamic investigations (Fig. 15.2). Micelle stability plays a vital role in scheduling sampling for PK studies; similarly, the fate of the drug and micelles and route of

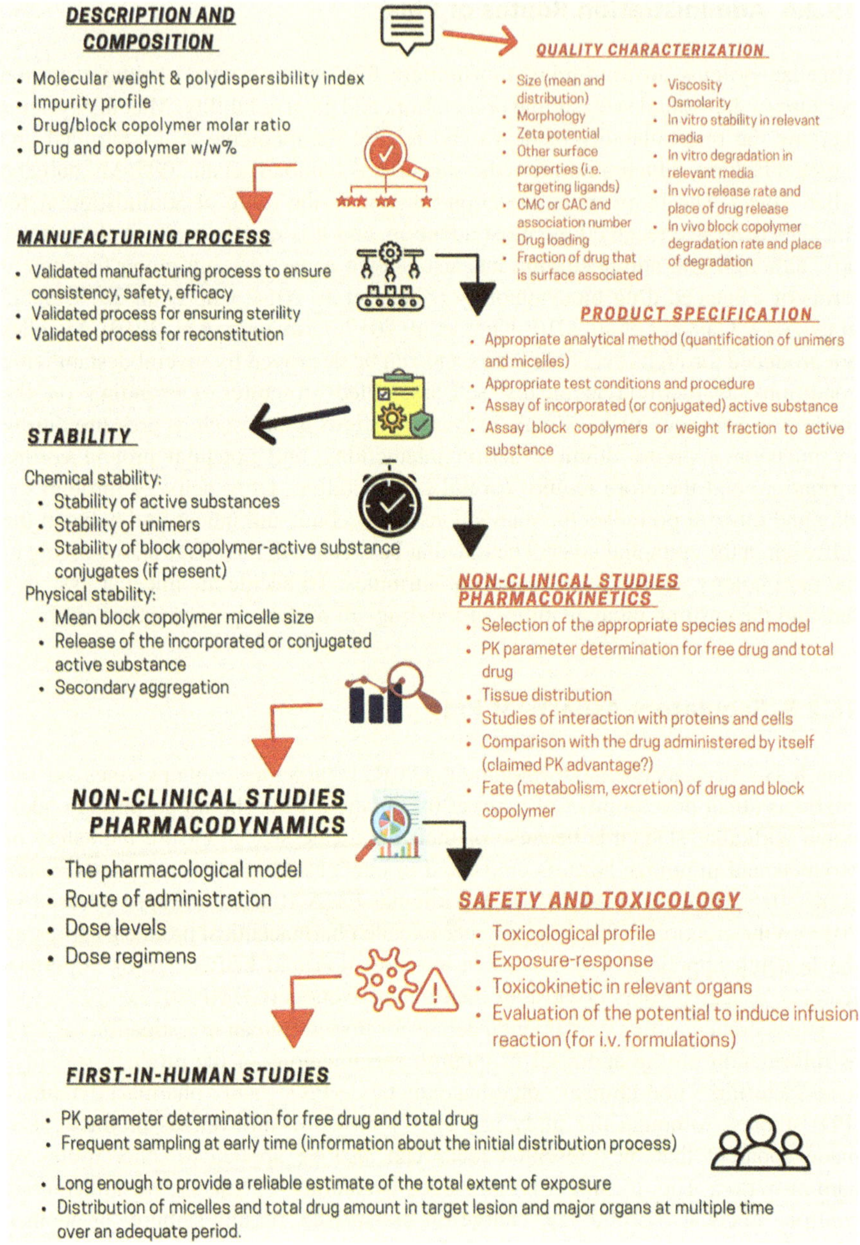

Fig. 15.2 European Medicines Agency (EMA) regulation for PMs

administration in cellular levels are considered utmost during designing PD studies. The lack of compendial (pharmacopeia) methodologies to assess the release of drugs and the integrity of micelles necessitates the development of internal procedures that must be validated with their repeatability and discriminatory capacity. The proposed release medium should thus accurately represent the physiological milieu of the micellar formulation while in use. These authorized assessments, to be periodically exercised in the system, are capable to foresee the release into circulatory paths and at the desired site of action. Conversely, the technique must also be delicate enough to guarantee uniformity from batch to batch. To construct the polymeric micellar systems in vivo and in vivo, it is important to have an early conversation with regulators about both the crucial attributes of the product and cutting-edge procedures that could be employed. This is due to the complicated nature of the PMs and the difficulties associated with their construe. However, the in vivo tests' predictability is a crucial concern, and as much as in the reflection piece, it is admitted that it would only be sporadically probable to associate between in vivo and in vivo results. There are currently very few publications in the literature that discuss the relationship between the release and kinetics. A reasonable match between in vivo and in vivo data was obtained by Zhang et al. when they used mouse serum for their in vivo PEG-PCL micelles stability investigations. After 72 h, about 60% of the micelles were intact in the bloodstream in vivo, while 70% of the micelles were recorded in vivo. Liu et al. (Zhang et al. 2020) using a fluorescent probe linked with AF4 and DLS investigated the intercommunication of the PMs with diverse plasma proteins. The data indicated identical stability of the PMs in both in vivo and in vivo. Nonetheless, it is imperative that a concept obtained from a few specific cases need not be true in all.

15.3 Physicochemical and Kinetic-Related Stability Aspects of PMs

Upon intravenous injection, micelles experience a variety of environmental changes, including considerable concoction, changes in pH and salt fluctuations, and interaction with extensive peptides and cells. Ruptured micelles are avoided as drug delivery vehicles because this would cause the drug cargo to leak before reaching the target cells. Micelles must maintain their structural integrity throughout preparation and till reach the target sites for medication solubilization or local drug delivery. Micelle stability can be conceptualized generally in terms of kinetic and thermodynamic stability. As micelles develop and attain equilibrium, the system behaves according to the concept of thermodynamic stability. Kinetic stability describes the rate at which the exchange of polymers occurs and micellar deconstruction and indicates how the system behaves throughout the circulation period.

15.3.1 Physicochemical Stability

15.3.1.1 Thermodynamic Stability

A crucial variable for assessing the micelle's thermodynamic stability is the CMC. It is connected to the term k_BT, which stands for thermal energy, the energy produced by efficient interactions between the bulk and polymers is denoted as ε_h. Greater thermodynamic stability is indicated by lower values. In the equation $\Delta G_{mic} = RT \ln (CMC) \ln$, the Gibbs energy of micellization, or ΔG_{mic}, is also directly connected to $CMC = \exp(-n\varepsilon_h/k_BT)$ (Zana 1996). Because the site of interaction of polymers is not only limited to tiny molecules, they display decreased CMC values than surfactant micelles with lower molar mass. Above and below the CMC, polymer solutions display various physical characteristics. Commonly, micromolar quantities of polymeric micelle CMCs are used (Maysinger 2007; Diezi et al. 2010). Stability and hydrophobic segment length are directly correlated (Van Domeselaar et al. 2003; Adams and Kwon 2002). The nature and cohesiveness of the hydrophobic core affect micelles' propensity to separate (Gaucher et al. 2005). Lower CMC is obtained by reducing the copolymer's hydrophobicity and increasing the hydrophobic core's coherence (Van Domeselaar et al. 2003; Ranger et al. 2001). The drug–core interaction can potentially have an impact on the micellar system stability employed in the transportation of the drugs. The substantiality of the micelle can increase with supplementary hydrophobic interactions between the core and enclosed drug molecule (Lee et al. 2004). Interactions between the surrounding aqueous environment and the corona encompassing the polymer chains and each other also have an impact on thermodynamic stability. PEG is used as the hydrophilic element in the majority of micelles for drug delivery. Van der Waals forces act between individual PEG chains, while the PEG chains' interactions with H_2O in a bulk solution are hydrophilic forces like hydrogen bonds and dipole–dipole forces (Owens III and Peppas 2006). The polymers will be forced to take up the stern and stretched, brush-like conformations when the surface density and the length of the PEG chain are increased. PEG's lower molecular weight and surface density lead to the destabilization of the micelles and the disclosure of the hydrophobic crux to aqueous solutions, on the other hand. To permit fluid mobility of surface chains and shield the hydrophobic core from exposure, sufficient coverage of the hydrophilic polymers is essential (see the section on destabilization mechanisms for further information on these mechanisms). A solution's optical clarity, surface tension readings, viscosity, and other physical properties that undergo abrupt changes at the CMC can all be used to calculate the CMC. The most used techniques include chromatography, particle size analysis, and spectroscopy, which are all chosen for their great sensitivity (Torchilin 2001).

15.3.1.2 Impact of a Hydrophobic Segment on Micelles Stability

Increased length hydrophobic chain is associated with greater stability for block copolymer micelles (and therefore reduced CMC) (Adams and Kwon 2002, 2003). For the copolymer to be as stable as possible, the chain lengths of both hydrophilic and hydrophobic must be balanced. After a particular threshold, an increase in the

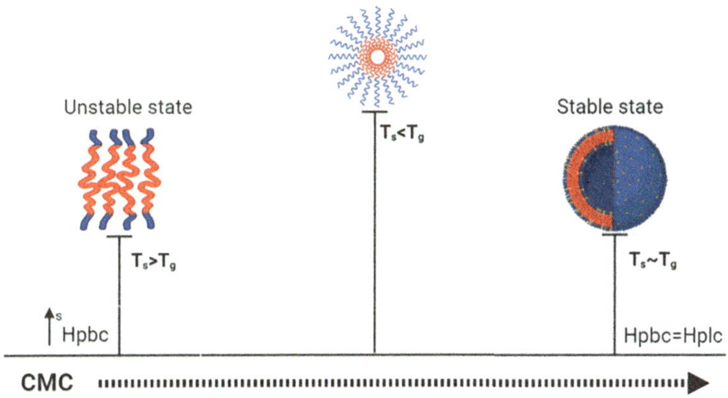

Fig. 15.3 Impact of hydrophobicity on the stability of micelles. *Hpbc* hydrophobic chains, *Hplc* hydrophilic chains, *Ts* surface temperature, *Tg* glass transition temperature

length of the hydrophobic chains produces micellar systems with reduced homogeneity, which eventually gives rise to non-spherical aggregates (Gädt et al. 2009). The CMC was reduced with an increase in hydrophobicity and stacking interactions in the core (Fig. 15.3). In contrast, the CMCs of micelles produced by combining two hydrophobic copolymer chains were significantly greater. The scientists hypothesized that interactions between the molecules of micellar core, such as stacking interactions, produced a "glassy" form in the crux that affects stableness and that for predicting stability of micelles hydrophobicity alone is insufficient.

15.3.1.3 Impact of Encapsulated and Conjugated Drug

Micelle stability is also impacted by drug–core interactions. Coherence of the drug and polymer core was another indicator of how well the drug was released after being enclosed. The efficiency of drug loading increased if the carboxylic acid was higher in the block copolymer for ionic intercommunication with the encapsulated pharmaceuticals. This improved relationship between the medication and micelle core reduced the CMC and slowed down the rate drug of release (Lee et al. 2004). It is believed that the encapsulated drug in the core and the conjugates had increased hydrophobic interactions which might have caused the beneficial rise in solubilization of the drug and more homogenous morphology. It increases the solubilization of poorly soluble drugs drastically. It is unclear if these traits are shared by other conjugated/encapsulated therapeutic micelles or if a comparable result might be obtained with a different tiny hydrophobic molecule (Mikhail and Allen 2010).

15.3.1.4 Environmental Impact on the Stability

The development and stability of micelles are greatly influenced by the microenvironment. Dialysis and co-solvent evaporation are two frequently used techniques for micellization (De Villiers et al. 2008). In contrast to dialysis, which created significantly larger and more scattered micelles, co-solvent evaporation produced smaller, more uniform micelles with a polydispersity index (PDI). According to scientists, the variance in the size of the micelle is dependent on the path and might be brought on by modifications in the rates required to establish the equilibrium. It is interesting to see the variations in micelle size caused by preparation techniques. According to Okano and colleagues, even minor adjustments to a single preparation process, such as changing the temperature or the molecular weight cut-off of a dialysis membrane, can have a significant impact on micelle size (Chung et al. 1999).

The CMC and size of the produced micelles are both dramatically affected by changes in solvent conditions (Chen et al. 1999; Greenall et al. 2011; Cui et al. 2007; Aliabadi et al. 2007). THF, DMF, acetone, and acetonitrile, which are four popular solvents with various degrees of water miscibility, were examined, which indicated a general relationship between micelle size and solvent/water miscibility, whereby a rise in miscibility was followed by a fall in micelle size. The mobility of the inter-micelle chain is also influenced by temperature. For instance, PEG-PDLLA micelles' PDLLA core was more mobile at temperatures higher than the glass transition temperature (Tg). Because of this, the CMC rises at higher temperatures, illuminating the direct relationship between temperature and thermodynamic stability (Yamamoto et al. 2002). Micelle production is a very delicate process that is vulnerable to a variety of internal and external permeation. Though the formation of a stable micellar system is present lyophilization issues can still arise during insufficient PEG or another hydrophilic component coverage.

15.3.2 Kinetic Stability

The dynamics between individual micelles, their surroundings, and one another are the focus of the description of kinetic stability, which deals with the behavior of the PMs with time in an aqueous solution. A micelle's stability may be impacted by any change in its surroundings. Micelles are subjected to significant and abrupt environmental changes because they are used as drug delivery systems, as previously mentioned. Individual chains continue to be dynamic after micellization and move back and forth between the micellar system and the bulk solution. Finally, micelles will start to disintegrate after being subjected to environmental changes or by simple dilution. To characterize the dynamics of micelles over the period and during dismantling, kinetic stability is used.

The following diagram illustrates how the concentration of the micellar system changes in equilibrium depending on the individual polymer chain's concentration:

$$K_{\mathrm{M}} = \frac{[A]^n}{[\text{micelle}]}$$

K_{M} is the micelle dissociation constant and has the units of concentration; n is the aggregation number of the micelle.

According to Mattice et al., three different mechanisms make up the dynamic equilibrium for copolymer exchange in micelles: chain insertion/expulsion, micelle merging/splitting, and micelle spanning (Haliloğlu et al. 1996). Chain expulsion and insertion is the process of inserting a polymer chain into a second micelle after the chain has been expelled from the first micelle and returned to the bulk. When two micelles briefly merge and their micellar cores touch, chains can be swapped. In this instance, the chain moves from the first to the second micelle. Micellar spanning serves as the ultimate mechanism for copolymer exchange. One prolonged chain, which never fully returns to the bulk, connects the exteriors of two micelles. This allows the chain to migrate across micelles without being expelled from one and without coming into touch with the core of the two micelles. By labeling singular chains and tracking the progress, one can predict or see empirically such complex kinetic investigations. Aniansson and Wall's earlier research outlined the dynamics of the relaxation of the micellar systems through the interchange of polymers and separation (Aniansson 1978; Aniansson and Wall 1974). Recently, Diamant et al. revised the kinetic modeling that defined the micellization process in several stages (Hadgiivanova et al. 2011). In general, kinetic stability can be measured using the same analytical methods as thermodynamic stability. The most popular of them are dynamic light scattering (DLS) and size exclusion chromatography (SEC).

15.4 Conclusion

PMs are developing significant drug carriers due to their distinctive characteristics. The ability of block copolymer micelles to generate distinct core-shell structures is their most important characteristic for therapeutic administration. Drugs that are poorly soluble in water are easily loaded into the polymeric micelle's hydrophobic core, giving the chance to increase their bioavailability. The two biggest challenges to the clinical development of PMs are stability and regulatory requirements. By addressing these problems, it may be possible to close the gap between formulation development and academic research and may hasten the clinical translation of those findings. There is no international regulatory body in this field, and the regulation of nanocarriers (including PMs) is still in its early phase. However, regulations on nanomaterials have been created and implemented by the European Union and Switzerland. It is still a need for regulatory bodies to create reliable and consistent processes for evaluating the safety of nanomaterials globally. The construction of these laws will make sure that the boon of nanotechnology may be secured without harming consumer and environmental health. There are many ways to improve the chemistry of the micelles and their stability. The complication of the system increases with each alteration, and to produce insightful findings, the appraisal of these structures needs to be equitably practical. The ultimate clinical success will be facilitated by increased localized therapeutic efficacy and decreased systemic toxicity, which will be made possible by a deeper understanding of polymeric micelle stability and regulations.

References

Adams ML, Kwon GS (2002) The effects of acyl chain length on the micelle properties of poly (ethylene oxide)-block-poly (N-hexylL-aspartamide)-acyl conjugates. J Biomater Sci Polym Ed 13(9):991–1006

Adams ML, Kwon GS (2003) Relative aggregation state and hemolytic activity of amphotericin B encapsulated by poly (ethylene oxide)-block–poly (N-hexyl-l-aspartamide)-acyl conjugate micelles: effects of acyl chain length. J Control Release 87(1–3):23–32

Aliabadi HM, Lavasanifar A (2006) Polymeric micelles for drug delivery. Expert Opin Drug Deliv 3(1):139–162

Aliabadi HM, Elhasi S, Mahmud A, Gulamhusein R, Mahdipoor P, Lavasanifar A (2007) Encapsulation of hydrophobic drugs in polymeric micelles through co-solvent evaporation: the effect of solvent composition on micellar properties and drug loading. Int J Pharm 329(1–2):158–165

Ambade AV, Savariar EN, Thayumanavan S (2005) Dendrimeric micelles for controlled drug release and targeted delivery. Mol Pharm 2(4):264–272

Aniansson EA (1978) Theory of micelle formation kinetics. Ber Bunsenges Phys Chem 82(9): 981–988

Aniansson EA, Wall SN (1974) Kinetics of step-wise micelle association. J Phys Chem 78(10): 1024–1030

Bandi SP, Kumbhar YS, Venuganti VV (2020) Effect of particle size and surface charge of nanoparticles in penetration through intestinal mucus barrier. J Nanopart Res 22(3):1–1

Cabral H, Matsumoto Y, Mizuno K, Chen Q, Murakami M, Kimura M, Terada Y, Kano MR, Miyazono K, Uesaka MJ, Nishiyama N (2011) Accumulation of sub-100 nm polymeric micelles in poorly permeable tumors depends on size. Nat Nanotechnol 6(12):815–823

Chen L, Shen H, Eisenberg A (1999) Kinetics and mechanism of the rod-to-vesicle transition of block copolymer aggregates in dilute solution. J Phys Chem B 103(44):9488–9497

Chen T, He B, Tao J, He Y, Deng H, Wang X, Zheng Y (2019) Application of Förster resonance energy transfer (FRET) technique to elucidate intracellular and in vivo biofate of nanomedicines. Adv Drug Deliv Rev 143:177–205

CHMP (2013) Joint MHLW/EMA reflection paper on the development of block copolymer micelle medicinal products. CHMP, Amsterdam

Cho H, Lai TC, Tomoda K, Kwon GS (2015) Polymeric micelles for multi-drug delivery in cancer. AAPS PharmSciTech 16(1):10–20

Choi YH, Han HK (2018) Nanomedicines: current status and future perspectives in aspect of drug delivery and pharmacokinetics. J Pharm Investig 48(1):43–60

Christian DA, Cai S, Garbuzenko OB, Harada T, Zajac AL, Minko T, Discher DE (2009) Flexible filaments for in vivo imaging and delivery: persistent circulation of filomicelles opens the dosage window for sustained tumor shrinkage. Mol Pharm 6(5):1343–1352

Chung JE, Yokoyama M, Yamato M, Aoyagi T, Sakurai Y, Okano T (1999) Thermo-responsive drug delivery from polymeric micelles constructed using block copolymers of poly (N-isopropylacrylamide) and poly (butylmethacrylate). J Control Release 62(1–2):115–127

Cui H, Chen Z, Zhong S, Wooley KL, Pochan DJ (2007) Block copolymer assembly via kinetic control. Science 317(5838):647–650

Cui C, Xue YN, Wu M, Zhang Y, Yu P, Liu L, Zhuo RX, Huang SW (2013) Cellular uptake, intracellular trafficking, and antitumor efficacy of doxorubicin-loaded reduction-sensitive micelles. Biomaterials 34(15):3858–3869

Dabholkar RD, Sawant RM, Mongayt DA, Devarajan PV, Torchilin VP (2006) Polyethylene glycol–phosphatidylethanolamine conjugate (PEG–PE)-based mixed micelles: some properties, loading with paclitaxel, and modulation of P-glycoprotein-mediated efflux. Int J Pharm 315(1–2):148–157

De Villiers MM, Aramwit P, Kwon GS, editors. (2008) Nanotechnology in drug delivery. Springer, New York

Diezi TA, Bae Y, Kwon GS (2010) Enhanced stability of PEG-block-poly (N-hexyl stearate l-aspartamide) micelles in the presence of serum proteins. Mol Pharm 7(4):1355–1360
Discher DE (2014) Shape effects of filaments versus spherical particles in flow and drug delivery. In: Summer bioengineering conference 2008 Jun 25 (vol. 43215). American Society of Mechanical Engineers, New York, pp. 739
Gädt T, Ieong N, Cambridge G et al (2009) Complex and hierarchical micelle architectures from diblock copolymers using living, crystallization-driven polymerizations. Nat Mater 8:144–150. https://doi.org/10.1038/nmat2356
Gaucher G, Dufresne MH, Sant VP, Kang N, Maysinger D, Leroux JC (2005) Block copolymer micelles: preparation, characterization and application in drug delivery. J Control Release 109(1–3):169–188
Gaucher G, Satturwar P, Jones MC, Furtos A, Leroux JC (2010) Polymeric micelles for oral drug delivery. Eur J Pharm Biopharm 76(2):147–158
Greenall MJ, Schuetz P, Furzeland S, Atkins D, Buzza DM, Butler MF, McLeish TC (2011) Controlling the self-assembly of binary copolymer mixtures in solution through molecular architecture. Macromolecules 44(13):5510–5519
Grimaudo MA, Pescina S, Padula C, Santi P, Concheiro A, Alvarez-Lorenzo C, Nicoli S (2019) Topical application of polymeric nanomicelles in ophthalmology: a review on research efforts for the noninvasive delivery of ocular therapeutics. Expert Opin Drug Deliv 16(4):397–413
Hadgiivanova R, Diamant H, Andelman D (2011) Kinetics of surfactant micellization: a free energy approach. J Phys Chem B 115(22):7268–7280
Haliloğlu T, Bahar I, Erman B, Mattice WL (1996) Mechanisms of the exchange of diblock copolymers between micelles at dynamic equilibrium. Macromolecules 29(13):4764–4771
Halperin A, Alexander S (1989) Polymeric micelles: their relaxation kinetics. Macromolecules 22(5):2403–2412
Hoeller S, Sperger A, Valenta C (2009) Lecithin based nanoemulsions: a comparative study of the influence of non-ionic surfactants and the cationic phytosphingosine on physicochemical behaviour and skin permeation. Int J Pharm 370(1–2):181–186
Honary S, Zahir F (2013) Effect of zeta potential on the properties of nano-drug delivery systems-a review (part 2). Trop J Pharm Res 12(2):265–273
ICH (1996) Stability testing of biotechnological/biological products Q5C. ICH, Geneva
ICH (1998) Duration of chronic toxicity testing in animals (rodent and non-rodent toxicity testing) S4. ICH, Geneva
ICH (2000) ICH (S7A) (ICH). Safety pharmacology studies for human pharmaceuticals S7A. ICH, Geneva
ICH (2003) Stability testing of new drug substances and drug products Q1A(R2). ICH, Geneva
ICH. Non-clinical evaluation of the potential for delayed ventricular repolarization (QT interval prolongation) by human pharmaceuticals S7B; ICH: Geneva, 2005
ICH (2008) Nonclinical evaluation for anticancer pharmaceuticals S9. ICH, Geneva
ICH (2011) Preclinical safety evaluation of biotechnology-derived pharmaceuticals S6(R1). ICH, Geneva
ICH (2013a) Development and manufacture of drug substances (chemical entities and biotechnological/biological entities) Q11. ICH, Geneva
ICH (2013b) Non-clinical safety studies for the conduct of human clinical trials for pharmaceuticals M3(R2). ICH, Geneva
Jachak SM, Saklani A (2007) Challenges and opportunities in drug discovery from plants. Curr Sci 10:1251–1257
Jiang GB, Quan D, Liao K, Wang H (2006) Preparation of polymeric micelles based on chitosan bearing a small amount of highly hydrophobic groups. Carbohydr Polym 66(4):514–520
Jin X, Zhou B, Xue L, San W (2015) Soluplus® micelles as a potential drug delivery system for reversal of resistant tumor. Biomed Pharmacother 69:388–395
Jubeh TT, Barenholz Y, Rubinstein A (2004) Differential adhesion of normal and inflamed rat colonic mucosa by charged liposomes. Pharm Res 21(3):447–453

Kabanov AV, Batrakova EV, Melik-Nubarov NS, Fedoseev NA, Dorodnich TY, Alakhov VY, Chekhonin VP, Nazarova IR, Kabanov VA (1992) A new class of drug carriers: micelles of poly (oxyethylene)-poly (oxypropylene) block copolymers as microcontainers for drug targeting from blood in brain. J Control Release 22(2):141–157

Ke W, Lu N, Japir AA, Zhou Q, Xi L, Wang Y, Dutta D, Zhou M, Pan Y, Ge Z (2020) Length effect of stimuli-responsive block copolymer prodrug filomicelles on drug delivery efficiency. J Control Release 318:67–77

Khan AR, Liu M, Khan MW, Zhai G (2017) Progress in brain targeting drug delivery system by nasal route. J Control Release 268:364–389

Kim S, Park K (2010) Polymer micelles for drug delivery. In: Targeted delivery of small and macromolecular drugs. CRC Press/Taylor & Francis Group, pp 513–551

Kondiah PP, Choonara YE, Kondiah PJ, Marimuthu T, Kumar P, du Toit LC, Modi G, Pillay V (2018) Nanocomposites for therapeutic application in multiple sclerosis. In: Applications of nanocomposite materials in drug delivery. Woodhead Publishing, Sawston, Cambridge, pp 391–408

Kulthe SS, Choudhari YM, Inamdar NN, Mourya V (2012) Polymeric micelles: authoritative aspects for drug delivery. Des Monomers Polym 15(5):465–521

Kuntsche J, Horst JC, Bunjes H (2011) Cryogenic transmission electron microscopy (cryo-TEM) for studying the morphology of colloidal drug delivery systems. Int J Pharm 417(1–2):120–137

Lee J, Cho EC, Cho K (2004) Incorporation and release behavior of hydrophobic drug in functionalized poly (D, L-lactide)-block–poly (ethylene oxide) micelles. J Control Release 94(2–3):323–335

Lee SY, Tyler JY, Kim S, Park K, Cheng JX (2013) FRET imaging reveals different cellular entry routes of self-assembled and disulfide bonded polymeric micelles. Mol Pharm 10(9):3497–3506

Li M, Coenjarts CA, Ober CK (2005) Patternable block copolymers. Block Copolym II:183–226

Li J, Li Z, Zhou T, Zhang J, Xia H, Li H, He J, He S, Wang L (2015) Positively charged micelles based on a triblock copolymer demonstrate enhanced corneal penetration. Int J Nanomedicine 10:6027

Lipinski CA (2000) Drug-like properties and the causes of poor solubility and poor permeability. J Pharmacol Toxicol Methods 44(1):235–249

Logie J, Owen SC, McLaughlin CK, Shoichet MS (2014) PEG-graft density controls polymeric nanoparticle micelle stability. Chem Mater 26(9):2847–2855

Mandal A, Bisht R, Rupenthal ID, Mitra AK (2017) Polymeric micelles for ocular drug delivery: from structural frameworks to recent preclinical studies. J Control Release 248:96–116

Masayuki Y, Mizue M, Noriko Y, Teruo O, Yasuhisa S, Kazunori K, Shohei I (1990) Polymer micelles as novel drug carrier: adriamycin-conjugated poly (ethylene glycol)-poly (aspartic acid) block copolymer. J Control Release 11(1–3):269–278

Maysinger D (2007) Nanoparticles and cells: good companions and doomed partnerships. Org Biomol Chem 5(15):2335–2342

Maysinger D, Lovrić J, Eisenberg A, Savić R (2007) Fate of micelles and quantum dots in cells. Eur J Pharm Biopharm 65(3):270–281

Mi P, Cabral H, Kataoka K (2020) Ligand-installed nanocarriers toward precision therapy. Adv Mater 32(13):1902604

Mikhail AS, Allen C (2009) Block copolymer micelles for delivery of cancer therapy: transport at the whole body, tissue and cellular levels. J Control Release 138(3):214–223

Mikhail AS, Allen C (2010) Poly (ethylene glycol)-b-poly (ε-caprolactone) micelles containing chemically conjugated and physically entrapped docetaxel: synthesis, characterization, and the influence of the drug on micelle morphology. Biomacromolecules 11(5):1273–1280

Miller DW, Batrakova EV, Kabanov AV (1999) Inhibition of multidrug resistance-associated protein (MRP) functional activity with pluronic block copolymers. Pharm Res 16(3):396–401

Mu Y, Fu Y, Li J, Yu X, Li Y, Wang Y, Wu X, Zhang K, Kong M, Feng C, Chen X (2019) Multifunctional quercetin conjugated chitosan nano-micelles with P-gp inhibition and permeation enhancement of anticancer drug. Carbohydr Polym 203:10–18

Murgia X, Pawelzyk P, Schaefer UF, Wagner C, Willenbacher N, Lehr CM (2016) Size-limited penetration of nanoparticles into porcine respiratory mucus after aerosol deposition. Biomacromolecules 17(4):1536–1542

Nelemans LC, Gurevich L (2020) Drug delivery with polymeric nanocarriers-cellular uptake mechanisms. Materials 13(2):366

Oltra NS, Swift J, Mahmud A, Rajagopal K, Loverde SM, Discher DE (2013) Filomicelles in nanomedicine—from flexible, fragmentable, and ligand-targetable drug carrier designs to combination therapy for brain tumors. J Mater Chem B 1(39):5177–5185

Oltra NS, Nair P, Discher DE (2014) From stealthy polymersomes and filomicelles to "self" peptide-nanoparticles for cancer therapy. Annu Rev Chem Biomol Eng 5:281

Owen SC, Chan DP, Shoichet MS (2012) Polymeric micelle stability. Nano Today 7(1):53–65

Owens DE III, Peppas NA (2006) Opsonization, biodistribution, and pharmacokinetics of polymeric nanoparticles. Int J Pharm 307(1):93–102

Pepić I, Lovrić J, Filipović-Grčić J (2013) How do polymeric micelles cross epithelial barriers? Eur J Pharm Sci 50(1):42–55

Ranger M, Jones MC, Yessine MA, Leroux JC (2001) From well-defined diblock copolymers prepared by a versatile atom transfer radical polymerization method to supramolecular assemblies. J Polym Sci A Polym Chem 39(22):3861–3874

Rapoport N, Marin A, Luo Y, Prestwich GD, Muniruzzaman MD (2002) Intracellular uptake and trafficking of Pluronic micelles in drug-sensitive and MDR cells: effect on the intracellular drug localization. J Pharm Sci 91(1):157–170

Rossi S, Vigani B, Sandri G, Bonferoni MC, Caramella CM, Ferrari F (2019) Recent advances in the mucus-interacting approach for vaginal drug delivery: from mucoadhesive to mucus-penetrating nanoparticles. Expert Opin Drug Deliv 16(8):777–781

Sabra S, Abdelmoneem M, Abdelwakil M, Mabrouk MT, Anwar D, Mohamed R, Khattab S, Bekhit A, Elkhodairy K, Freag M, Elzoghby A (2017) Self-assembled nanocarriers based on amphiphilic natural polymers for anti-cancer drug delivery applications. Curr Pharm Des 23(35):5213–5229

Sainz V, Conniot J, Matos AI, Peres C, Zupan�i� E, Moura L, Silva LC, Florindo HF, Gaspar RS (2015) Regulatory aspects on nanomedicines. Biochem Biophys Res Commun 468(3):504–510

Salatin S, Maleki Dizaj S, Yari KA (2015) Effect of the surface modification, size, and shape on cellular uptake of nanoparticles. Cell Biol Int 39(8):881–890

Taipaleenmäki EM, Mouritzen SA, Schattling PS, Zhang Y, Städler B (2017) Mucopenetrating micelles with a PEG corona. Nanoscale 9(46):18438–18448

Taipaleenmäki E, Brodszkij E, Städler B (2020) Mucopenetrating zwitterionic micelles. ChemNanoMat 6(5):744–750

Torchilin VP (2001) Structure and design of polymeric surfactant-based drug delivery systems. J Control Release 73(2–3):137–172

Torchilin VP (2006) Nanoparticulates as drug carriers. Imperial College Press, London

Truong NP, Whittaker MR, Mak CW, Davis TP (2015) The importance of nanoparticle shape in cancer drug delivery. Expert Opin Drug Deliv 12(1):129–142

Van De Waterbeemd H, Smith DA, Beaumont K, Walker DK (2001) Property-based design: optimization of drug absorption and pharmacokinetics. J Med Chem 44(9):1313–1333

Van Domeselaar GH, Kwon GS, Andrew LC, Wishart DS (2003) Application of solid phase peptide synthesis to engineering PEO–peptide block copolymers for drug delivery. Colloids Surf B Biointerfaces 30(4):323–334

Vllasaliu D, Fowler R, Stolnik S (2014) PEGylated nanomedicines: recent progress and remaining concerns. Expert Opin Drug Deliv 11(1):139–154

Wang Y, van Steenbergen MJ, Beztsinna N, Shi Y, Lammers T, van Nostrum CF, Hennink WE (2020) Biotin-decorated all-HPMA polymeric micelles for paclitaxel delivery. J Control Release 328:970–984

Yadav HK, Almokdad AA, Sumia IM, Debe MS (2019) Polymer-based nanomaterials for drug-delivery carriers. In: Nanocarriers for drug delivery. Elsevier, Amsterdam, pp 531–556

Yamamoto Y, Yasugi K, Harada A, Nagasaki Y, Kataoka K (2002) Temperature-related change in the properties relevant to drug delivery of poly (ethylene glycol)–poly (d, l-lactide) block copolymer micelles in the aqueous milieu. J Control Release 82(2–3):359–371

Yokoyama M, Kwon GS, Okano T, Sakurai Y, Seto T, Kataoka K (1992) Preparation of micelle-forming polymer-drug conjugates. Bioconjug Chem 3:295–301

Zana R (1996) Critical micellization concentration of surfactants in aqueous solution and free energy of micellization. Langmuir 12(5):1208–1211

Zhang Y, Huang Y, Li S (2014) Polymeric micelles: nanocarriers for cancer-targeted drug delivery. AAPS PharmSciTech 15(4):862–871

Zhang H, Li H, Cao Z, Du J, Yan L, Wang J (2020) Investigation of the in vivo integrity of polymeric micelles via large stokes shift fluorophore-based FRET. J Control Release 324:47–54

Zhao B, Wang XQ, Wang XY, Zhang H, Dai WB, Wang J, Zhong ZL, Wu HN, Zhang Q (2013) Nanotoxicity comparison of four amphiphilic polymeric micelles with similar hydrophilic or hydrophobic structure. Part Fibre Toxicol 10(1):1–6

Zhong S, Pochan DJ (2010) Cryogenic transmission electron microscopy for direct observation of polymer and small-molecule materials and structures in solution. Polym Rev 50(3):287–320

Zhu Y, Meng T, Tan Y, Yang X, Liu Y, Liu X, Yu F, Wen L, Dai S, Yuan H, Hu F (2018) Negative surface shielded polymeric micelles with colloidal stability for intracellular endosomal/lysosomal escape. Mol Pharm 15(11):5374–5386

Zhulina EB, Borisov OV (2012) Theory of block polymer micelles: recent advances and current challenges. Macromolecules 45(11):4429–4440

Correction to: Polymeric Micelles: Principles, Perspectives and Practices

Sachin Kumar Singh, Monica Gulati, Srinivas Mutalik, Muralikrishnan Dhanasekaran, and Kamal Dua

Correction to:
S. K. Singh et al. (eds.), *Polymeric Micelles: Principles, Perspectives and Practices*,
https://doi.org/10.1007/978-981-99-0361-0

This book was inadvertently published with incorrect affiliation of the third editor, which has been corrected as follows:

Srinivas Mutalik
Department of Pharmaceutics
Manipal College of Pharmaceutical Sciences
Manipal Academy of Higher Education
Manipal, Karnataka, India

The updated version of the book can be found at
https://doi.org/10.1007/978-981-99-0361-0

S. K. Singh et al. (eds.), *Polymeric Micelles: Principles, Perspectives and Practices*,
https://doi.org/10.1007/978-981-99-0361-0_16

The manufacturer's authorised representative in the EU is Springer Nature Customer Service Centre GmbH, Europaplatz 3, 69115 Heidelberg, Germany. If you have any concerns regarding our products, please contact ProductSafety@springernature.com

Printed and bound by CPI Group (UK) Ltd, Croydon, CR0 4YY
30/09/2026
02225303-0012